Energy Materials

Inorganic Materials Series

Editors:

Professor Duncan W. Bruce
Department of Chemistry, University of York, UK

Professor Dermot O'Hare
Chemistry Research Laboratory, University of Oxford, UK

Dr Richard I. Walton
Department of Chemistry, University of Warwick, UK

Series Titles

Functional Oxides
Molecular Materials
Porous Materials
Low-Dimensional Solids
Energy Materials

Energy Materials

Edited by

Duncan W. Bruce
University of York, UK

Dermot O'Hare
University of Oxford, UK

Richard I. Walton
University of Warwick, UK

WILEY

A John Wiley and Sons, Ltd, Publication

Library of Congress Cataloging-in-Publication Data

Energy materials / [edited by] Duncan W. Bruce, Dermot O'Hare, Richard I. Walton.
    p. cm. — (Inorganic materials series ; 4)
    ISBN 978-0-470-99752-9 (hardback)
    1. Energy storage—Materials.   2. Electric batteries—Materials.   3. Power electronics—Materials.
  I. Bruce, Duncan W.   II. Walton, Richard I.   III. O'Hare, Dermot.
  TK2896.E524 2011
  620.1′1297—dc22

                                                 2010042192

A catalogue record for this book is available from the British Library.

Print ISBN: 978-0-470-99752-9 (Cloth)
e-book ISBN: 978-0-470-97778-1
o-book ISBN: 978-0-470-97779-8
e-pub ISBN: 978-0-470-97806-1

Set in 10.5/13 Sabon by Integra Software Services Pvt. Ltd, Pondicherry, India.

FSC
www.fsc.org
MIX
Paper from
responsible sources
FSC® C013604

Contents

Inorganic Materials Series Preface

Back in 1992, two of us (DWB and DO'H) edited the first edition of *Inorganic Materials* in response the growing emphasis and interest in materials chemistry. The second edition, which contained updated chapters, appeared in 1996 and was reprinted in paperback. The aim had always been to provide the reader with chapters that while not necessarily comprehensive, nonetheless gave a first-rate and well-referenced introduction to the subject for the first-time reader. As such, the target audience was from first-year postgraduate student upwards. Authors were carefully selected who were experts in their field and actively researching their topic, so were able to provide an up-to-date review of key aspects of a particular subject, whilst providing some historical perspective. In these two editions, we believe our authors achieved this admirably.

In the intervening years, materials chemistry has grown hugely and now finds itself central to many of the major challenges that face global society. We felt, therefore, that there was a need for more extensive coverage of the area and so Richard Walton joined the team and, with Wiley, we set about a new and larger project. *The Inorganic Materials Series* is the result and our aim is to provide chapters with a similar pedagogical flavour but now with much wider subject coverage. As such, the work will be contained in several themed volumes. Many of the early volumes concentrate on materials derived from continuous inorganic solids, but later volumes will also emphasise molecular and soft matter systems as we aim for a much more comprehensive coverage of the area than was possible with *Inorganic Materials*.

We approached a completely new set of authors for the new project with the same philiosophy in choosing actively researching experts, but also with the aim of providing an international perspective, so to reflect the diversity and interdisciplinarity of the now very broad area of inorganic materials chemistry. We are delighted with the calibre of authors who have agreed to write for us and we thank them all for their efforts and cooperation. We believe they have done a splendid job and that their work will make these volumes a valuable reference and teaching resource.

DWB, York
DO'H, Oxford
RIW, Warwick
January 2010

Preface

In an age of global industrialisation and population growth, and with concerns about current consumption of dwindling traditional fuels by a society that has high demands, the area of *energy* is one that is very much in the public consciousness. Fundamental scientific research is recognised as being crucial to delivering solutions to these issues, particularly to yield novel means of providing efficient, ideally recyclable, ways of converting, transporting and delivering energy. Although, the area of inorganic materials has long been associated with the topic of energy (consider the now ubiquitous rechargeable lithium batteries based on layered, transition-metal oxides), with the current challenges faced in energy, it is now particularly timely to publish a volume of reviews that considers some of the state-of-the-art materials that are being designed to meet some of the very specific challenges.

As with earlier volumes in this series, we approached authors who are at the forefront of research in their field. The topic of energy is tremendously broad and spans synthetic chemistry, solid-state physics and device fabrication, but we have chosen topics carefully that show how the skill of the synthetic chemist can be applied to allow the targeted preparation of inorganic materials with properties optimised for a specific application. We feel that the authors have risen to this challenge and in so doing have produced clearly written chapters that summarise the current status of research, but with an eye to how future research may develop materials' properties further. These chapters cover several important aspects of energy, from efficient conversion of natural resources (solar cells and solid-oxide fuel cells), through recyclability (electrolytes for batteries) to transport of fuels (hydrogen storage). We hope that this will give the reader a taste for the high level of activity and excitement in this topical field.

DWB, York
DO'H, Oxford
RIW, Warwick
December 2010

List of Contributors

Michel B. Armand LRCS, Université de Picardie Jules Verne, Amiens, France

Peter G. Bruce EaStCHEM, School of Chemistry, University of St Andrews, Fife, Scotland

Maria Forsyth Institute of Technology and Research Innovation (ITRI), Deakin University, Burwood, Victoria, Australia

Elizabeth A. Gibson School of Chemistry, University of Nottingham, University Park, Nottingham NG7 2RD, UK

Anders Hagfeldt Centre for Molecular Devices, Department for Physical and Analytical Chemistry, Uppsala University, Uppsala, Sweden

Miguel A. Laguna-Bercero Department of Materials, Imperial College London, London, UK

Bruno Scrosati Dipartimento di Chimica, Università di Roma La Sapienza, Italy

Stephen J. Skinner Department of Materials, Imperial College London, London, UK

K. Mark Thomas Sir Joseph Swan Institute for Energy Research and School of Chemical Engineering and Advanced Materials, Newcastle University, Newcastle upon Tyne, UK

Władysław Wieczorek Polymer Ionics Research Group, Faculty of Chemistry, Warsaw University of Technology, Warsaw, Poland

1

Polymer Electrolytes

Michel B. Armand[1], Peter G. Bruce[2], Maria Forsyth[3], Bruno Scrosati[4] and Władysław Wieczorek[5]

[1]LRCS, Université de Picardie Jules Verne, Amiens, France
[2]EaStCHEM, School of Chemistry, University of St Andrews, Fife, Scotland
[3]Institute of Technology and Research Innovation (ITRI), Deakin University, Burwood, Victoria, Australia
[4]Dipartimento di Chimica, Università di Roma La Sapienza, Italy
[5]Polymer Ionics Research Group, Faculty of Chemistry, Warsaw University of Technology, Warsaw, Poland

1.1 INTRODUCTION

1.1.1 Context

The discovery of polymer electrolytes (ionically conducting polymers) in the 1970s by Peter Wright and Michel Armand introduced the first new class of solid ionic conductors since the phenomenon of ionic conductivity in the solid state was first identified by Michael Faraday in the 1800s.[1–3] Faraday's materials were solids such as the F^- ionic conductor PbF_2. Polymer electrolytes are distinguished from such materials in that they combine ionic conductivity in the solid state with mechanical flexibility, making them ideal replacements for liquid electrolytes in electrochemical cells because of their ability to form good interfaces with solid electrodes. All solid state electrochemical devices, such as lithium batteries, electrochromic displays and smart windows are much sought after.[4,5] Although the major focus of attention remains on Li^+ conducting polymer electrolytes, because of their potential applications, salts of

Energy Materials   Edited by Duncan W. Bruce, Dermot O'Hare and Richard I. Walton
© 2011 John Wiley & Sons, Ltd.

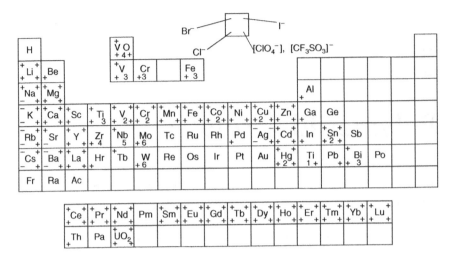

Figure 1.1 Complex formation between poly(ethylene oxide) (PEO) and various metal salts. +, Complex formed; −, no evidence of complex. Reprinted from *High Conductivity Solid Ionic Conductors. Recent Trends and Applications*, T. Takahashi (Ed.), World Scientific, Singapore, 1989, p. 117. With permission from World Scientific

almost every element in the periodic table have been incorporated into polymers to form electrolytes (Figure 1.1).

Today, the field embraces high molecular weight amorphous polymers, gels, hybrid composite materials and crystalline polymers. The work carried out over the last thirty years is too extensive to be described in detail within the constraints of space available here. Instead we shall begin by summarising the key developments in the early years, then focusing, for the rest of the chapter, on only three key areas of recent development, nanofillers, ionic liquids and crystalline polymer electrolytes. The choice of topics reflects the expertise of the authors and the desire to concentrate on a few areas rather than all in a very superficial manner. As a result we have not been able to include recent results on new amorphous polymer and gels, or the elegant work on polymer electrolytes as solid solvents and in medical applications.[6–8]

1.1.2 Polymer Electrolytes – The Early Years

The earliest polymer electrolytes, which remain one of the most important classes of polymer electrolytes to this day, consist of a salt dissolved in a high molecular weight polymer. The latter must contain donor atoms capable of acting as ligands coordinating the cations of the salt and hence providing the key solvation enthalpy to promote formation of the polymer electrolyte.[9,10] In a classic example of $LiCF_3SO_3$ in poly(ethylene oxide) (PEO) the polymer wraps around the cation in a fashion that is reminiscent of crown ether or cryptand based coordination compounds, so familiar in molecular inorganic chemistry (Figure 1.2).[11,12] The anion is invariably singly charged and often polyatomic, and is barely solvated. Although strong cation solvation is important for promoting complex formation in polymer electrolytes, if it is too strong it inhibits ion transport which, unlike motion in liquid electrolytes cannot occur by the transport of an ion along with its solvation sheath. In polymer electrolytes the cation must dissociate, at least in part, from its coordination site in order to move. Therefore, the cation–polymer interaction must be sufficiently strong to promote dissolution but not so strong as to inhibit ion exchange. If the interaction is too strong to permit cation transport,

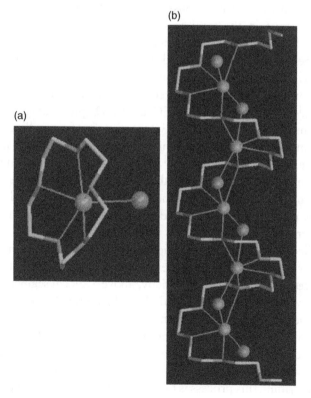

Figure 1.2 Comparison between the 15-crown-5:NaI complex (a) and the corresponding poly(ethylene oxide)$_3$:NaI solid state complex (b)

the resulting material will be an anion conductor. The cation–polymer interactions may be classified according to the hard-soft acid-base theory of Pearson, where polymers such as the ubiquitous PEO [$(CH_2\text{-}CH_2\text{-}O)_n$] which contains ether oxygens, a hard base, will complex strongly to hard cations such as Mg^{2+}, hence PEO:Mg(ClO$_4$)$_2$ exhibits immobile cations, whereas soft bases such as $(CH_2CH_2\text{-}S)_n$ will complex strongly soft cations such as Ag^+.[12] Although the mobility can be related to the hard-soft acid-base principle it has also be correlated to the Eigen values for the kinetics of exchange of a ligand such as H_2O, where fast H_2O exchange accords with mobility and slow exchange with cation immobilisation.[13] More extensive discussion of the thermodynamics of complex formation and the mobility/immobility of the cations is given in the literature.[9,10,14]

Although the above considerations help us to understand whether cations are likely to be mobile or not in the polymer host they do not of course provide a model for ion transport of either the cations or the anions. The earliest theories considered that the ordered helical polymer chains that exist in the semi-crystalline PEO were important for ion transport. However, this was quickly dispelled by elegant solid state NMR experiments on LiCF$_3$SO$_3$ dissolved in PEO.[15] A range of compositions were studied including the crystalline complex which exists at a composition corresponding to three ether oxygens per lithium, PEO$_3$:LiCF$_3$SO$_3$.[16] According to the PEO-LiCF$_3$SO$_3$ phase diagram, at compositions more dilute in salt than 3:1, a mixture of crystalline PEO plus the 3:1 complex or, at sufficiently high temperatures a viscous liquid plus 3:1 complex exists (Figure 1.3).[17] The authors noted that ion transport occurred in the amorphous state above the glass transition temperature T_g. Since that time until recently such thinking has dominated the synthesis of new polymer electrolytes and the understanding of ion transport in these materials.

A detailed discussion on the various theories of ion transport in amorphous polymers above T_g and the equations that describe it are given in references [9,15] and [18–24]. Briefly, above T_g, local segmental motion of the polymer chains occurs and these facilitate the motion of ions through the polymer, in a fashion that is somewhat analogous to the transport of gases through amorphous polymers above T_g. The polymer chains are constantly creating suitable coordination sites adjacent to the ions, into which the ions can then hop. One of the most sophisticated models describing this process is known as the dynamic bond percolation model.[20] The term 'bond' here should not be confused with a chemical bond but refers to the dynamic generation of suitable coordination sites

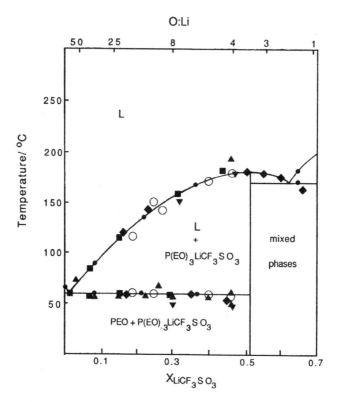

Figure 1.3 Phase diagram of the PEO-LiCF$_3$SO$_3$. X, Salt mole fraction; L, liquid phase. Reprinted with permission from F. M. Gray, *Solid Polymer Electrolytes: Fundamentals and Technological Applications*, VCH, New York, Weinheim, Cambridge, 1991, p. 77. Copyright (1991) Wiley-VCH Verlag GmbH & Co. KGaA

which provide a 'bond' or temporary bridge for the ion to hop from one site to the next. The temperature dependence of the conductivity is given by the Vogel–Tamman–Fulcher (VTF) equation $\sigma = \sigma_0 \cdot \exp[-B/(T-T_0)]$,[25] where σ_0 is the pre-exponential factor, B should not be confused with an activation energy in the Arrhenius expression and T_0 is related to the so-called thermodynamic T_g. Plots of $\log\sigma$ *vs* $1/T$ are curved because of the reduced temperature $(T-T_0)$.

The model implies that the highest conductivities will be obtained in amorphous polymers with the lowest T_g, resulting in the highest local segmental motion and therefore high diffusivity of the ions. By moving away from semi-crystalline polymers, such as PEO, to amorphous materials with low T_g, higher conductivities were indeed obtained, especially at room temperature. More recently it has been demonstrated that the ionic conductivity in amorphous, but not percolating at room temperature, regions of a semi-crystalline electrolyte can be four orders of magnitude higher.[26] Amorphous polymer electrolytes were obtained by a variety of elegant approaches including the formation of random and block copolymers, comb-branched polymers, cross-linked networks, *etc.* In most cases the CH_2CH_2-O repeat unit was retained because the C-C-O repeat provides an excellent ligand for cations, as it does in the crown ethers. Some of the main polymer architectures and examples of particular materials are given in Figure 1.4.[27–29] The anions of the salts were also designed to promote amorphicity and help plasticise high segmental motion. Anion design reached its zenith with $LiN(CF_3SO_2)_2$, which has a very low lattice energy, thus promoting dissolution in the polymer, as well as an anion architecture that promotes amorphicity and plasticises the polymer. In fact, such a salt is so effective that it can be combined even with pure PEO to produce, at certain compositions, an amorphous polymer electrolyte with a conductivity higher than 10^{-5} S cm^{-1} at 25 °C.[30] The conductivities of several polymer electrolytes, illustrating progress over the last thirty years are given in Figure 1.5.

Although ethylene oxide based polymers prove to be excellent ligands from the viewpoint of complexing a variety of cations they do suffer from

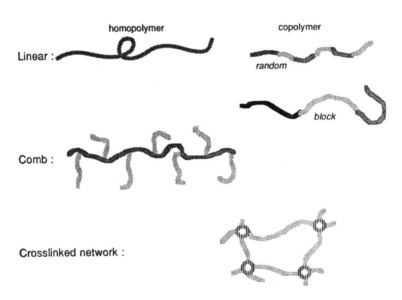

Figure 1.4   Examples of polymer architectures. Reprinted from *High Conductivity Solid Ionic Conductors. Recent Trends and Applications*, T. Takahashi (Ed.), World Scientific, Singapore, 1989, p. 125. With permission from World Scientific

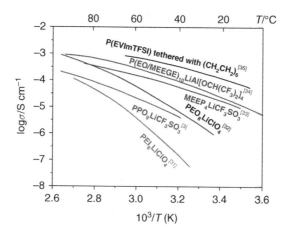

Figure 1.5 Temperature variation of the conductivity for selected amorphous polymer electrolytes. Superscript numbers are the literature references

one significant disadvantage, a low dielectric constant ($\varepsilon_r = 5 - 7$). The conductivity of a polymer electrolyte is given by the equation:

$$\sigma = nq\mu \tag{1.1}$$

where σ is the conductivity, n the concentration of charge carriers, q their charge and μ their mobility. Whereas low T_g amorphous polymers maximise the ionic mobility, μ, the low dielectric constant of ethylene oxide based polymers leads to strong ion–ion interactions and the formation of ion pairs and higher aggregates.[36–38] The formation of ion pairs reduces the concentration of charge carriers, n, and hence conductivity, σ, compared with a fully dissociated salt. The strength of association depends on the dielectric constant but it is also influenced by the charge density such that NMR studies of PEO:LiN(CF$_3$SO$_2$)$_2$ electrolytes indicate that the anion charge is sufficiently delocalised to mitigate the strength of the ion–ion interaction leading to an almost fully dissociated salt, even at relatively high salt concentrations. To further address the problem of ion association high dielectric constant polymers have been investigated such as polycarbonates. Unfortunately, they exhibit relatively high T_gs and therefore the higher dissociation of salt is purchased at the cost of lower mobility. At the other end of the concentration ratios, once the polymer:salt ratio exceeds a certain threshold and a polymer-in-salt compound is formed, which is a rubbery version of a glassy electrolyte, ionic conductivity shows a significant increase.[39,40]

1.2   NANOCOMPOSITE POLYMER ELECTROLYTES

Polymer electrolytes are a class of materials which play a key role in modern energy technology. In particular, they are presently widely studied for the development of high energy density batteries, with special interest in lithium metal and lithium ion batteries.

Conventional lithium batteries use liquid electrolytes.[41–43] An important step forward in this technology is the replacement of the latter with polymer electrolytes in order to achieve the production of advanced energy storage devices having a full plastic configuration. This is an important concept since it allows the combination of high energy and long life, *i.e.* the characteristics that are typical of liquid electrolyte cell configurations, with reliability, safety and easy manufacturing. These characteristics are typical of polymer-based, all-plastic structures. The practical development of this concept, however, requires the availability of polymer electrolytes having transport and interfacial properties approaching those of the conventional liquid solutions.

Classical examples of lithium polymer electrolytes are the previously discussed blends of a lithium salt, LiX, where X is preferably a large soft anion, *e.g.* $[ClO_4]^-$ or $[N(CF_3SO_2)_2]^-$, and a high molecular weight polymer containing Li^+-coordinating groups, *e.g.* PEO.[9,44,45] As for all conductors, the conductivity of the PEO-LiX polymer electrolytes depends on the number of the ionic carriers and on their mobility. The number of the Li^+ carriers increases as the LiX concentration increases, but their mobility is greatly depressed by the progressive occurrence of ion–ion association phenomena.[46] Due to their particular structural position, the Li^+ ions can be released to transport the current only upon unfolding of the coordinating PEO chains. In other words, this type of polymer electrolyte requires local relaxation and segmental motion of the solvent (*i.e.* PEO) chains to allow fast Li^+ ion transport. Despite major improvements in the conductivity of high molecular weight polymer electrolytes, the conductivities at room temperature and below are often not sufficient for use in applications where relatively high rates are required, *e.g.* lithium batteries. The conductivities at around 80–100 °C do exhibit values of practical interest, *i.e.* of the order of 10^{-3} S cm^{-1}. This implies that the use of the PEO-LiX electrolytes is mainly restricted to batteries for which a relatively high temperature of operation does not represent a major problem, *e.g.* batteries designed for electric vehicles (EVs). Indeed, various R&D projects aimed at the production of polymer lithium batteries for EV application are in progress worldwide. These polymer batteries typically use a lithium metal anode and a Li-intercalation cathode, such as V_2O_5 or $LiFePO_4$.[45,47]

As a result of the above limitations, various approaches to raising the conductivity have been considered, such as the addition of plasticisers, e.g. organic liquids, propylene carbonate or ethylene carbonate or low molecular weight ethylene glycols.[48–52] However, the gain in conductivity is adversely associated with a loss of the mechanical properties and by a loss of the compatibility with the lithium electrode, both effects resulting in serious problems since they affect the battery cycle life and increase the safety hazard.

A promising approach to circumvent the issue of the temperature dependence of the conductivity, which still ensures efficient cyclability of the lithium anode and a high safety level, is the use of 'solid plasticisers'; solid additives which promote amorphicity at ambient temperature without affecting the mechanical and the interfacial properties of the electrolyte. Examples of such additives are ceramic powders, e.g. TiO_2, Al_2O_3 and SiO_2, composed of nanoscale particles.[53–57]

The preparation of these 'nanocomposite' polymer electrolytes involves first the dispersion of the selected ceramic powder (e.g., TiO_2, SiO_2 or Al_2O_3) and of the lithium salt (e.g. $LiClO_4$ or $LiCF_3SO_3$) in a low boiling solvent, e.g. acetonitrile, followed by the addition of the PEO polymer component and thorough mixing of the resulting slurry. The slurry is then cast yielding homogenous and mechanically stable membranes.[58] Figure 1.6 illustrates the typical appearance of these ceramic-containing composite membranes.

The general concept of adding ceramic powders to PEO-LiX polymer electrolytes dates back to the early 1980s when this procedure was successfully employed to improve their mechanical properties,[59] their interface with the lithium electrode[60–65] and their ionic conductivity.[66,67] However, it is only recently that the role of the dispersed ceramics in influencing the transport and interfacial properties of the PEO-LiX polymer electrolytes has been clearly understood and demonstrated.[53]

Figure 1.7 shows the conductivity Arrhenius plot of PEO_8LiClO_4 and the same polymer with nanoparticles of TiO_2 or Al_2O_3 added. $PEO_8 LiClO_4$ without nanoparticles exhibits a break around 70 °C, reflecting the melting of crystalline PEO (the 8:1 composition is a mixture of PEO and a crystalline PEO:salt complex) to the amorphous state. When the temperature is reduced below 70 °C the conductivity decays to low values.

When ceramic nanoparticles are added, the conductivity is almost one order of magnitude higher over the entire temperature range; the break is

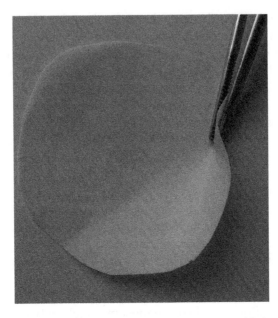

Figure 1.6   Typical appearance of a PEO-based, ceramic-added, nanocomposite membrane

much less prominent and occurs at a lower temperature. This difference is also seen in other nanocomposite polymer electrolytes using different types of ceramic fillers.[68–73] Stress–strain measurements have revealed a large enhancement of the Young's modulus and of the yield point stress when passing from ceramic-free to nanocomposite polymer electrolyte samples, thus demonstrating that the higher conductivity of the latter is not due to polymer degradation but is in fact accompanied by a substantial increase in the electrolyte's mechanical properties.[58] It is then reasonable to conclude that the favourable transport behaviour is an inherent feature of the nano-composite materials.

Various models have been proposed to account for the effects of the ceramic fillers. One common model assumes that the conductivity enhance-ment is due to the promotion of a large degree of amorphicity in the polymer. Accordingly, once the electrolytes are annealed at temperatures higher than the PEO melting temperature (*i.e.* above 70 °C), the ceramic additive, due to its large surface area, prevents local PEO chain reorganisation and hence crystallisation so that a high degree of amorphicity is preserved to ambient temperatures, consistent with enhancement of the ionic conductivity.

However, one may observe from Figure 1.7 that the conductivity enhancement in the composite polymer electrolytes occurs in the entire

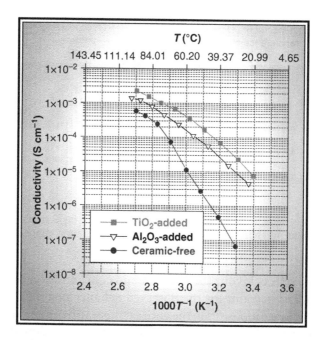

Figure 1.7 Conductivity Arrhenius plots of composite, PEO-based, polymer electrolytes. Also the plot of a ceramic-free sample is reported for comparison

temperature range, *i.e.* not only below but also *above* 70 °C where the PEO is amorphous in any case. Therefore, the model must be extended to assume that the role of the ceramic cannot be limited to the sole action of preventing crystallisation of the polymer chain, but must also favour the occurrence of specific interactions between the surface groups of the ceramic particles and both the PEO segments and the lithium salt.

One can anticipate that the Lewis acid groups on the surface of the ceramics (*e.g.* the –OH groups on the SiO_2 surface) may compete with the Lewis-acid lithium cations for the attentions of the PEO chains, as well as with the anions of the LiX salt (see Figure 1.8).

More specifically, the following may occur:

(i) The nanoparticles may act as cross-linking centres for segments of the PEO chains, which not only inhibits crystallisation but also destabilises the coordination around the cations easing migration of Li^+ ions from site to site in the vicinity of the fillers.

(ii) Lewis acid–base interactions between the ceramic surface and the anions will compete with interactions between the cations and anions promoting salt dissociation *via* a sort of 'ion-ceramic complex' formation.

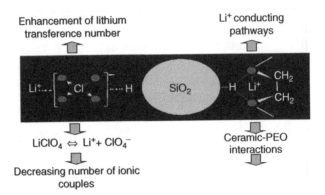

Figure 1.8 Schematic model of the surface interactions of the ceramic particle with the polymer chain and with the salt anion in PEO-LiClO$_4$ nanocomposite electrolytes

These two effects favour the mobility and concentration of 'free' ions and may indeed account for the observed enhancement of the conductivity of the nanocomposites in the entire temperature range. This model has been confirmed by a series of measurements; including determination of the conductivity and of the lithium ion transference number T_+ of various composite electrolyte samples differing in the type and the nature of the ceramic filler.[74]

Finally, the model has been confirmed by spectroscopic analysis. Raman results reported by Best *et al.* have demonstrated the specific interaction between nanometric TiO$_2$ powders and the salt.[56] In addition, NMR data have shown that the diffusion of Li$^+$ ions, and thus the related T_+ value, in the nanocomposite electrolytes, is considerably higher than that of the parent ceramic-free electrolytes.[75]

On the basis of the model described above, one would expect that the enhancement of the transport properties should depend upon the degree of acidity of the ceramic's surface states. This is indeed the case as demonstrated by the behaviour of PEO-based polymer electrolytes using ceramic fillers with a high surface acidity, *e.g.* the sulfate-promoted superacid zirconia, S-ZrO$_2$. The results show that this ceramic filler considerably enhances the transport properties of the electrolyte.[76,77]

The S-ZrO$_2$-composite polymer electrolytes are particularly suited as separators in advanced rechargeable lithium batteries. Recent results reporting the performance of battery prototypes using this type of polymer electrolyte confirms expectations demonstrating long cycle life and high safety.[76]

1.3   IONIC LIQUID BASED POLYMER ELECTROLYTES

An alternative method of achieving high ionic conductivity whilst retaining the useful properties of polymer electrolytes in device applications such as lithium batteries, fuel cells, actuators and dye sensitised solar cells, is to use ionic liquids (ILs) either as the main conductivity medium supported in a polymer membrane or as a plasticising component in polymer electrolytes. In this section we provide some background of ILs and their properties in the context of device applications followed by their use in polymer electrolyte systems.

1.3.1   Ionic Liquid Properties

ILs are fluids composed solely of ions and, and by consensus, have a melting point below 100 °C. ILs are the focus of interest of a growing community for their unique properties such as:[78,79]

- negligible vapour pressure in many cases;
- high conductivity up to 20 mS cm^{-1} at room temperature;
- non-flammability;
- high thermal stability;
- exceptional solvent behaviour;
- exceptional electrochemical behaviour with large electrochemical windows.

The lack of flammability and the high electrochemical stability are key attractions for battery electrochemists, solving the most urgent problems of a battery electrolyte in terms of safety and enabling the use of high voltage electrode materials (*e.g.* LiMn$_{1.5}$Ni$_{0.5}$O$_2$ at 4.5 V).

One of the largest family of ILs, now with innumerable representatives, stems from the early study of acidic chloroaluminates ([AlCl$_4$]$^-$, [Al$_2$Cl$_7$]$^-$) of the delocalised cations based on imidazolium derivatives:

Numerous ILs based on substituted imidazolium cations and classic anions such as [BF$_4$]$^-$, [CF$_3$SO$_3$]$^-$, [PF$_6$]$^-$ have been developed since the early

1990s.[80,81] The reduction potential of the imidazolium cation, however, limits the use this family of ILs in some electrochemical applications such as lithium batteries.

The acidity of the C2 proton (indicated in the scheme above) is estimated as $pK_a = 24$, and this corresponds to a reduction potential of 1.5 V vs $Li^+:Li^0$. The methylation of the C2 proton increases the reduction potential by ≈ 300 mV; however, this is still not sufficiently negative for lithium battery applications and thus such ILs require the use of additives such as vinylene carbonate (VC), which form a stable solid-electrolyte interphase (SEI) layer, in order to be viable.[82–84]

The introduction of the $[N(CF_3SO_2)]^-$ anion has considerably broadened the scope of ILs, as the delocalisation of charge within this anion leads to ILs with very low freezing point.[85–87] Similarly, the combination of delocalisation and the presence of N or C centres have even further broadened the scope and temperature of operation of ILs (see below). These anions tend to give lower melting points and/or viscosities as compared with conventional $[PF_6]^-$ or $[BF_4]^-$.

The bis(fluorosulfonyl) imide (FSI) anion is considered very promising for use in lithium battery technology in terms of conductivity and only a small viscosity increase with Li salt addition. In addition, FSI has been shown to provide the *ad hoc* SEI layer that allows dendrite-free cycling of lithium metal at the point of reconsidering this anode as a viable alternative for high energy density batteries.[88]

TFSI   FSI   DCI

TCM   TCB   DCTA

ILs are not limited to those based on imidazolium cation derivatives. In particular, quaternary ammonium salts and phosphonium salts, which are appreciably more resistant to reduction as compared with azoles, have considerably extended the variety of ILs.[89,90]

n = 0: MPPy; n = 1: BMPy   DEMME or 1,2,2,2O1

An important concept in the field of ILs is just how 'ionic' is the liquid? MacFarlane et al.,[90] Ueno et al.[91] and Schrödle et al.[92] in particular have used the concept of 'ionicity' to define the IL.

By using the Nernst–Einstein equation and the measured diffusion coefficients, a molar conductivity can be calculated for a given electrolyte material (Λ_{NMR}):

$$\Lambda_{NMR} = \frac{N_A e^2}{kT} (D_+ + D_-) \tag{1.2}$$

where N_A is the Avogadro constant, k is the Boltzmann constant, T is temperature, e is electron charge and D_+ and D_- are the diffusion coefficients of cations and anions, respectively.

The ratio of the measured molar conductivity (determined from the measured σ and the density of the IL) to the calculated molar conductivity quantifies the discrepancy between mass and charge transport and is often referred to as ionicity ($\Lambda_{meas}/\Lambda_{NMR}$). An ionicity value of unity implies that all ions are moving independently of one another and all contribute individually to the conductivity. An ionicity value less than unity suggests that some fraction of ions are in fact not contributing to the conductivity and has been interpreted as increased ion pairing or aggregation. One might expect that an IL being composed entirely of ions might have a high dielectric constant but in fact it has been shown that ε is generally between 9 and 15,[92,93] which is not too different to traditional polymer

electrolytes. The latter are also known to have significant ion aggregation, as discussed above, which leads to deviations of the conductivity
from that predicted by the Nernst–Einstein calculation for a given salt
concentration.[94,95] The Walden plot has also been used to describe how
'good' an IL is, in terms of the degree of ion association.[90,96]

In many practical applications of these materials, the target ions [e.g. Li^+
for lithium batteries, $I^-/[I_3]^-$ for dye-sensitised solar cells (DSSCs), H^+ for
fuel cells] need to be added to the IL. The issue of the appropriate choice of
lithium salt in the case of lithium batteries has been addressed in a recently
published review.[97] Generally the addition of lithium salt results in a
considerable decrease in specific ionic conductivity of an IL-based system
from 1–20 mS cm^{-1} for pure IL-based electrolytes to less than 1 mS cm^{-1}
for IL–lithium salt systems. This is due to the rise in electrolyte viscosity
resulting from strong Coulombic interaction between ionic components.
More critically, the lithium ion has been shown to strongly associate with
the anion, for example $[N(CF_3SO_2)]^-$. This leads to Li-X pairs that do not
contribute to the ion conductivity or $[LiX_2]^-$ or $[Li_2X]^+$ which would also
restrict Li^+ transport. Indeed, diffusion measurements of all components in
such IL electrolyte mixtures show that the Li^+ ion has the lowest diffusion
coefficient which is in contrast to its small size![98,99] The need to enhance
the lithium ion transport number has led to the use of low molecular weight
diluents[99] or even zwitterionic additives[100–102] to assist in reducing the
ion association between the lithium and the IL anion. Zwitterions are
another group of the IL family that have been extensively studied by
Yoshizawa et al.[103] as an alternative to traditional ILs where the ions
also contribute to conduction. In the case of zwitterions, the excellent
properties of low flammability, increased stability, etc., are retained whilst
the conductivity can be predominantly due to the addition of the desirables
species such as H^+ or Li^+.[103,104]

There are several reviews on IL chemistries and the use of ILs in various
electrochemical applications, as discussed above; indeed an entire issue of
the journal Physical Chemistry, Chemical Physics is devoted to the topic of
ILs (Volume 12, Issue 8, 2010). Therefore, the remainder of this section will
focus on ILs in polymer electrolyte applications.

1.3.2   Ion Gels

Polymer electrolytes incorporating an IL can be classified into two main
groups: (i) ion gels whereby the IL is the main conducting medium and the
polymer is the support; and (ii) polyelectrolytes prepared via the

polymerisation of an IL. The first class of materials has been pioneered by the groups of Watanabe[105,106] and Forsyth and MacFarlane[107–110] and have been termed either 'polymer in ionic liquid electrolytes (PILS)' or 'ion gels'. In these cases the materials can be prepared either by swelling a relatively inert polymer such as poly(1-vinyl pyrrolidone) (PVP), poly(N-dimethyl acrylamide) (PDMAA) and poly(1-vinyl pyrrolidone-co-vinyl acetate) [P(VP-co-VA)] with a variety of ILs or by in-situ polymerisation of the polymer such as poly(methyl methacrylate) in the IL. The fact that ILs readily support such free radical polymerisation opens great opportunities for material development in this field. For example Winther-Jensen et al. have recently reported a ion gel electrolyte with only 5% poly(hydroxy ethyl methacrylate) (PHEMA) and a eutectic mixture of phosphonium tosyslate ILs which greatly improves the performance of an electrochromic device.[109] Similarly, photo-initiated in-situ polymerised ion gels for solar cell applications may offer opportunities for simple assembly of DSSC devices.[110] In this latter case the PHEMA polymer component (>30%) was polymerised in an iodide/SCN IL mixture. A recent report of photopolymerised ternary systems consisting of dianol diacrylate resin, tetramethyl sulfone (TMS) as compatibilising solvent and an imidazolium cation based IL reports that free standing thin film foils could be obtained which surprisingly display ionic conductivities higher than liquid IL/TMS mixture.[89] Electrochemical stability of these films was also reported to be between 3.3 and 3.7 V against Ag/Ag^+.

Shobukawa et al. have also demonstrated the in-situ radical polymerisation of a copolymer of ethylene oxide and propylene oxide P(EO-co-PO) triacrylate in the presence of a lithium ion based IL.[105] In this case the IL is based on a large, diffuse trifluoroacetyl substituted borate anion coupled with a lithium cation. Ion gels prepared in this case displayed room temperature conductivities of the order of 10^{-4} S cm^{-1} and demonstrated efficient plating and stripping of lithium metal, thus being of interest in lithium battery applications. In the case of proton conducting ILs with potential application for PEM fuel cells, several groups have[106,111,112] reported the development of IL electrolytes, and ion gels based on these. For example, materials based on substituted imidazolium cations and either multivalent anion such as $[HSO_4]^-$ and $[H_2PO_4]^-$ or the ubiquitous $[N(CF_3SO_2)]^-$ anion have been shown to have sufficient conductivity and proton activity for such application. These protic ILs have been incorporated for example into cellulose[111] or poly(vinylidene)-co-hexafluoropropylene (PVdF-HFP)[112] substrates.

With respect to the conduction processes in this class of polymer electrolyte, of course some loss of conductivity is expected given that a rigid, high T_g polymer is combined with a fluid liquid component. Nonetheless these materials are shown to retain the high ionic conductivities of the IL component even up to 40% polymer. Furthermore, unlike the traditional PEO-based systems, where conductivity is practically suppressed at T_g, the ion transport in these ion gels is decoupled from the bulk polymer motions and hence conductivities as high as 10^{-7} S cm^{-1} are observed at T_g. This is more akin to the polymer-in-salt systems described by Xu et al.[113] and Forsyth et al.[114] when the fraction of lithium salt component in the polymer electrolyte is significantly greater than the polymer.

Finally, ILs can be used as simple, nonvolatile plasticisers as in the case of IL plasticised PEO-based polymeric electrolytes.[115–121] Studies on the effect of 1-ethyl-3-methylimidazolium tetrafluoraoborate (EMIBF$_4$) on the properties of PEO-EC-LiBF$_4$ electrolyte were reported,[115] showing a decrease in the degree of crystallinity of the polymeric component and a subsequent fivefold increase in conductivity for a material containing 0.2 mol of this IL. The increase in conductivity was combined with a substantial rise in lithium transference number (up to 0.65 for the sample with the highest conductivity) and extension of the electrolyte stability window from 3.7 V vs Li/Li$^+$ for IL-free sample to 4.5 V vs Li/Li$^+$ for the electrolyte with the highest conductivity at room temperature.

1.3.3 Polymer Electrolytes Based on Polymerisable Ionic Liquids

The idea of the preparation of IL polymers was originally introduced by Ohno[122] and is illustrated schematically in Figure 1.9. In this approach IL monomers bearing double bonds in the chain were prepared and subsequently polymerised via radical polymerisation. The obtained IL polymers exhibit however much lower conductivities (10^{-5} S cm^{-1}) compared with the monomers used in the polymerisation reaction (10^{-2} S cm^{-1}). The drop in conductivity was considerably reduced by tethering IL polymer chains with oligo (ethylene oxide), -(CH$_2$CH$_2$O)$_8$-, (PEO)$_8$ (see Figure 1.10). The modification of the chemical structure of the IL, i.e. by tethering the imidazlium functionality with (PEO)$_8$, resulted in the increase in ionic conductivity by almost two orders of magnitude. The presented procedure leads to the preparation of single ion conducting soft polymer electrolytes with conductivities promising for

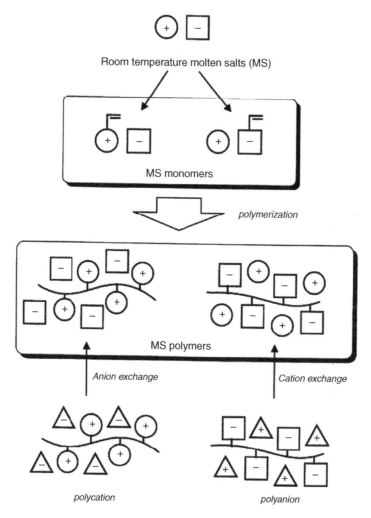

Figure 1.9 Scheme of preparation of molten salt polymer electrolytes. Reprinted from Ohno, 2001[122] with permission from Elsevier (license number 2404140702181)

application of these systems in electrochemical devices. Some of the monomers considered by Ohno et al. are given in Figure 1.11. Further exciting work undertaken by Ohno et al.,[123] and also explored by Batra et al.,[124] involves the formation of oriented ion conducting channels through the use of photopolymerisation of imidazolium functionalised liquid crystals. It was found that the conductivity depended on the degree of order and the relative alignment of the channels and this provides a unique method to tailor ion conducting materials on surfaces. An excellent review of polymerisable ILs based on imidazolium cations is found

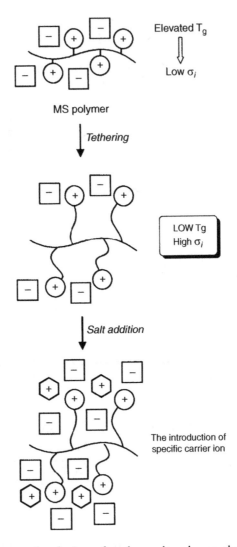

Figure 1.10 Scheme of tethering of molten salt polymer electrolytes with low molecular weight polyglycols. Reprinted from Ohno, 2001[122] with permission from Elsevier (license number 2404140702181)

in a recent publication by Green and Long[125] which discusses, amongst other things, the design of IL monomers for emerging technologies. In this review, they indicate that such imidazolium IL-based polymer electrolytes are promising candidates for ionomeric polymer transducers (actuators) and solar cells.

In conclusion, polymer electrolytes incorporating ILs have been shown to be promising materials in energy applications. The new chemistries afforded

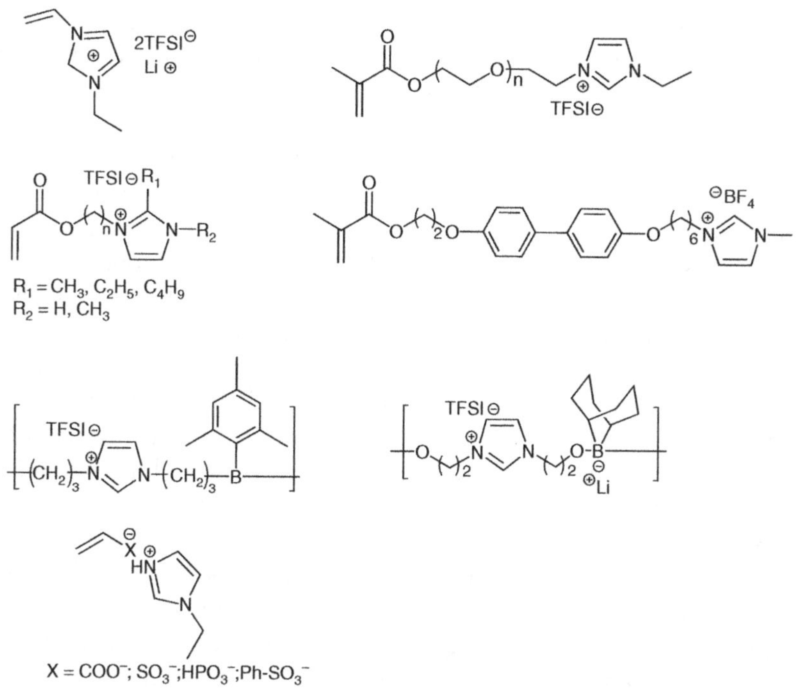

Figure 1.11 Monomer structures for several polymerisable ILs developed by Ohno *et al.* Reprinted with permission from Green and Long, 2009[125]. Copyright (2009) Taylor and Francis

by the enormous variety of ILs and subsequent ion-gels and the families of electrochemically, chemically and thermally stable materials will open up new possibilities of IL-based polymer electrolytes in devices from advanced batteries to solar cells, actuators and fuel cells. The transport processes in these polymer electrolytes tend to be decoupled from the polymer, in contrast to traditional polymer electrolytes, although ion aggregation is still a limiting factor which has to be addressed in order to maximise the performance in these materials.

1.4   CRYSTALLINE POLYMER ELECTROLYTES

For some 25 years since their discovery, the term polymer electrolytes was synonymous with amorphous polymers above T_g because only such materials were considered to support ionic conductivity. This belief has its origin in the elegant NMR experiments carried out on

PEO:LiCF$_3$SO$_3$ where, as described above, compositions giving rise to amorphous phases were observed to support higher conductivity than the crystalline 3:1 complex, PEO$_3$:LiCF$_3$SO$_3$.[15] However, this result only demonstrates that PEO$_3$:LiF$_3$SO$_3$ is a poor conductor not that all crystalline polymer electrolytes are necessarily insulators. The situation is analogous to believing that because NaCl is a relatively poor ionic conductor then all sodium compounds will demonstrate low conductivity, which is not of course the case, for example Naβ–Al$_2$O$_3$ exhibits a conductivity of 1.4×10^{-2} S cm^{-1} at 25 °C.[126] Indeed the solid that exhibits the highest Li$^+$ conductivity at room temperature is Li$_3$N; its conductivity is around one order of magnitude higher than the best amorphous polymer electrolyte (1.2×10^{-3} S cm^{-1} at room temperature).[127] The material is, of course, a ceramic so it lacks the desirable mechanical properties of a polymer but even more significantly it is highly crystalline, reminding us that ion transport can be high in crystalline (organised) structures. The study of amorphous polymer electrolytes continues to be a major topic embracing gels and inorganic fillers, as described above. However, given the observations of high conductivity in ceramic materials and the recognition that within biology ion transport occurs ubiquitously in organised membranes, why should crystalline polymer electrolytes inevitably be insulators?

Investigation of ionic conductivity in crystalline soft solids has followed two paths; crystalline polymer salt complexes and plastic crystals. Although the latter are not strictly polymer electrolytes they have similar mechanical properties and help to provide a more complete view of ion transport in non-amorphous soft solids.

1.4.1 Crystalline Polymer: Salt Complexes

A lack of knowledge concerning the structures of crystalline polymer electrolytes represented a major hurdle to their investigation as ionic conductors; analogous structure is the foundation on which much of our understanding of materials in general has been based. Very few polymer electrolytes were susceptible to fibre diffraction (single crystal studies). The development of powerful methods by which complete crystal structures could be determined *ab initio* from powder diffraction data unlocked the door to the structure of crystalline polymer electrolytes.[128,129] Amongst a number of crystal structures that were solved, the structure of PEO$_6$:LiXF$_6$, X=P, As, Sb was shown to possess tunnels formed by

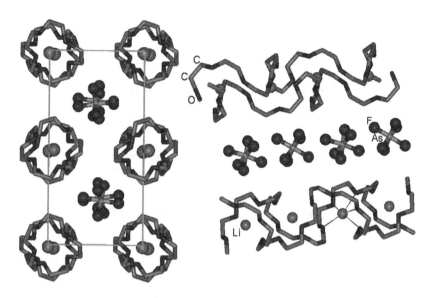

Figure 1.12 The structure of PEO$_6$:LiAsF$_6$. (left) View of the structure along *a* showing rows of Li$^+$ ions perpendicular to the page. (right) View of the structure showing the relative position of the chains and their conformation (hydrogens not shown). Thin lines indicate coordination around the Li$^+$ cation

pairs of PEO chains within which Li$^+$ ions are located and along which they could migrate (Figure 1.12).[130] The temperature dependent conductivity of the 6:1 crystal structure is shown in Figure 1.12.[131,132] All three 6:1 compounds had similar crystal structures and all conduct.[133]

The conductivities in Figure 1.13 are too low to be of technological interest; their significance is that they demonstrate ion transport can occur in crystalline polymers (*i.e.* not enabled by segmental motion), they open up a new direction in which to explore ion transport in polymers and represent a new class of solid electrolytes distinct from amorphous polymers in that ion transport occurs in a rigid 'environment' not enabled by segmental motion, and distinct from crystalline ceramics in that they are soft solids. They also offer a completely different set of approaches to raising conductivity, that are quite distinct from amorphous polymers above T_g. The fundamentally different nature of ion transport in crystalline and amorphous polymers is highlighted by comparing the temperature dependent conductivities in Figures 1.13 and 1.5; in the former the Arrhenius plot is linear due to hopping in a rigid or semi-rigid environment whereas the latter is curved, typical of transport enabled by segmental motion. Concerning the mechanism of ion

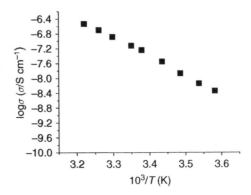

Figure 1.13 Ionic conductivity σ (S cm^{-1}) of crystalline complexes PEO$_6$:LiAsF$_6$

transport in crystalline polymers, it should be noted that low molecular weights (short chain lengths in the case of PEO typically 22 ethylene oxide units, 1000 Da) are employed to maximise crystallinity. Such a chain length is significantly smaller than the average particle size so that, unlike the conventional picture of a semi-crystalline polymer where chains meander from crystalline to amorphous regions, in crystalline polymer electrolytes there are many chains and therefore many chain ends within a crystallite. The chain ends cannot wrap around the cations in the same fashion as is demonstrated by the crystal structure in Figure 1.12, therefore such chain ends represent natural sources of point defects essential for ion transport by hopping in a rigid 'environment'.

The results of various approaches to modifying the crystalline 6:1 complexes in order to increase conductivity are presented in Figure 1.14. Reducing the chain length and thus increasing the density of chain ends increases the conductivity.[132] Replacing, in part, the polymer chains with low molecular weight molecules that are chemically identical such as CH$_3$-(CH$_2$CH$_2$O)$_4$-CH$_3$ also results in an increase in chain ends and hence conductivity.[134] Substituting polydispersed PEO with close to mono-dispersed material of the same chain length reduces the conductivity.[135] This may be explained by coalescence of the chain ends in the case of monodispersed material whereas polydispersed does not permit chain end alignment and therefore induces a higher density of chain ends and hence higher conductivity. The polymer chains are normally end-capped by CH$_3$ units; replacing these by larger end groups such as C$_2$H$_5$ increases conductivity, however further increasing the size of the chain ends, *e.g.* to C$_3$H$_7$ or beyond reduces the conductivity.[135] Clearly there is a balance to be obtained between increasing chain disruption leading to more

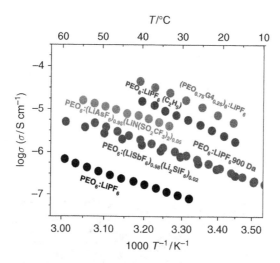

Figure 1.14 Temperature variation of the conductivity for selected crystalline polymer electrolytes. Unless stated otherwise, all complexes were prepared with 1000 Da PEO end-capped with $-CH_3$. G4 is $CH_3-(CH_2CH_2O)_4-CH_3$

defects and such disruption interrupting the migration pathways and destroying crystallinity. Replacing the $[XF_6]^-$ ions with bulkier ions of the same charge such $[N(CF_3SO_3)_2]^-$ or having ions of different charge results in an increase in the number of defects and hence conductivity, in the former case due to increased local strain and in the latter case due to the need to maintain electroneutrality.[136,137] Modelling studies, especially based on molecular dynamics, have been used to probe ion transport in the crystalline state.[138–143] These highlight the significance of the chain ends and doping in the conduction process.

The modifications described above have, in a few years, raised the conductivity of crystalline complexes by two orders of magnitude, to levels comparable with the best amorphous polymer electrolytes that have been reported over the last few decades. Further advances in crystalline polymer electrolytes will depend on a better understanding of the conduction mechanism, especially how the chain ends influence the creation and migration of defects. There is considerable scope for investigation of other dopants and further modification of the polymers that could lead to yet higher conductivities, especially recalling the high level of conductivity exhibited by lithium conducting crystalline ceramics. One approach is to increase the dimensionality of the ion transport above the one-dimensional conduction exhibited by the 6:1 complexes. Wright *et al.* achieved this by forming polymer electrolytes where insulating blocks were separated by planes containing PEO chains and a lithium salt. High levels of conductivity have been reported.[144] Introducing order into predominantly amorphous polymer electrolytes can also increase conductivity.[3] The ambient-temperature conductivity of a film of PEO$_{20}$:LiI increases fivefold on application of an uniaxial stress which aligns the crystalline regions within the polymer.[145]

Crystalline polymer electrolytes have been incorporated into lithium ion cells which are being shown to sustain cycling thus demonstrating potential for applications.[146]

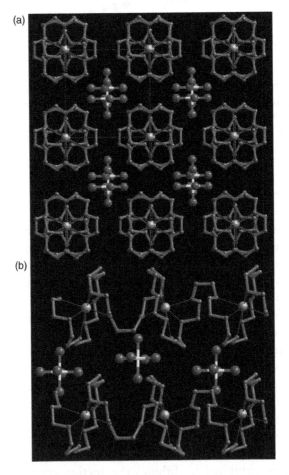

Figure 1.15 The structure of PEO$_8$:NaAsF$_6$. (a) View of complete structure (hydrogens are not shown). (b) Fragment of the structure showing neighbouring tunnels with dedicated anions. Thin dotted lines indicate coordination around the Na$^+$ cation

There is increasing interest in sodium-based batteries, because of the greater abundance and lower costs of sodium compared with lithium. Recently the first examples of crystalline polymer electrolytes that support Na^+ conduction have been reported.[147] Potassium and rubidium based crystalline polymer electrolytes have also been described.[147] This result is significant not only because these are the first Na^+, K^+, Rb^+, crystalline polymer electrolytes but also because they are the first example with a crystal structure that differs from the $PEO_6:LiXF_6$ complexes. In the former case, the ratio of ether oxygens to cations is 8:1 rather than 6:1 and only one polymer chain wraps around the cations (Figure 1.15).

REFERENCES

[1] M. Faraday, *Experimental Researches in Electricity*, Vol. 1, R. Taylor and W. Francis, London, 1849.

[2] D. E. Fenton, J. M. Parker and P. V. Wright, *Polymer*, 14, 589 (1973).

[3] M. B. Armand, J. M. Chabango and M. J. Duclot, in *Fast Ion Transport in Solids*, P. Vashishta, J. N. Mundy and G. K. Shenoy (Eds), North-Holland, Amsterdam, 1979, p. 131.

[4] J.-M. Tarascon and M.Armand, *Nature*, 414, 359 (2001).

[5] A. Barnes, A. Despotakis, T. C. P. Wong, A. P. Anderson, B. Chambers and P. V. Wright, *Smart Mater. Struct.*, 7, 752 (1998).

[6] H. Tokuda, S. Tabata, S. Seki and M. Watanabe, *Kobunshi Ronbunshu (Jap. J. Polym. Sci. Tech.)*, 63, 1 (2006).

[7] R. J. Latham, R. G. Linford and W. S. Schlindwein, *Ionics*, 9, 41 (2003).

[8] D. M. F. Santos and C. A. C. Sequeira (Eds), *Polymer Electrolytes: Fundamentals and Applications*, CRC Press, Taylor & Francis Group, London, 2010.

[9] F. M. Gray, *Polymer Electrolytes. RSC Materials Monographs*, The Royal Society of Chemistry, Cambridge, 1997.

[10] P. G. Bruce (Ed.), *Solid State Electrochemistry*, Cambridge University Press, Cambridge, 1995.

[11] J. M. Lehn, *Supramolecular Chemistry: Concepts and Perspectives*, VCH, Weinheim, 1995.

[12] P. G. Bruce, *Phil Trans. R. Soc. London. A*, 354, 415 (1996).

[13] M. B. Armand, *Solid State Ionics*, 69, 309 (1994).

[14] M. Armand and M. Gauthier, in *High Conductivity Solid Ionic Conductors. Recent Trends and Applications*, T. Takahashi (Ed.), World Scientific, Singapore, 1989, p. 114.

[15] C. Berthier, W. Gorecki, M. Minier, M.B. Armand, J.M. Chabagno and P. Rigaud, *Solid State Ionics*, 11, 91 (1983).

[16] P. Lightfoot, M. A. Mehta and P. G. Bruce, *Science*, 262, 883 (1993).

[17] F. M. Gray, *Solid Polymer Electrolytes: Fundamentals and Technological Applications*, VCH, New York, 1991.

[18] B. Scrosati (Ed.), *Applications of Electroactive Polymers*, Chapman & Hall, London, 1993.

[19] M. A. Ratner, in *Polymer Electrolytes Reviews – 1*, J. R. MacCallum and C. A. Vincent (Eds), Elsevier Applied Science, London, 1987, p. 173.

[20] S. D. Druger, M. A. Ratner and A. Nitzan, *Solid State Ionics*, 9–10, 1115 (1983).

[21] J. J. Fontanella, M. C. Wintersgill, J. P. Calame and C. G. Andeen, *Solid State Ionics*, 8, 333 (1983).

[22] M. C. Wintersgill, J. J. Fontanella, J. P. Calame, D. R. Figueroa and C. G. Andeen, *Solid State Ionics*, 11, 151 (1983).

[23] Z. Stoeva, C. T. Imrie and M. D. Ingram, *Phys. Chem. Chem. Phys.*, 5, 395 (2003).

[24] M. D. Ingram, C. T. Imrie, Z. Stoeva, S. J. Pas, K. Funke and H. W. Chandler, *J. Phys. Chem. B*, 109, 16567 (2005).

[25] H. Vogel, *Phys. Z.*, 22, 645 (1921); G. Tammann and W. Hesse, *Z. Anorg. Allg. Chem.*, 156, 245 (1926); G.S. Fulcher, *J. Am. Ceram. Soc.*, 8, 339 (1925).

[26] A. J. Bhattacharyya, J. Fleig, Y. G. Guo and J. Maier, *Adv. Mater.*, 17, 2630 (2005).

[27] D. G. H. Ballard, P. Cheshire, T. S. Mann and J. E. Przeworski, *Macromolecules*, 23, 1256 (1990).

[28] J. M. G. Cowie and A. C. S. Martin, *Polym. Commun.*, 26, 298 (1985).

[29] M. Watanabe, S. Oohashi, K. Sanui, N. Ogata, T. Kobayashi and Z. Ohtaki, *Macromolecules*, 18, 1945 (1985).

[30] S. Sylla, J.-Y. Sanchez and M. Armand, *Electrochim. Acta*, 37, 1699 (1992).

[31] C. K. Chiang, G. T. Davis, C. A. Harding and T. Takahashi, *Macromolecules*, 18, 825 (1985).

[32] J. R. MacCallum, M. J. Smith and C. A. Vincent, *Solid State Ionics*, 11, 307 (1984).

[33] J. S. Tonge and D. F. Shriver, *J. Electrochem. Soc.*, 134, 269 (1987).

[34] H. Tokuda and M. Watanabe, *Electrochim. Acta*, 48, 2085 (2003).

[35] H. Ohno, *Electrochim. Acta*, 46, 1407 (2001).

[36] R. Dupon, B. L. Papke, M. A. Ratner, D. H. Whitmore and D. F. Shriver, *J. Am. Chem. Soc.*, 104, 6247 (1982).

[37] M. Armand, *Solid State Ionics*, 9–10, 745 (1983).

[38] R. Frech and J. P. Manning, *Electrochim. Acta*, 37, 1499 (1992).

[39] C. A. Angell, C. Liu and E. Sanchez, *Nature*, 362, 137 (1993).

[40] M. Forsyth, J. Z. Sun and D. R. MacFarlane, *Solid State Ionics*, 112, 161 (1998).

[41] S. Megahed and.B. Scrosati, *J. Power Sources*, 51, 79 (1994).

[42] C. A. Vincent and B. Scrosati, *Modern Batteries. An Introduction to Electrochemical Power Sources*, Arnold, London, 1997.

[43] M. Wakihara and O. Yamamoto (Eds), *Lithium Ion Batteries*, Kodansha & Wiley-VCH, Weinheim, 1998.

[44] F. M. Gray and M. Armand, *in Handbook of Battery Materials*, J. O. Besenhard (Ed.), Wiley-VCH, Weinheim, 1999, p. 499.

[45] B. Scrosati and R. J. Neat, *in Applications of Electroactive Polymers*, B. Scrosati (Ed.), Chapman & Hall, London, 1993, p. 182.

[46] C. A. Vincent and B. Scrosati, *MRS Bull.*, 25, 28 (2000).

[47] M. Armand and J. M. Tarascon, *Nature*, 451, 652 (2008).

[48] M. Z. A. Munshi and B. B. Owens, *Solid State Ionics*, 26, 41 (1988).

[49] S. Chintapalli and R. Frech, *Solid State Ionics*, 86–88, 341 (1996).

[50] G. B. Appetecchi, G. Dautzenberg and B. Scrosati, *J. Electrochem. Soc.*, 143, 6 (1996).

[51] I. Kelly, J. R. Owen and B. C. H. Steele, *J. Electroanal. Chem.*, 168, 467 (1984).

[52] I. E. Kelly, J. R. Owen and B. C. H. Steele, *J. Power Sources*, 14, 13 (1985).

[53] F. Croce, G. B. Appetecchi, L. Persi and B. Scrosati, *B., Nature*, 394, 456 (1998).

[54] B. Kumar and L. G. Scanlon, *Solid State Ionics*, **124**, 239 (1999).

[55] C. Capiglia, P. Mustarelli, E. Quartarone, C. Tomasi and A. Magistris, *Solid State Ionics*, **118**, 73 (1999).

[56] A. S. Best, A. Ferry, D. R. MacFarlane and M. Forsyth, *Solid State Ionics*, **126**, 269 (1999).

[57] F. Croce and B. Scrosati, *Ann. N. Y. Acad. Sci.*, **984**, 194 (2003).

[58] F. Croce, R. Curini, L. Persi, F. Ronci, B. Scrosati and R. Caminiti, *J. Phys. Chem B*, **103**, 10 632 (1999).

[59] J. E. Weston and B. C. H. Steele, *Solid State Ionics*, **7**, 75 (1982).

[60] B. Scrosati, *J. Electrochem. Soc.*, **136**, 2774 (1989).

[61] F. Croce and B. Scrosati, *J. Power Sources*, **43**, 9 (1993).

[62] M. C. Borghini, M. Mastragostino, S. Passerini and B. Scrosati, *J. Electrochem. Soc.*, **142**, 2118 (1995).

[63] E. Peled, D. Golodinitsky, G. Ardel and V. Eshkenazy, *Electrochim. Acta*, **40**, 2197 (1995).

[64] B. Kumar and L. G. Scanlon, *J. Power Sources*, **52**, 261 (1994).

[65] E. Quartarone, P. Mustarelli and A. Magistris, *Solid State Ionics*, **110**, 1 (1998).

[66] W. Wieczorek, Z. Florjancyk and J. R. Stevens, *Electrochim. Acta*, **40**, 2251 (1995).

[67] J. Przyluski, M. Siekierski and W. Wieczorek, *Electrochim. Acta*, **40**, 2101 (1995).

[68] F. Croce, L. Persi, F. Ronci and B. Scrosati, *Solid State Ionics*, **135**, 47 (2000).

[69] G. B. Appetecchi, F. Croce, L. Persi, F. Ronci and B. Scrosati, *Electrochim. Acta*, **45**, 1481 (2000).

[70] D. R. MacFarlane, P. J. Newman, K. M. Nairn and M. Forsyth, *Electrochim. Acta*, **43**, 1333 (1998).

[71] B. Kumar and L. G. Scanlon, *Solid State Ionics*, **124**, 239 (1999).

[72] J. Plocharski and W. Wieczorek, *Solid State Ionics*, **28**, 979 (1988).

[73] C. C. Tambelli, A. C. Bloise, A. V. Rosario, E. C. Pereira, C. J. Magon and J. P. Donoso, *Electrochim. Acta*, **47**, 1677 (2002).

[74] F. Croce, L. Persi, B. Scrosati, F. Serraino-Fiory, E. Plichta and M. A. Hendrickson, *Electrochim. Acta*, **46**, 2457 (2001).

[75] S. H. Chung, Y.Wang, L. Persi, F. Croce, S. G. Greenbaum, B. Scrosati, B. and E. Plichta, *J. Power Sources*, **97–98**, 644 (2001).

[76] F.Croce, S. Sacchetti and B. Scrosati, *J. Power Sources*, **161**, 560 (2006).

[77] F.Croce, F. Settimi and B. Scrosati, *Electrochem. Commun.*, **8**, 364 (2006).

[78] M. Galiński, A. Lewandowski and I. Stępniak, *Electrochim. Acta*, **51**, 5567 (2006).

[79] M. Armand, F. Endres, D. R. MacFarlane, H. Ohno and B. Scrosati, *Nat. Mater*, **8**, 621 (2009).

[80] J. S. Wilkes and M. J. Zaworotko, *J. Chem. Soc, Chem. Commun.*, 965 (1992).

[81] J. Fuller, R. T. Carlin and R. A. Osteryoung, *J. Electrochem. Soc.*, **144**, 3881 (1997).

[82] S. R. Sivakkumar, D. R. MacFarlane, M. Forsyth and D. W. Kim, *J. Electrochem. Soc*, **154**, A834 (2007).

[83] M. Egashira, H. Shimomura, N. Yoshimoto, M. Morita and J. Yamaki, *Electrochemistry*, **73**, 585 (2005).

[84] G. H. Lane, A. Best, M. Forsyth, D. R. MacFarlane and A. Hollenkamp, *Electrochim. Acta*, **55**, 2210 (2010).

[85] J. Sun, M. Forsyth and D. R. MacFarlane, *J. Phys. Chem. B*, **102**, 8858 (1998).

[86] D. R. MacFarlane, P. Meakin, J. Sun, N. Amini and M. Forsyth,. *J. Phys. Chem. B*, **103**, 4164 (1999).

[87] D. R. MacFarlane, M. Forsyth, P. C. Howlett, J. M. Pringle, J. Sun, G. Annat, W. Neil and E. I. Izgorodina, *Acc. Chem. Res.*, **40**, 1165 (2007).

[88] A. Guerfi, M. Dontigny, P. Charest, M. Petitclerc, M. Lagace, A. Vijh and K. Zaghib, *J. Power Sources*, **195**, 845 (2010).

[89] K. J. Fraser, E. I. Izgorodina, M. Forsyth, J. L. Scott and D. R. MacFarlane, *Chem. Commun.*, **37**, 3817 (2007).

[90] D.R. MacFarlane, M. Forsyth, E. I. Izgorodina, A. P. Abbott, G. Annat and K. Fraser, *Phys. Chem., Chem. Phys*, **11**, 4962 (2009).

[91] K. Ueno, H. Tokuda and M. Watanabe,. *Phys. Chem., Chem. Phys*, **12**, 1649 (2010).

[92] S. Schrödle, G. Annat, D. R. MacFarlane, M. Forsyth, R. Buchner and G. Hefter, *Aust. J. Chem.*, **60**, 6 (2007).

[93] S. Schrodle, G. Annat, D. R. MacFarlane, M. Forsyth, R. Buchner and G. Hefter, *Chem. Commun.*, 1748 (2006).

[94] N. Boden, S. A. Leng and I. M. Ward, *Solid State Ionics*, **45**, 261 (1991).

[95] M. Forsyth, V.A. Payne, M.A. Ratner, S. W. DeLeeuw and D. F. Shriver, *Solid State Ionics*, **53**, 1011 (1992).

[96] W. Xu, E. I. Cooper and C. A. Angell, *J. Phys. Chem. B*, **107**, 6170 (2003).

[97] A. Lewandowski and A. Świerczek-Mocek, *J. Power Sources*, **194**, 601 (2009).

[98] K. Hayamizu, S. Tsuzuki, S. Seki, Y. Ohno, H. Miyashiro and Y. Kobayashi, *J. Phys. Chem. B*, **112**, 1189 (2008).

[99] P. Bayley, G. Lane, N. Rocher, A. Best, D. R. MacFarlane and M. Forsyth, *Phys. Chem., Chem. Phys.*, **11**, 7207 (2009).

[100] N. Byrne, P. C. Howlett, D. R. MacFarlane, M. E. Smith, A. Howes, A. F. Hollenkamp, T. Bastow, P. Hale and M. Forsyth, *J. Power Sources*, **184**, 288 (2008).

[101] N. Byrne, P. C. Howlett, D. R. MacFarlane and M. Forsyth, *Adv. Mater.*, **17**, 2497 (2005).

[102] C. Tiyapiboonchaiya, J. M. Pringle, J. Sun, N. Byrne, P.C. Howlett, D.R. MacFarlane and M. Forsyth, *Nat. Mater.*, **3**, 29 (2004).

[103] M. Yoshizawa, A. Narita and H. Ohno, *Aust. J. Chem.*, **57**, 139 (2004).

[104] A. Narita, W. Shibayama, K. Sakamoto, T. Mizumo, N. Matsumi and H. Ohno, *Chem. Commun.*, 1926 (2006).

[105] H. Shobukawa, H. Tokuda, M. A. H. Susan and M. Watanabe, *Electrochim. Acta*, **50**, 3872 (2005).

[106] T. Ueki and M. Watanabe, *Macromolecules*, **41**, 3739 (2008).

[107] C. Tiyapiboonchaiya, D. R. MacFarlane, J. Sun and M. Forsyth, *Macromol. Chem. Phys.*, **203**, 1906 (2002).

[108] J. Sun, D. R. MacFarlane, N. Byrne and M. Forsyth, *Electrochim. Acta*, **51**, 4033 (2006).

[109] O. Winther-Jensen, R. Vijayaraghavan, J. Sun, B. Winther-Jensen and D. R. MacFarlane, *Chem. Commun.*, 3041, (2009).

[110] O. Winther-Jensen, V. Armel, M.Forsyth and D. R. MacFarlane, *Macromol. Rapid Commun.*, **31**, 479 (2010).

[111] W. Ogihara, H. Kosukegawa and H. Ohno, *Chem. Commun.*, 3637 (2006).

[112] L. A. Boor Singh and S. S. Sekhon, *Chem. Phys. Lett.*, **425**, 294 (2006).

[113] W. Xu, L. M. Wang and C. A. Angell, *Electrochim. Acta*, **48**, 2037 (2003).

[114] M. Forsyth, J. Sun and D. R. MacFarlane, *Electrochim. Acta*, **45**, 1249 (2000).

[115] S. Kim and S. J. Park, *Electrochim. Acta*, **54**, 3775 (2009).

[116] G. T. Kim, G. B. Appettecchi, F. Alessandrini and S. Passerini, *J. Power Sources*, **171**, 861 (2007).

[117] H. Ye, J. Huang, J. J. Xu, A. Khalfan and S. G. Greenbaum, *J. Electrochem. Soc.*, **154**, A1048 (2007).

[118] A. Fernicola, F. Croce, B. Scrosati, T. Watanabe and H. Ohno, *J. Power Sources*, **174**, 342 (2007).

[119] C. Sirisopanaporn, A. Fernicola and B. Scrosati, *J. Power Sources*, **186**, 490 (2009).

[120] J. Reiter, J. Vondrak, J. Michalek and Z. Micka, *Electrochim. Acta*, **52**, 1398 (2006).

[121] I. Stępniak and E. Andrzejewska, *Electrochim. Acta*, **54**, 5660 (2009).

[122] H. Ohno, *Electrochim. Acta*, **46**, 1407 (2001).

[123] M. Yoshio, T. Kagata, K., Hoshino, T. Mukai, H. Ohno and T. Kato, *J. Am. Chem. Soc.*, **128**, 5570 (2006).

[124] D. Batra, S. Seifert and M. A. Firestone, *Macromol. Chem. Phys.*, **208**, 1416 (2007).

[125] M. D. Green and T. E. Long, *J. Macromol. Sci., C*, **49**, 291 (2009).

[126] A. R West, *in Solid State Electrochemistry*, P. G. Bruce (Ed.), Cambridge University Press, Cambridge, 1995, p. 7.

[127] H. Aono, N. Imanaka and G. Adachi, *Acc. Chem. Res.*, **27**, 265 (1994).

[128] Y. G. Andreev, P. Lightfoot and P. G. Bruce, *Chem. Commun.*, 2169 (1996).

[129] Y. G. Andreev, P. Lightfoot and P. G. Bruce, *J. Appl. Crystallogr.*, **18**, 294 (1997).

[130] G. S. MacGlashan, Y. G. Andreev and P. G. Bruce, *Nature*, **398**, 792 (1999).

[131] Z. Zadjourova, Y. G. Andreev, D. P. Tunstall and P. G. Bruce, *Nature*, **412**, 520 (2001).

[132] Z. Stoeva, I. Martin-Litas, E. Staunton, Y. G. Andreev and P. G. Bruce, *J. Am. Chem. Soc.*, **125**, 4619 (2003).

[133] Z. Zadjourova, D. Martin y Marero, K. H. Andersen, Y. G. Andreev and P. G. Bruce, *Chem. Mater.*, **13**, 1282 (2001).

[134] C. Zhang, E. Staunton, Y. G. Andreev and P. G. Bruce, *J. Mater. Chem.*, **17**, 3222 (2007).

[135] E. Staunton, Y. G. Andreev and P. G. Bruce, *Faraday Discuss.*, **134**, 143 (2007).

[136] A. M. Christie, S. Lilley, E. Staunton, Y. G. Andreev and P. G. Bruce, *Nature*, **433**, 50 (2005).

[137] C. Zhang, E. Staunton, Y. G. Andreev and P. G. Bruce, *J. Am. Chem. Soc.*, **127**, 18305 (2005).

[138] D. Brandell, A. Liivat, H. Kasemagi, A. Aabloo and J. O. Thomas, *J. Mater. Chem.*, **15**, 1422 (2005).

[139] D. Brandell, A. Liivat, A. Aabloo and J. O. Thomas, *Chem. Mater.*, **17**, 3673 (2005).

[140] D. Brandell, A. Liivat, A. Aabloo and J. O. Thomas, *J. Mater. Chem.*, **15**, 4338 (2005).

[141] A. Liivat, D. Brandell and J. O. Thomas, *J. Mater. Chem.*, **17**, 3938 (2007).

[142] A. Liivat, D. Brandell, A. Aabloo and J. O. Thomas, *Polymer*, **48**, 6448 (2007).

[143] P. Johansson and P. Jacobsson, *Electrochim. Acta*, **48**, 2279 (2003).

[144] P. V. Wright, Y. Zheng, D. Bhatt, T. Richardson and G. Ungar, *Polym. Int.*, **47**, 34 (1998).

[145] S. H. Chung, Y. Wang, S. G. Greenbaum, D. Golodnitsky and E. Peled, *Electrochem. Solid-State Lett.*, **2**, 553 (1999).

[146] P. G. Bruce, unpublished results, 2010.

[147] C. Zhang, S. Gamble, D. Ainsworth, A. M. Z. Slawin, Y. G. Andreev and P. G. Bruce, *Nat. Mater.*, **8**, 580 (2009).

2

Advanced Inorganic Materials for Solid Oxide Fuel Cells

Stephen J. Skinner and Miguel A. Laguna-Bercero
Department of Materials, Imperial College London, London, UK

2.1   INTRODUCTION

Fuel cells are widely regarded as a next generation technology that will contribute to reductions in emissions of gases responsible for climate change such as CO_2. There are several modifications to the basic concept of a fuel cell that allow operation over a wide variety of temperatures and with differing fuels. One of the most promising types of fuel cell for large scale power generation and combined heat and power applications is the solid oxide fuel cell (SOFC). The SOFC offers many advantages over conventional combustion-based power generation technologies and has been the subject of intensive investigation for many years. As with all other fuel cells the SOFC is an electrochemical energy conversion device that converts the chemical energy contained within a fuel (for example, H_2) to useful electrical energy through an electrochemical oxidation process. Using the simplest fuel as an example the overall chemical reaction occurring is:

$$H_{2(g)} + \frac{1}{2}O_{2(g)} \rightarrow H_2O_{(g)} \qquad (2.1)$$

Energy Materials   Edited by Duncan W. Bruce, Dermot O'Hare and Richard I. Walton
© 2011 John Wiley & Sons, Ltd.

To achieve this overall reaction it is essential that we consider the basic components of the fuel cell. A SOFC consists of three ceramic functional components (anode 'fuel side', cathode 'air side' and electrolyte) plus an electrical interconnect that is either ceramic or metallic depending upon the operating environment. These components are referred to as the fuel cell and to achieve suitable power outputs these individual cells are connected to form a fuel cell stack. Schematics of two alternative designs of a single fuel cell are shown in Figure 2.1.

At the anode the fuel is oxidised with the oxidant (oxygen) reduced at the cathode. These two components are separated by a gas tight electrolyte membrane that is a pure ionic conductor, transporting the oxide ion species from the cathode to the fuel side. This process releases electrons to an external circuit, providing the useful electrical power. These reactions are summarised as:

$$\text{cathode reaction} \qquad \frac{1}{2}O_{2(g)} + 2e^- \rightarrow O^{2-} \qquad (2.2)$$

$$\text{anode reaction} \qquad H_{2(g)} + O^{2-} \rightarrow H_2O_{(g)} + 2e^- \qquad (2.3)$$

Evidently these electrode reactions are significantly more complex than shown in Equations 2.2 and 2.3 and fuller discussions of the proposed fuel oxidation and oxygen reduction electrode reactions are given in the literature.[1–4]

It is therefore clear that in a SOFC there are many processes that require optimised materials, and consequently the SOFC is a complex solid state device. As a summary, it is essential that the electrolyte is a gas tight, pure ionic conductor stable over a wide pO_2 range ($>10^{-24}$ atm). Anodes have to possess stability in reducing conditions with both high ionic and electronic conductivity, and chemical compatibility with the electrolyte material. Similar properties are required for the cathode with the exception that stability is now in oxidising environments.

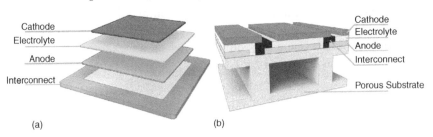

Figure 2.1   Schematic representation of two solid oxide fuel cell designs: (a) planar; and (b) integrated planar

Additionally the cathode has to be catalytically active towards oxygen reduction, which is often a rate limiting process, particularly at lower temperatures of operation (<650 °C). These are demanding properties for a set of materials and have led to a relatively small number of materials finding application in SOFCs. Several excellent reviews[5–11] have reported on the development of conventional SOFC materials and these details will only be briefly summarised here.

2.1.1 Conventional SOFC Electrolytes

A SOFC was proposed by Baur and Preis[12] as far back as 1937 based upon an electrolyte of stabilised zirconia with metallic electrodes. Since then stabilised zirconia has been the electrolyte that has received most attention by fuel cell developers. Most zirconia electrolytes are based upon either yttria or scandia stabilisation of the tetragonal polymorph,[6,7,13] commonly referred to as YSZ and ScSZ, respectively, although a number of alternative dopants have been investigated (Tables 2.1 and 2.2). Conventionally the substitution level is between 3 mol% and 8 mol% for the yttria-based materials and at 10–12 mol% for the scandia-based materials. The choice of the dopant level is dictated by a compromise between mechanical robustness and overall conductivity, as summarised in Table 2.1. Substitution of zirconia results in the stabilisation of either the tetragonal or cubic polymorphs adopting the fluorite type structure as shown in Figure 2.2. This substitution

Table 2.1 Conductivity of typical zirconia-based fluorite electrolytes

Electrolyte	Conductivity at 1000 °C (S cm^{-1})	
	As-sintered	After annealing
ZrO_2-3 mol% Y_2O_3	0.059	0.050
ZrO_2-3 mol% Yb_2O_3	0.063	0.090
ZrO_2-2.9 mol% Sc_2O_3	0.090	0.063
ZrO_2-8 mol% Y_2O_3	0.130	0.090
ZrO_2-9 mol% Y_2O_3	0.130	0.120
ZrO_2-8 mol% Yb_2O_3	0.200	0.150
ZrO_2-10 mol% Yb_2O_3	0.150	0.150
ZrO_2-8 mol% Sc_2O_3	0.300	0.120
ZrO_2-11 mol% Sc_2O_3	0.300	0.300
ZrO_2-11 mol% Sc_2O_3-1 wt% Al_2O_3	0.260	0.260

After Singhal and Kendal.[326]

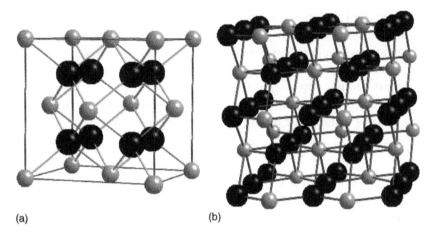

(a)   (b)

Figure 2.2   Schematic representation of the (a) cubic and (b) tetragonal polymorphs of the fluorite structured yttria stabilised zirconia. Metal ions in grey, oxygen in black

also provides an increased concentration of oxygen vacancies within the lattice facilitating increased ionic conductivity, as illustrated by the incorporation reaction:

$$Y_2O_3 \xrightarrow{ZrO_2} Y'_{Zr} + V_O^{\bullet\bullet} + 3O_O^x \qquad (2.4)$$

One significant disadvantage with the ZrO_2-based SOFCs is the temperature of operation at which ionic conductivity is sufficiently high to support a device which for 8YSZ is typically 800–1000 °C, depending upon the thickness of the electrolyte. This leads to a further consideration for fuel cell operation – that of electrolyte-supported or electrode-supported designs. Again each design has advantages but to increase performance it is generally accepted that lower operating temperatures are required and thus thinner electrolytes. As electrolyte thickness decreases to below 100 μm self-supporting electrolytes are no longer mechanically feasible. Hence the deposition of layers of dense electrolytes on porous electrodes becomes a significant processing issue. On moving to a scandia-based system it is possible to achieve greater conductivities at lower temperatures than with YSZ thus reducing the operating temperature of the fuel cell.

Despite the relatively low levels of conductivity displayed by zirconia-based electrolytes these materials have the major advantage of exhibiting considerable chemical stability as a function of oxygen partial pressure, presenting no degradation at a pO_2 as low as 10^{-24} atm. However there are some further considerations when using ZrO_2-based cells, including

Table 2.2 Conductivity of typical ceria-based fluorite electrolytes (CeO_2-Ln_2O_3)

Ln_2O_3	Mol%	Conductivity (S cm^{-1})	
		700 °C	500 °C
Sm_2O_3	10	3.5×10^{-2}	2.9×10^{-3}
Sm_2O_3	10	4.0×10^{-2}	5.0×10^{-3}
Gd_2O_3	10	3.6×10^{-2}	3.8×10^{-3}
Y_2O_3	10	1.0×10^{-2}	0.21×10^{-3}
CaO	5	2.0×10^{-2}	1.5×10^{-3}

After Singhal and Kendal.[326]

the reactivity of the electrolyte with cathodes and the necessity to use exotic alloys or high cost interconnect ceramics, combined with issues surrounding cell sealing. For these reasons two other types of electrolyte have been developed – lanthanum gallate based perovskites and ceria-based fluorites.

Ceria-based electrolytes have been proven to operate at much reduced temperatures (500–700 °C) with ionic conductivity comparable with YSZ at much higher temperatures (see Table 2.2).[14,15] Commonly the CeO_2 host is substituted with either Sm or Gd ($Ce_{1-x}Sm_xO_{2-\delta}$, CSO, and $Ce_{1-x}Gd_xO_{2-\delta}$, CGO), creating significant vacancy concentrations. Use of these ceria-based materials is limited by the redox characteristics of the $Ce^{3+/4+}$ couple, with reduction occurring at temperatures above about 650 °C leading to a reduction of the ionic transport number. This in turn can lead to short circuits within the cell and hence a loss of performance. However, as conductivity in ceria-based compounds is sufficient at temperatures below 650 °C for fuel cell electrolytes, the issue is then one of suitably active cathodes, addressed in Section 2.1.3.

The final class of 'conventional' electrolyte is based upon the perovskite structured (ABO_3) lanthanum gallate (Figure 2.3). Ishihara et al.[16–20] and Huang et al.[21–24] both reported in the 1990s fast oxide ion conduction in doubly substituted gallates, giving a composition of $La_{1-x}Sr_x$ $Ga_yMg_{1-y}O_{3-\delta}$ (LSGM) where $x = 0.1$ or 0.2 and $y = 0.2$. Levels of conductivity reported were higher than the YSZ materials in the 600–750 °C temperature range (Figure 2.4) and consequently received significant attention. Further reports indicated that a minimal Ni or Co content on the B site enhanced the ionic conductivity further.[25,26] Many authors have attempted to optimise the LSGM type electrolytes and further details of these novel electrolytes will be discussed in Section 2.2.1.1. Given the optimal operating temperatures of the three electrolytes detailed above it is clear that there are three main operating regimes for

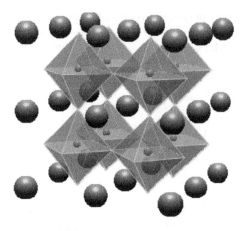

Figure 2.3   Representation of the perovskite structure adopted by $LaGaO_3$-based electrolytes. Ga ions are located at the centre of the shaded octahedra

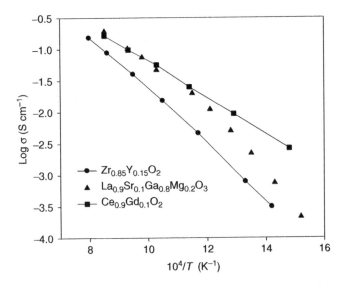

Figure 2.4   Conductivity of conventional oxide ion conducting electrolytes. Data obtained from Tsipis and Kharton[2]

SOFCs: 500–650 °C, 650–800 °C and 800–1000 °C, with regimes corresponding to ceria-based, LSGM-based and ZrO_2-based electrolytes, respectively. Each of these regimes required the development of specific electrodes optimised (or compromised?) for the operating temperature and these electrodes are discussed in the subsequent sections.

2.1.2 Conventional Anodes

In SOFCs the composition of the anode is limited to a large extent by the stringent requirements placed upon the material, including oxide ion mobility, high electronic conductivity, catalytic activity for hydrogen oxidation and stability at very low pO_2. Typically anodes in SOFCs will operate under a gas mixture containing CH_4/H_2O rather than a pure H_2 fuel stream. This is possible as the typical anode of Ni/YSZ exploits the catalytic activity of Ni for steam methane reforming, forming a hydrogen rich syngas that can then be oxidised electrochemically. As a consequence of this there are currently very few 'conventional' anode materials to choose from. In general the anode is a composite material consisting of a metallic component mixed with the electrolyte composition. Hence for YSZ-based cells the anode material would typically be of Ni-YSZ composite type.[27,28] These composites are often referred to as 'cermets'. Similarly for CGO- and LSGM-based cells the composite materials are of the Ni-CGO and Ni-LSGM type.

Preparation of the anode structure is usually achieved through the mixing of NiO with the appropriate electrolyte component, followed by a reduction step to produce the Ni cermet. Volume changes associated with the reduction step do not appear to be detrimental to the overall durability of the cell but there have been observations of serious degradation of the anode when reoxidation occurs. These are significant concerns for thermal cycling and it is clear that over relatively modest timescales there is a significant degradation rate of $10 \, m\Omega \, cm^{-2}$ per $1000 \, h$ at $1000 \, °C$ for YSZ-based devices.[29] Clearly these degradation rates and the evolution of reaction products is too great for long term operation. Redox stability of the electrode structure is therefore a common problem for Ni-based anodes, and significant volume changes upon oxidation of Ni to NiO have been demonstrated to impact significantly upon the mechanical integrity of the cell.[30]

One further limitation of the anode component in these devices is the degradation of performance with the increase of sulfur content in the fuel stream and also with the risk of coking if carbon rich fuel streams are used, such as with natural gas.[31] Evidently if a pure hydrogen fuel were used in the SOFC these issues of anode operation would not arise, however one of the advantages of the SOFC is the potential fuel flexibility offered by the use of oxide anodes, such as the reforming of natural gas in the anode.

2.1.3 Conventional Cathodes

Cathode materials for SOFCs based on any of the electrolytes described in Section 2.1.1 are of the perovskite structure type, generally La-based with transition metals located on the B site. Several authors[4,5,32–35] have summarised the range of perovskites investigated to date, concentrating on the conductivity, ion transport and compatibility of these materials. As such it is superfluous to continue the discussion in detail here. Instead we will refer to the main cathode types only, leaving the reader to consult the relevant literature for further details.

The majority of SOFC cathodes are based upon either $LaMnO_3$ or $LaCoO_3$ substituted with Sr on the A site and Fe on the B site. The rationale for using these materials is a result of the best compromise in properties for the cathodic reactions. In general, with higher temperature cells, such as those based on YSZ, the cathode functions as a triple phase boundary (TPB) material, where the chemical reactions between the electrode, electrolyte and gas phase take place at a limited area within the cathode structure (Figure 2.5). In this case the electrode material is functioning as an electronic conductor and oxygen reduction catalyst. For this case the Sr doped $LaMnO_3$ (LSM) is a suitable choice with conductivity of the order of 320 S cm^{-1} at 800 °C[36] for the $La_{0.6}Sr_{0.4}MnO_3$ composition. Whilst LSM acts as a suitable electrode at these elevated temperatures, there are some fundamental issues with this cathode material that compromise its performance, including the reactivity of LSM with YSZ to form insulating interfacial reaction products of the pyrochlore type, $La_2Zr_2O_7$.[37,38] This is clearly detrimental to the overall performance of the cell as this reaction product acts as a blocking phase for the cathode reactions.

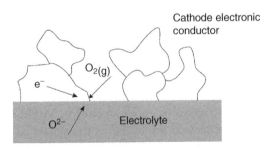

Figure 2.5 Schematic illustration of electrolyte-cathode three phase boundary, typical of high temperature YSZ-based SOFCs

On reducing the temperature of operation we find that the performance of the LSM cathode with respect to oxygen reduction and surface exchange is no longer sufficient. Indeed at the lower temperatures where CGO is used as the electrolyte it becomes necessary to move away from the TPB model and use materials that are both ionic and electronic conductors (MIEC, mixed ionic electronic conductor) effectively extending the TPB over the entire cathode surface. On replacing Mn with Co it is noted that a material with both high electronic conductivity and fast ion conductivity is produced, $La_{1-x}Sr_xCoO_{3-\delta}$ (LSC). However, the thermal properties of this composition are such that a significant mismatch in the thermal expansion coefficients proves detrimental to performance as thermal cycling results in cracking and delamination of the cathode. To overcome this issue a substitution of Fe for Co was performed and a resulting decrease in thermal expansion coefficient observed.[39,40] Unfortunately this substitution also resulted in a reduction of the MIEC properties of the cathode. Hence, to promote cathode performance at lower temperatures, composite cathodes of $La_{1-x}Sr_xCo_{1-y}Fe_yO_{3-\delta}$ (LSCF) with CGO have been produced.[41]

With LSGM electrolytes similar cathodes have been developed with perhaps some of the most exciting recent results focusing on the $Ba_{1-x}Sr_xCo_{1-y}Fe_yO_{3-\delta}$ (BSCF)[42–50] and $Sm_{1-x}Sr_xCoO_{3-\delta}$ (SSC)[51–55] compositions. The development of these materials has resulted from the continued interest in substitution of both the lanthanide and transition metal components in the perovskite lattice. A more detailed discussion of these new cathode materials in operation with LSGM, YSZ and CGO electrolytes is given in Section 2.2.2.

2.1.4 Summary

In the preceding sections the main materials currently used in SOFCs have been detailed along with the background regarding the materials requirements controlling application of these oxides. It is clear that the SOFC is a complex all ceramic device with many challenges to overcome before a truly robust device can be marketed. In the subsequent sections we outline the main materials developments in terms of new compositions and structure types that will compete for inclusion in next generation devices with enhanced performance and durability. In this we include the potential for development of proton conducting ceramic fuel cells.

2.2   NEXT GENERATION SOFC MATERIALS

Whilst SOFCs have developed significantly over the past few decades the main focus of materials development has been on the fluorite- and perovskite-based functional components. Each of the materials selections discussed above has deficiencies and there is now considerable effort directed towards the development of novel materials and structures with better performance for next generation SOFCs. In the following sections the potential choices available for low temperature electrolytes and cathodes and sulfur tolerant anodes will be discussed.

2.2.1   Novel Electrolyte Materials

2.2.1.1   Electrolytes Based on ABO_3 Perovskite

Initial studies of the oxide ion conductivity of materials with the ABO_3 perovskite structure were reported by Takahashi and Iwahara[56] in solid solutions such as $CaTi_{1-x}\text{-}Al_xO_{3-\delta}$, $CaTi_{1-x}\text{-}Mg_xO_{3-\delta}$ and $SrTi_{1-x}\text{-}Al_xO_{3-\delta}$. Relatively recently, perovskite and related crystal structures have been investigated and proposed as good oxide ion conductors, as presented in Table 2.3. The highest overall bulk oxide ion conductivity was reported for the $La_{0.8}Sr_{0.2}Ga_{0.83}Mg_{0.17}O_{3-\delta}$ (LSGM) composition. At 800 °C this material presents a conductivity of 0.166 S cm^{-1}, which is approximately four times the conductivity of YSZ at the same

Table 2.3 Conductivity data reported for some novel perovskite oxides as potential fuel cell electrolytes

Composition	Temperature (°C)	Conductivity (S cm^{-1})	Ref
$Nd_{0.9}Ca_{0.1}Al_{0.5}Ga_{0.5}O_{3-\delta}$	700	$\sim 10^{-2}$	[327]
$KNb_{0.5}Al_{0.5}O_{2.5}$	700	5.1×10^{-3}	[328]
$KTa_{0.5}Al_{0.5}O_{2.5}$	700	2.0×10^{-3}	[328]
$NaNb_{0.5}Al_{0.5}O_{2.5}$	700	1.5×10^{-3}	[328]
$NaTa_{0.5}Al_{0.5}O_{2.5}$	700	1.3×10^{-3}	[328]
$La_{0.8}Sr_{0.2}Ga_{0.9}Mg_{0.1}O_{3-\delta}$	700	7.14×10^{-2}	[329]
$La_{0.8}Sr_{0.2}Ga_{0.85}Mg_{0.15}O_{3-\delta}$	800	0.14	[329]
$La_{0.8}Sr_{0.2}Ga_{0.83}Mg_{0.17}O_{2.815}$	700	0.079	[22]
$La_{0.8}Sr_{0.2}Ga_{0.83}Mg_{0.17}O_{2.815}$	800	0.166	[22]
$La_{0.8}Sr_{0.2}Ga_{0.85}Mg_{0.15}O_{3-\delta}$	700	0.066	[330]
$La_{0.8}Sr_{0.2}Ga_{0.85}Mg_{0.15}O_{3-\delta}$	800	0.146	[330]

temperature,[22] and is therefore considered as a potential candidate for intermediate temperature solid oxide fuel cell (IT-SOFC) applications.

Apart from its good conductivity, higher than that of both YSZ and ScSZ and similar to that of CGO, Sr and Mg doped $LaGaO_3$ (LSGM) is stable in both fuel and air, and especially at low oxygen partial pressures, where GDC is easily reduced as redox of the Ce^{4+}/Ce^{3+} couple is facile. It was also noted that the conductivity of LSGM depends strongly on the dopant concentration as shown in Figure 2.6.

In order to improve the conductivity of LSGM several approaches involving the addition of dopants on both A and B sites have been suggested. Partial substitution (10 mol%) of the La site by other rare earth ions such as Nd, Sm, Gd, Y and Yb in the $La_{0.9}Sr_{0.1}Ga_{0.8}Mg_{0.2}O_{2.85}$ composition was however found to decrease the total electrical conductivity.[17,57] The lanthanum site can also be substituted with barium[58] as an alternative to Sr and on substitution with barium it was observed that the octahedral tilt angle was affected reducing the activation energy for conduction. $La_{0.9}Ba_{0.1}Ga_{0.8}Mg_{0.2}O_{2.85}$ (LBGM) presents lower conductivity than LSGM at high temperatures but conversely presents higher conductivity values at lower temperatures,[59] thus making the material of interest for lower operating temperature devices. It has also been shown that replacement of Sr^{2+} ions by K^+ ions in LSGM decreases the electrical conductivity of LSGM and increases the activation energy

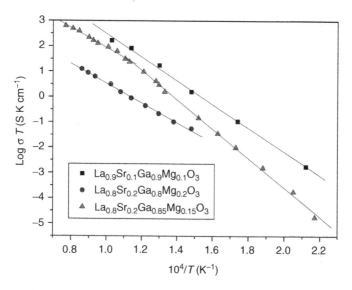

Figure 2.6 Illustration of the conductivity of LSGM type electrolytes as a function of composition. Data from Fergus[7]

whereas the substitution of La by substantial amounts of up to 50 atom% of Pr in LSGM does not affect the electrical conductivity.[60] Hence A-site substitution appears to be detrimental to overall conductivity in the LSGM type electrolytes.

The effect of B-site substitution has been also widely studied. Partial substitution of Ga by Al or In also decreases the electrical conductivity of LSGM. This decrease may be attributed to changes in the crystal structure and lattice parameter.[61] However, Pr and In co-substitutions forming the composition $La_{0.4}Pr_{0.4}Sr_{0.2}In_{0.8}Mg_{0.2}O_{2.8}$ results in high electrical conductivity at low temperatures with a remarkably low activation energy of only 0.44 eV. This increase in the electrical conductivity has been suggested to be due to the higher number of oxygen vacancies or an increase in the electronic conductivity due to the mixed valence of Pr.[61] The addition of transition metals, such as cobalt[25, 62–64] and iron,[62,65] were found to increase the conductivity of LSGM, especially at low temperatures as shown in Figure 2.7. Small additions of Fe or Co increase the electrical conductivity at low temperatures due to the introduction of electronic charge carriers into the lattice. However, at high temperatures the total conductivity is dominated by ionic conduction, being similar to that of LSGM. Therefore, the addition of transition metals such as Fe or Co into LSGM does not have a significant effect on the ionic conductivity. On the contrary, high doping levels might yield LSGM-based

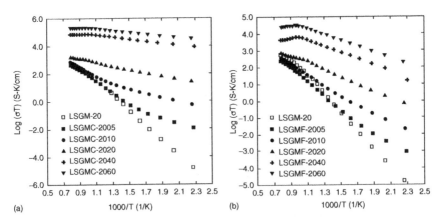

Figure 2.7 Illustration of the ionic conductivity data for different compositions of cobalt (a) and iron (b) doped LSGM electrolytes. Reprinted with permission from Stevenson et al., 2000[25]. Copyright (2000) *J. Electrochem. Soc.* – American Institute of Physics

compositions for use as SOFC cathodes. The use of nickel as a dopant for LSGM has also been reported, presenting similar results to those observed with cobalt substitution.[66]

In order to produce high quality electrolyte powders interest has been directed towards novel preparation methods of the LSGM compositions to ensure high homogeneity, good sinterability and enhanced performance. For example, $La_{0.9}Sr_{0.1}Ga_{0.8}Mg_{0.1}Co_{0.1}O_{2.85}$ (LSGMC) powders have been synthesised by carbonate co-precipitation resulting in the ionic conductivity of the LSGMC samples increasing to 1.13 S cm^{-1} at 800 °C.[67] Further preparation methods recently developed for LSGM and LSGMC are sol-gel synthesis[68] and spark plasma sintering,[69] with the latter technique used to reduce the sintering temperature and time taken to produce gas tight, high density membranes.

2.2.1.2   Other Perovskite-based Derivatives

Other potential electrolytes with the perovskite-type structure are the $SrSn_{1-x}Fe_xO_{3-\delta}$ $(0<x<1)$ compositions or Ga and Sc doped $CaTiO_3$. It has been found that solid solutions of $SrSn_{1-x}Fe_xO_{3-\delta}$ with low Fe contents (<0.3) exhibit regimes of predominant ionic and electronic conductivity. For example, $SrSn_{0.9}Fe_{0.1}O_3$ exhibits p- and n-type electronic conductivity at high and low oxygen partial pressures, respectively.[70] Ga and Al doped $CaTiO_3$ samples also present p- and n-type electronic conductivity at both high and low oxygen partial pressures, while $CaTi_{0.9}Sc_{0.1}O_{3-\delta}$ shows predominantly ionic conductivity over a wide range of oxygen partial pressures at a temperature of 800 °C.[71]

2.2.1.3   Electrolytes Based on La$_2$Mo$_2$O$_9$ (LAMOX)

Lanthanum molybdate, $La_2Mo_2O_9$, has been reported to exhibit fast oxide ion conducting properties comparable with the conventional zirconia and ceria compositions.[72] This compound presents a different crystal structure from all known oxide electrolytes, and consists of isolated $[MoO_4]^{2-}$ units in a three-dimensional matrix of $[La_2O]^{4+}$. $La_2Mo_2O_9$ undergoes a reversible phase transition from the non-conductive monoclinic α-form to the highly conductive cubic β-form at approximately 580 °C. Powder X-ray diffraction (XRD) of the phases α and β are practically identical, because the structural phase transition $\alpha \rightarrow \beta$ is actually a transition from a static to dynamic distribution of the oxygen defects without any shift of the heavy atoms.[73] High-resolution XRD is however required to reveal the structure of the monoclinic α-phase, a complex structure suggested to comprise 312 nonequivalent atoms.[74] The complex structure of the monoclinic phase is viewed as impeding the oxygen ordering underlying the $\beta \rightarrow \alpha$ phase transition.

In Figure 2.8 ionic conductivity data are presented and compared with those of stabilized zirconia. The high-temperature phase, β-La$_2$Mo$_2$O$_9$, has a cubic structure which is derived from that of β-SnWO$_4$ (Figure 2.9). Partial site occupation by oxygen atoms, strongly anisotropic thermal factors, and short-range order with a distance characteristic of O-O pairs has been evidenced.[75] A structural model for the origin of the oxygen ion conduction in β-La$_2$Mo$_2$O$_9$ based on the β-SnWO$_4$ compound has been proposed. It was suggested that the substitution of a lone pair cation (Sn^{2+}) by a non lone pair one with higher oxidation state (e.g. La^{3+}) could be used as a model to design new fast oxide ion conductors.[76]

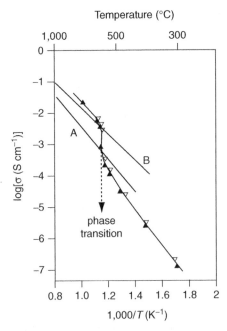

Figure 2.8 Total conductivity of the ionic conductor La$_2$Mo$_2$O$_9$ highlighting the drop in conductivity associated with an α–β phase change. For comparison, A and B represent the conductivity of the stabilized zirconias (ZrO$_2$)$_{0.87}$(CaO)$_{0.13}$ and (ZrO$_2$)$_{0.9}$(Y$_2$O$_3$)$_{0.13}$, respectively. Reprinted with permission from Lacorre et al., 2000[72]. Copyright (2000) Macmillan

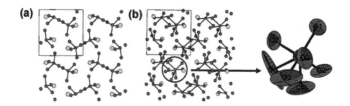

Figure 2.9 Representation of the structure of (a) β–$SnWO_4$ and (b) β–$La_2Mo_2O_9$. The coordination environment of the Mo site and atomic displacement parameters of oxygens in (b) are highlighted. Reprinted with permission from Evans *et al.*, 2005[74]. Copyright (2005). American Chemical Society

In order to stabilise the β-structure at lower temperatures, suppressing the α–β phase transition, minor doping of $La_2Mo_2O_9$ has been found to be effective.[77] For example, substitution of W for Mo has been shown to stabilise the β-phase at room temperature with Collado *et al.*[78] reporting that $La_2W_{2-x}Mo_xO_9$ can be prepared as single β-phase at any cooling rate for compositions where $x \geq 0.7$. Small additions of niobium were found to increase the conductivity when compared with the undoped $La_2Mo_2O_9$ phase.[79,80]

Substitution of Nd, Gd and Y for La has also been studied with gadolinium and yttrium both found to stabilise the cubic β-phase at room temperature. Gd and Y substitutions also seem to increase anion conductivity of the β-phase.[81] Small amounts of Ba doping were also found to increase the overall conductivity of the parent compound $La_2Mo_2O_9$ to a notable extent both at low and high temperatures.[82] Other dopants substituting for lanthanum (Ca^{2+}, Sr^{2+}, Ba^{2+} or K^+) have also been studied.[83] Although these substitutions do not improve the ionic conductivity compared with the intrinsic vacancies in the structure, an increase of the redox stability, essential for SOFC applications in the low temperature range for these materials was observed.

A series of optimised $La_2Mo_2O_9$ materials have been studied using the isotope exchange depth profile technique and secondary ion mass spectrometry (SIMS) analysis. Oxide ion diffusion coefficients (D^*) in both the parent compound and optimised materials were found to be significantly higher than those reported for any of the fluorite structured electrolyte materials, obtaining the best value for the $La_{1.7}Gd_{0.3}Mo_2O_9$ composition of $D^* = 1.41 \times 10^{-6}\,cm^2\,s^{-1}$ at 800 °C.[84] These diffusion coefficients obtained for stabilised compositions indicate the potential for use as intermediate temperature SOFC electrolytes, assuming that the durability and compatibility with electrodes are proven competitive with existing zirconia- and ceria-based systems.

Recently, different preparation methods, such as spark-plasma sintering,[85] thermal processing methods[86] or polyaspartate precursor methods,[87] have been shown as potential synthesis methods in order to improve the ionic conductivity of $La_2Mo_2O_9$-based electrolytes. As previously, these techniques optimise the particle morphology and homogeneity resulting in improved performance.

2.2.1.4   Brownmillerite Structured Oxides and Derivatives

The $A_2B_2O_5$ brownmillerite structure is derived from the ABO_3 cubic perovskite structure in which half of the anions have been removed from alternate BO_2 layers. This gives the stacking sequence $AO–BO_2–AO–BO–AO–$ with alternating layers of apex-linked BO_6 octahedra and BO_4 tetrahedra as shown in Figure 2.10. The oxygen vacancies are ordered along (010) planes and they may contribute to ionic transport forming

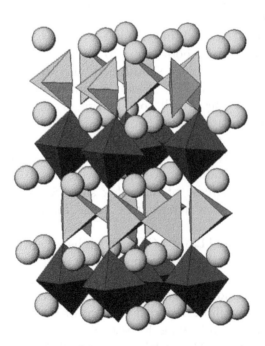

Figure 2.10   Representation of the Brownmillerite structure adopted by $Ba_2In_2O_5$. Ba atoms are shown as spheres and In atoms are located at the centre of both tetrahedra and octahedra

one-dimensional diffusion pathways for oxygen ion migration in the tetrahedral layers.[88] Electrical conductivity is typically oxygen-ionic in dry atmospheres with moderate oxygen partial pressures, mixed ionic and p-type electronic under oxidising conditions, and protonic under gas mixtures containing H_2O. The parent compound, $Ba_2In_2O_5$, presents mixed conductivity with dominant oxygen transport in dry air and at room temperature has orthorhombic symmetry. A transition into the tetragonal polymorph disordered perovskite phase at above 925 °C is reported, which leads to a dramatic increase of the ionic conduction.[88] At 1040 °C, $Ba_2In_2O_5$ transforms into a disordered cubic phase which is a pure oxide ion conductor. In order to stabilise the disordered cubic perovskite structure and thus to increase the ionic transport in the intermediate temperature range, substitution of indium with higher valence cations, such as Zr^{4+}, Ce^{4+} or Sn^{4+} has been studied.[89,89] Total conductivity of $Ba_2In_2O_5$ and other related compounds is presented in Figure 2.11, and compared with 8 mol% YSZ.[91]

Doping of $Ba_2In_2O_5$ with Co has been investigated forming the $BaIn_{0.9}Co_{0.1}O_{3-\delta}$ composition (cubic perovskite structure) and shows high oxide ionic conductivity comparable with the original $Ba_2In_2O_5$ and high electronic conductivity. For example, at 850 °C the sample presents a total conductivity of 4.7×10^{-1} S cm^{-1} and an oxide ion transport number of 0.52 at 850 °C[92] disqualifying this composition

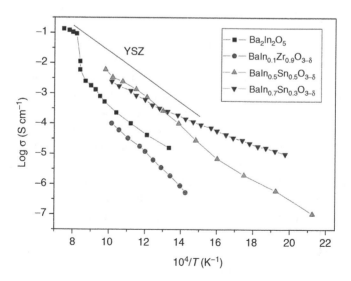

Figure 2.11 Total conductivity of $Ba_2In_2O_5$ and other related compounds. Data from Kharton et al.[91] YSZ is included as a solid line for comparison

from consideration as an electrolyte. Partial substitution for In in $Ba_2In_2O_5$ with V, Mo and W also led to the stabilisation of the fast oxide ion conducting cubic form at lower temperatures. As for substitution by cations with a smaller radius than that of In^{3+}, a decrease of the order–disorder transition temperature with the increased level of substitution is reported, and it is associated with a decrease in the conductivity.[93] Below 400 °C, an uptake of water is observed for all of the materials, which makes them promising as potential proton conductors. Further partial substitutions on the In site including Cu,[94] Ga,[95] Y,[96] Ti,[97], Sc, Ta, Nb or Si,[98] and Sr[99] or La,[99,100] as a substitution for Ba, also stabilise the cubic perovskite phase at low temperatures.

2.2.1.5 Bi_2O_3-based Electrolytes

δ-Bi_2O_3 is renowned as the best oxide ion conductor with conductivity one to two orders of magnitude higher than the best performing YSZ. A phase transition from the monoclinic α-phase to the highly conductive cubic δ-phase at about 730 °C was first reported by Gattow and Schröder,[101] who noted that the δ-phase was stable up to its melting point of approximately 825 °C. They also performed high temperature powder XRD, showing that the δ-phase crystallised in a fluorite-type cubic cell (space group $Fm\overline{3}m$) with a ¼ defective anionic $8c$ Wyckoff site. Neutron diffraction studies at 778 °C[102] allowed precise determination of the structural disorder, with the oxygen ions in $8c$ and $32f$ sites and partial occupancy factors fitting to the total oxygen content. This disorder of oxide ions is viewed as being responsible for the observed fast oxide ion conduction. Unfortunately δ-Bi_2O_3 suffers from significant instability in reducing environments leading to electronic conduction, and hence this discounts the parent phase from serious consideration as an electrolyte. In an effort to instigate stabilisation of the δ-phase at lower temperatures substitution of bismuth with other dopants has been attempted by many authors. Of the species used for stabilisation Y_2O_3[103–105] and rare earth oxides Er_2O_3,[106,107] Dy_2O_3,[108] Sm_2O_3[109] or Yb_2O_3[110] have been most prevalent. It has also been found that stabilisation of the face-centred cubic phase is possible by substituting with higher valent dopants such as M_2O_5 (M = V, Nb, Ta)[111] or WO_3.[112]

Another interesting system is Bi_2O_3-Gd_2O_3,[103,109] which adopts both the cubic and rhombohedral phases depending upon composition, both being good oxygen ionic conductors. In general, it is found that the rhombohedral phase is formed in the case of relatively large M^{3+} ions,

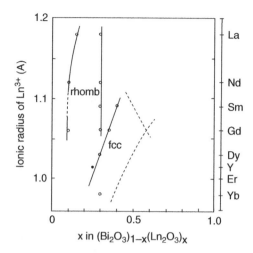

Figure 2.12 Illustration showing the formation range of the cubic and rhombohedral phases as a function of the ionic radius in the Bi_2O_3-M_2O_3 system. Reprinted with permission from Iwahara et al., 1981[114]. Copyright (1981) Elsevier

and the cubic structures are usually formed in the case of a relatively small cationic radius.[113] Iwahara et al.[114] showed the formation range of the cubic and rhombohedral phases as a function of the ionic radius in the Bi_2O_3-M_2O_3 system, as presented in Figure 2.12. It is clear from this that using dopants with small cationic radius facilitates the stabilisation of the cubic phase.

Further studies of La_2O_3 doping of Bi_2O_3 indicated that a rhombohedral ε-structure exists and that this presents the highest conductivity of all bismuth oxide based systems.[115,116] In Figure 2.13 we present the relative performance of different compositions of doped bismuth oxide samples compared with a standard YSZ electrolyte[117] indicating the attractive performance of this class of materials if purely ionic conductivity is considered.

Moreover, Watanabe reported that all δ-phases reported at room temperature were metastable,[118,119] claiming that all of the stabilised δ-phases reported were nothing but the quenched high-temperature phases which form solid solutions based on δ-Bi_2O_3. These quenched phases are obtained in the bismuth-rich region of a system with Ln_2O_3, Y_2O_3, TeO_2, Nb_2O_5 or Ta_2O_5. On annealing at low temperatures (below approximately 700 °C) they transform gradually to the low-temperature stable modification in the systems with Ln_2O_3, or decompose into two other phases in the other systems, to yield far lower electrical conductivities.[120,121]

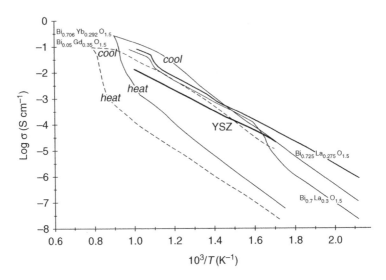

Figure 2.13   Conductivity data of some typical $Bi_{1-x}Ln_xO_{1.5}$ compounds compared with YSZ and BiMeVOX families. Reprinted with permission from Drache *et al.*, 2007[117]. Copyright (2007) American Chemical Society

More recently, it has been demonstrated that stabilisation of the δ-phase in ternary systems, such as Bi_2O_3-Er_2O_3-WO_3,[122] Bi_2O_3-ZrO_2-Nb_2O_5,[123] Bi_2O_3-ZrO_2-Y_2O_3 or Bi_2O_3-Nb_2O_5-Ho_2O_3 solid solutions[124] is possible, leading to the formation of new oxide ion conductors. These ternary systems appear to be more stable in comparison with the metastable binary systems. For example, the $(Bi_2O_3)_{0.705}$ $(Er_2O_3)_{0.245}(WO_3)_{0.05}$ system shows no degradation up to 1100 h at 600 °C[122] and has led to consideration of a new class of ion conductors discussed in the following section.

To improve performance as an electrolyte, thin films of δ-Bi_2O_3 have been synthesised by Laurent *et al.*[125] These authors believe that the small crystallite size (~10–15 nm) achieved through this preparation route might explain the stability of the metastable phase at low temperature, as they achieve a conductivity of about 0.39 S cm^{-1} at 440 °C. Although this high conductivity value is very promising, further work is required in order to probe the stability of these thin films.

2.2.1.6   Novel Materials Based upon Bi_2O_3 and $Bi_4V_2O_{11}$

As previously described, bismuth oxide has high ionic conductivity but also presents several disadvantages, including thermodynamic instability at low oxygen partial pressures, volatilisation at moderate temperatures, high corrosion activity and low mechanical strength.

Alternative bismuth-based materials have been investigated with the aim of exploiting the high oxide ion conductivities of the defect fluorites whilst stabilising the material to lower temperatures and wider pO_2 ranges. Mauvy et al.[126] studied compounds of the general formula $Bi_{18-4m}M_{4m}O_{27+4m}$ where the variable m ($0 \leq m \leq 1$) defines the ratio of the number of $[Bi_{14}M_4O_{31}]$ over the total number of layers in the sequence. Two new compounds were reported: $Bi_{14}P_4O_{31}$ and $Bi_{50}V_4O_{85}$. These materials present monoclinic symmetry and are characterised by cationic ordering of bismuth and the M atoms (M = P, V) in the framework. $Bi_{14}P_4O_{31}$ unfortunately is not a good ionic conductor but the related $Bi_{50}V_4O_{85}$ presents a conductivity of 0.001 S cm^{-1} at 650 °C and has an activation energy of 1.05 eV, close to that of YSZ ($\sim$1 eV).

By double doping with rhenium and a rare earth cation (La, Nd, Eu, Er, and Y), the δ-Bi_2O_3 can be stabilised using even the largest lanthanides, a phenomenon not previously reported, and some compositions show significantly higher conductivities at low temperatures (<400 °C) than other δ-Bi_2O_3 phases.[127] For example, the conductivity of $Bi_{12.5}La_{1.5}ReO_{24.5}$ at 300 °C is about one order of magnitude higher than for phases stabilised by joint Dy/W substitutions, which were previously reported to exhibit the highest conductivities,[128] and also approaches that of the best low-temperature conductor, the two-dimensional $Bi_2V_{0.9}Cu_{0.1}O_{5.35}$ (BICUVOX),[129] as presented in Figure 2.14.

Of further interest are also the solid solution compositions $Bi_{5-x}Ln_xNbO_{10}$.[130] These composition exist up to a value of approximately $x = 0.25$ for La, and to $x = 0.5$ for Gd and Y. Except for Bi_5NbO_{10}, which appears to be a pure ionic conductor, most of the solid solutions exhibit mixed conductivity with predominant ionic transport. Partial Gd and Y substitutions favour electron conductivity in the low temperature range, probably due to bismuth partial oxidation, indicating that electrode applications may be possible, but that again these materials are unsuitable for electrolytes without substantial optimisation.

Another important family of materials based on Bi_2O_3 are those that adopt the Aurivillius type structure (perovskite-related layered structure) with the chemical formula $Bi_4V_2O_{11}$, first reported by Abraham et al.[131] The low temperature phase is α-$Bi_4V_2O_{11}$, and it presents two reversible phase transitions at 450 °C (β-phase) and 570 °C (γ-phase). The high temperature γ-$Bi_4V_2O_{11}$ exhibits fast oxide ionic conductivity of about 0.1 S cm^{-1} at 600 °C, where the oxygen vacancies are disordered.[89] The crystal lattice of the doped γ-vanadate (BIMEVOX) family

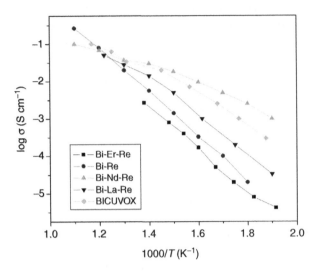

Figure 2.14   Conductivity data of $Bi_{12.5}La_{1.5}ReO_{24.5}$ and other related compounds. Data from Punn et al.[127]

consists of alternating $[Bi_2O_2]^{2+}$ and perovskite-like $[VO_{3.5}]^{2-}$ layers, with oxygen vacancies in the perovskite layers providing ion migration. Aliovalent cations have been used to stabilise the γ-phase at room temperature. Solid solutions of $Bi_2V_{1-x}MO_{5.5-\delta}$ (M = Cu, Ni and $0.07 \le x \le 0.12$) have some of the highest oxygen ionic conductivities reported for this class of materials[132] and, for example, $Bi_4V_{1.8}Cu_{0.2}O_{10.7}$ exhibits a conductivity of 10^{-3} S cm^{-1} at 300 °C, which is approximately two orders of magnitude higher than that of the pure $Bi_4V_2O_{11}$.[133] Despite the reasonably high conduction, the use of Aurivillius phases for electrochemical applications is restricted due to their high chemical reactivity and low mechanical strength.[134]

2.2.1.7   Apatite Structure Oxides

Oxyapatite materials adopt the general formula $A_{10}(MO_4)_6O_{2\pm\delta}$, where A is a rare earth or alkaline earth cation and M is a p-block element such as P, Si or Ge. Their structure consists of isolated MO_4 tetrahedra arranged so as to form oxide-ion and A channels parallel to the c-axis, with the oxide-ion channels being central to the oxide ion conductivity, as shown in Figure 2.15.[135]

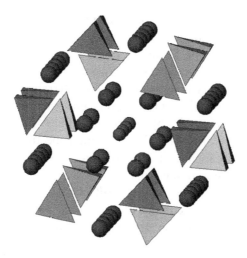

Figure 2.15 Representation of the apatite structure adopted by $A_{10}(MO_4)_6O_{2\pm\delta}$. A atoms are shown as large spheres, oxygen atoms are shown as the small spheres in the middle of the figure and M atoms are located at the centre of the tetrahedra

White and Dong[136] have proposed an alternative description of the apatite structure. They have shown that apatites can be simply described as zeolite-like microporous frameworks, $A_4(MO_4)_6$, composed of face-sharing AO_6 trigonal prismatic columns, that are corner connected to the MO_4 tetrahedra. These frameworks allow flexibility to accommodate the $A_6O_{2\pm\delta}$ units and indicate how the oxygen interstitial excess can easily be accommodated in the structure.

Alternative rare earth apatite materials have been proposed recently as solid electrolyte materials due to the discovery of fast oxide ion conductivity in the silicate-based apatite systems.[137-141] A recent computer modelling study by Tolchard *et al.*[142] indicates that an unusually broad range of dopant ions (in both size and charge state) can substitute for La in the $La_{9.33}Si_6O_{26}$ apatite. The range of dopants is much wider than that observed for doping on a single cation site in most other ionic conductors for fuel cell applications, such as the perovskite-type oxides. The observed conductivity is very sensitive to the doping regime and cation–anion nonstoichiometry. However, the highest conductivities are found for the oxygen-excess samples, with oxide-ion conduction occurring mainly along the oxide-ion channels. There is a wide range of doping possibilities in both the lanthanum site (Mg, Ca, Sr, Ba, Co, Ni, Cu, Mn, Bi) and the silicon site (B, Al, Ga, Zn, Mg, Ti, Ge, Fe, Co, Ni, Cu, Mn, P), as reported by Kendrick *et al.*[135] and references therein, giving considerable scope for future development.

In order to study the conduction mechanism of oxygen ion migration, atomistic modelling studies have been performed,[143] suggesting that the high ionic conductivity in the apatites of nonstoichiometric composition is mediated by oxygen interstitial migration along the c-axis and involving cooperative displacements of the SiO_4 tetrahedra. The predicted location of the interstitial next to the SiO_4 tetrahedra has been confirmed in the oxygen excess samples by structural studies[144] and spectroscopic techniques such as ^{29}Si NMR[145] and Raman spectroscopy.[146] In samples with cation vacancies, the conduction is found to be mediated by Frenkel-type interstitial oxide ions.[146–148]

Water incorporation in $La_{9.33+x}(Si/GeO_4)_6O_{2+3x/2}$ has also been reported leading to an enhancement of the total conductivity.[149] According to the authors, they attributed this increment to proton conduction at temperatures below 600 K in wet atmospheres. Little further work has been reported on the nature of this proton conduction, but it is evidently of significance when contemplating applications.

One of the drawbacks with the apatite ceramics is that despite their good conductivity, synthesis of gas tight membranes has generally required temperatures in the range of 1600–1750 °C. Evidently to implement these materials a significant reduction in sintering temperature is required. To achieve this Chesnaud et al. have used novel sintering and synthesis routes to produce fully dense apatite silicate ceramics through a combination of freeze-drying and spark plasma sintering.[150] This results in a reduction in synthesis temperature of fully dense membranes to temperatures as low as 1200 °C which is then competitive with processing conditions of conventional electrolytes.

High oxide-ion conductivity has also been reported in the germanium-based systems, showing higher activation energies than the analogous silicates.[151] Although higher levels of oxygen excess can be achieved in the germanates, the samples with high hyperstoichiometry show the additional complexity of a change in the crystal symmetry from hexagonal to triclinic, with an associated decrease in the conductivity at low temperatures.[141,152]

Regarding doping strategies for the Ge-based materials, substitution on the La site with divalent cations such as Ba, Sr, Ca, Mg, Cu, Ni, Mn or Co, helps to stabilise the more conductive hexagonal lattice, whereas doping on the Ge site with Co, Al or Ga tends to stabilise the triclinic symmetry.[153] However, the fact that the conductivities in the hexagonal germanate systems appear to be less sensitive than the silicates to the effect of the dopants or the dopant site may indicate a range of conduction pathways. In this trend, Kendrick et al.[154] have reported some atomistic simulation work on $La_{9.33+x}(GeO_4)_6O_{2+3x/2}$, showing the migration of interstitial oxide ions *via* the GeO_4 tetrahedra both parallel and perpendicular to the *c*-axis. These interstitial sites lead to the consequent formation of Ge_2O_9 units with the presence of five-coordinated germanium, also proposed by Pramana *et al.*[155]

Finally, Tsipis *et al.* have studied the integration of the $La_{10}Si_5AlO_{26.5}$ electrolyte with different cathode materials. Problems with surface diffusion of silica from the electrolyte were encountered, leading to a blocking of the electrode–electrolyte interface.[156]

2.2.1.8 LaBaGaO₄ Structured Materials

$LaBaGaO_4$ contains gallium in a distorted tetrahedral environment and ordered alternating layers of lanthanum and barium. The gallate tetrahedra are clearly seen to be isolated moieties within the lattice (Figure 2.16), as reported by Kendrick *et al.*[157] They performed density functional theory (DFT) calculations and found that oxide-ion vacancies were accommodated in the structure through the formation of Ga_2O_7 defects, which are also integral to the mechanism of oxide-ion conduction. The authors describe the oxide-ion conduction mechanism as an unusual cooperative 'cog-wheel'-type process involving the breaking and re-formation of Ga_2O_7 units, as well as the relatively facile rotation of the GaO_4 units. Although it presents good oxygen ionic conductivity at high temperatures, experiments in wet atmospheres demonstrate that this

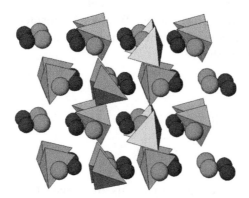

Figure 2.16 Representation of the structure adopted by $LaBaGaO_4$. La atoms are shown in dark grey, Ba atoms are shown in light grey and Ga atoms are located at the centre of the tetrahedra

system has high proton conductivity ($\approx 1 \times 10^{-4}$ S cm^{-1} at 500 °C) and hence may find application in an alternative device.[158]

2.2.1.9 Melilite Structured Electrolytes (LaSrGa$_3$O$_7$)

LaSrGa$_3$O$_7$ adopts the melilite structure, which consists of alternating cationic (La/Sr)$_2$ and corner-sharing tetrahedral anionic Ga$_3$O$_7$ layers and features fivefold tunnels that accommodate the eight-coordinate La/Sr as chains of cations.[159] This composition is an insulator and was found as a secondary phase in the synthesis of the Sr, Mg-doped lanthanum gallate (LSGM) electrolyte.[22] Rozumek et al. showed that the La/Sr ratio can be extended to La$_{1.6}$Sr$_{0.4}$, which presented a conductivity of $\sim$0.1 S cm^{-1} at 950 °C and an oxygen transport number of 0.80–0.95 above 600 °C.[160] However, Raj et al. reported that the La/Sr ratio can only be extended to La$_{1.05}$Sr$_{0.95}$ without the formation of a Sr-doped LaGaO$_3$ perovskite impurity (LSG), with an oxide ion conductivity of $\sim$10^{-3} S cm^{-1} at 950 °C.[161]

Recently, Kuang et al. reported the single phase melilite La$_{1.54}$Sr$_{0.46}$Ga$_3$O$_{7.27}$, obtained by reaction at 1200 °C for 12 h and 1400 °C for 12 h of a starting material with cation ratios La$_{1.54}$Sr$_{0.46}$Ga$_{3.05}$.[162] This material presents a high total conductivity of 0.02 S cm^{-1} at 600 °C and 0.1 S cm^{-1} at 900 °C. Their structural investigation by neutron powder diffraction shows that the interstitial oxygen located within the A cation tunnels at the level of the gallate sheets is responsible for the high oxide ion conductivity at high temperatures, and the decrease of the activation energy from $\sim$0.85 eV below 400 °C to $\sim$0.42 eV over the 600–1000 °C temperature range. Whilst of considerable interest this development requires significant further work before being adopted as a viable solid electrolyte.

2.2.2 Novel Cathodes

2.2.2.1 Perovskite-type Materials

Most of the current cathode materials are based on the perovskite-type ABO$_3$ structure with one of the conventional materials being LSM (La$_{1-x}$Sr$_x$MnO$_3$) as introduced earlier in Section 2.1.3. Other materials currently under development are based on the analogous system

$La_{1-x}Sr_xCoO_3$. These are viewed as traditional cathode materials but through substitution, primarily of the La site, several new cathode compositions have been proposed. These are discussed in subsequent sections.

2.2.2.2   The $Ln_{1-x}Sr_xMnO_3$ System

Introducing A-site substituents into the manganite perovskites has been reported to generate cation vacancies and the effect of this on the ionic and electronic transport properties of these materials has been widely studied.[34,36] For example, the electrical conductivity of $Ln_{1-x}A_x$ $MnO_{3\pm\delta}$ at moderate A^{2+} (A = alkaline earth) concentrations increases with x as the Mn^{4+} fraction increases, as described by Tsipis and Kharton.[1] Complementing these experimental studies De Souza et al.[163] have used atomistic simulation to model the energetics of cation migration and formation in substituted perosvkites. They found that in both rhombohedral and orthorhombic phases oxidative nonstoichiometry leads to cation vacancy formation on *both* cation sites with a tendency for La vacancies to form preferentially. These details are essential in understanding the behaviour of the material in a device and also in understanding the degradation processes that can lead to the formation of insulating phases which affect cathode performance detrimentally, as it is well known that LSM reacts with the YSZ electrolyte forming the insulating pyrochlore phase $La_2Zr_2O_7$. In order to avoid this reaction, Sakaki et al. examined the Pr, Nd and Sm analogues,[164] as the reactivity with YSZ to form $Ln_2Zr_2O_7$ becomes lower on decreasing Ln^{3+} radius. They analysed thermal expansion coefficient, reactivity and electrical conductivity data, and determined that the most promising electrode materials were $Pr_{0.7}Sr_{0.3}MnO_3$ and $Nd_{0.7}Sr_{0.3}MnO_3$. An improvement of between a half and one order of magnitude in the electrical conductivity was also found by Wen et al. for the $Ln_{0.7}Sr_{0.3}MnO_3$ composition ($Ln =$ Pr, Nd, Sm).[165] The superior properties of praseodymium-containing solid solutions is clearly observed in Figure 2.17,[1,166–168] and may be associated with a non-negligible contribution of the $Pr^{3+/4+}$ redox couple at the electrode surface and with a less pronounced interaction with zirconia during cell fabrication and operation. For example, $Pr_{0.7}Ca_{0.3}$ MnO_3 showed a maximum conductivity of 266 S cm^{-1} at 1000 °C (about three times higher than that of LSM) and no evidence of reaction with the electrolyte up to 1200 °C.[169]

Further improvements in the performance of cathode materials can be achieved by co-doping on both A and B sites. Phillipps et al. found that

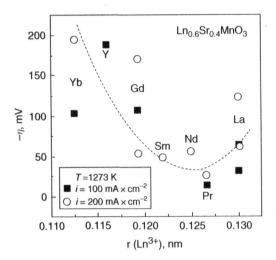

Figure 2.17   Illustration of the cathodic overpotentials of porous $Ln_{0.6}Sr_{0.4}MnO_3/YSZ$ electrodes as a function of the Ln^{3+} cationic radius. Reprinted with permission from Tsipis and Kharton, 2008[1]. Copyright (2008) Springer Science+Business Media

doping with Co on the B site improved the conductivity of Gd_{1-x} $(Sr,Ca)_xMnO_3$ by up to one order of magnitude in comparison with LSM.[170] Partial substitution on the B sites with Al or other transition metal oxides has also been shown to improve the total conductivity.[171,172]

2.2.2.3 $Ln_{1-x}A_xFe_{1-y}M_yO_{3-\delta}$ and $Ln_{1-x}A_xCo_{1-y}M_yO_{3-\delta}$ System

Iron-containing oxide phases are stable under SOFC cathodic conditions and exhibit, in general, larger electronic or mixed conductivity than in the LSM system. The highest level of electronic and ionic transport has been found for the $La_{0.5}Sr_{0.5}FeO_{3-\delta}$[173,174] composition. An important disadvantage of these perovskite ferrites is their high thermal expansion coefficient (TEC) (α values typically between 15 and 25 $\times$ 10^{-6} K^{-1}) in comparison with the common solid electrolytes (of about 10–11 $\times$ 10^{-6} K^{-1}).[175,176] Another important issue is the reactivity of the ferrite with the electrolyte material. A significant advantage with the ferrites is that the rate of formation of the zirconate phase ($La_2Zr_2O_7$) is reported to be much lower in comparison with lanthanum manganites, and can be further reduced by incorporation of Al^{3+}.[177] Double substituted materials with a slight A-site deficiency were investigated by Kindermann *et al.*[178]

showing that the reactivity of $(La_{0.6}Ca_{0.4})_xFe_{0.8}M_{0.2}O_3$ (M = Cr, Mn, Co, Ni) compositions with yttria-stabilised zirconia is minimised when M = Mn, highlighting the complexity of designing a suitable cathode candidate for high temperature operation. For lower temperatures of operation where ceria electrolytes would be used the reactivity with the oxide ferrites is negligible for these compositions. The introduction of doped ceria buffer layers between the cathode and the YSZ electrolyte for high operating temperatures significantly improves the SOFC performance.[179]

Perovskite related cobaltites present considerably better cathodic and transport properties than the analogous manganites and ferrites, however they also present higher thermal and chemical expansion.[180,181] Initially, of most interest was the $La_{1-x}Sr_xCoO_{3-\delta}$ (LSC) composition, possessing significant oxide ion conductivity and high electronic conductivity.[182] In considering the use of cobaltite cathodes the choice of the electrolyte is of importance. LSC was found to react with zirconia but presents superior performance when tested with $Ce_{0.8}Sm_{0.2}O_{1.9}$ (CSO) and $Ce_{0.8}Gd_{0.2}O_{1.9}$ (CGO).[183]

Adler[184] has investigated the behaviour of LSC porous electrodes by AC impedance spectroscopy and has found that the oxygen reduction reaction is limited by surface chemical exchange and solid state diffusion, contrary to the commonly accepted view that the electrode reactions are charge transfer limited. Furthermore he has also found that the reduction reaction extends over several micrometres and under certain conditions may approach the thickness of the electrode. Isotope exchange measurements on dense LSC deposited on a $Ce_{0.9}Ca_{0.1}O_{1.9}$ electrolyte were performed by Kawada et al.[185] These experiments showed that the isotope exchange rate was controlled by the surface exchange process, supporting the findings of Adler. van Doorn and Burggraff[186] have also focused on the effect of structure on the ionic conductivity. A variety of compositions ($0 < x < 0.8$) were studied and they have found evidence for the existence of oxygen vacancy ordering in these materials where $x > 0.5$. At this composition LSC adopts a cubic structure, whereas at $x < 0.5$ it is rhombohedral. High resolution transmission electron microscopy (HRTEM) and selected area electron diffraction (SAED) measurements at the $x = 0.7$ composition confirmed the presence of superstructure ordering, where it was proposed that the ionic conductivity would be lower than in disordered regions.

Replacement of La with Gd, Sm or Dy in LSC has been investigated in an attempt to produce materials of greater chemical stability, but it was established that at temperatures of greater than 900 °C $SrZrO_3$ phases formed, and therefore these materials are only attractive candidates for low temperature SOFC applications.[170,187,188]

To overcome the technological problems associated with the LSC materials at high temperatures, B-site substitutions have been developed. One interesting example refers to the $La_{1-x}Sr_xCo_{1-y}Fe_yO_3$ (LSCF) composition, initially studied by Tai et al.[189] In this work they identified the rhombohedral/orthorhombic transition at $y = 0.8$ for the $x = 0.2$ composition and also studied the effect this had on thermal expansion. Details of TECs and total conductivity values for different compositions can be found in Table 2.4. As can be seen in Table 2.4, moderate dopant additions provide a significant enhancement in the total conductivity and electrochemical activity, but also increase apparent TECs. Kostogloudis and Ftikos[190] reported that, on the creation of La-site cation vacancies, the conductivity and thermal expansion both decrease owing to the dominant charge compensation mechanism *via* oxygen vacancy formation. The charge compensation mechanism in these A-site deficient perovskites is then the formation of oxygen vacancies, rather that the oxidation of Mn^{3+} to Mn^{4+} as for LSM.

Tu et al.[191] have also studied a series of LSCF materials varying the type of lanthanide ion. They found that all compositions had a high electrical conductivity with $Nd_{0.6}Sr_{0.4}Co_{0.8}Fe_{0.2}O_{3-\delta}$ having the highest conductivity ($\sim$600 S cm^{-1} at 400 °C), and catalytic activity for oxygen reduction.

Table 2.4 Total conductivity and thermal expansion coefficients for $La_{1-x}Sr_x$ $Fe_{1-y}Co_yO_{3-\delta}$ ceramics in air. Data from Tsipis and Kharton[1]

x	y	Total conductivity (S cm^{-1})	Total conductivity (S cm^{-1})	Average TEC	Average TEC	Ref.
		873 K	1073 K	T (K)	$\alpha \times 10^{-6}$ (K^{-1})	
0	0.2	0.8	4.5	573–1173	17.5	[189]
0.2	0	93	1.09×10^2	573–1173	12.6	[189]
0.2	0.2	1.75×10^2	1.91×10^2	373–1173	15.4	[189]
0.2	0.2	1.27×10^2	1.49×10^2	373–1173	15.4	[189]
0.2	0.4	3.39×10^2	2.87×10^2	373–1173	17.6	[189]
0.2	0.6	4.14×10^2	4.55×10^2	373–1173	20	[189]
0.2	0.8	1.05×10^3	9.95×10^2	373–1173	20.7	[189]
0.2	1.0	1.69×10^3	1.52×10^3	303–1273	18.5	[331]
0.4	0.2	3.35×10^2	2.79×10^2	373–873	15.3	[189]
0.4	0.2	2.75×10^2	3.33×10^2	303–1273	17.5	[331]
0.4	0.2	4.60×10^2	3.30×10^2	973	15.3	[190]
0.6	1.0	1.81×10^3	1.16×10^3	303–1273	25.1	[331]
0.7	0.6	1.73×10^2	1.29×10^2	303–1273	24.1	[331]
0.7	0.8	4.80×10^2	3.88×10^2	303–1273	21.0	[331]
0.8	1.0	8.10×10^2	5.78×10^2	303–1273	25.6	[331]

2.2.2.4 Ruddlesden–Popper Type Compounds

Ruddlesden–Popper type materials have the general formula $A_{n+1}M_nO_{3n+1}$, and adopt the structures observed in the $Sr_{n+1}Ti_nO_{3n+1}$ titanates, first reported by Ruddlesden and Popper.[192] The structure of the Ruddlesden–Popper phases consists of n consecutive perovskite layers $(AMO_3)_n$ alternating with rock salt layers (AO), with AO along the crystallographic c direction. The well-known perovskite AMO_3 and K_2NiF_4-type (A_2MO_4) structures correspond to $n = \infty$ and $n = 1$, respectively (Figure 2.18). Ruddlesden–Popper phases present attractive electrochemical and transport properties, but chemical phase stability in the range of temperatures and oxygen chemical potentials necessary for the SOFC applications is still an issue.[193–195] Several K_2NiF_4-type compounds have been proposed as SOFC cathodes, including those with A = La, Sr, Ba, Pr, Nd and M = Ni, Cu, Co, Fe.[196–199] The total conductivity

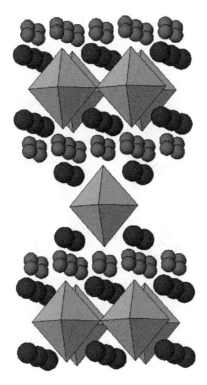

Figure 2.18 Representation of the K_2NiF_4-type (A_2MO_4) structure. A atoms are shown as larger spheres, oxygen atoms are shown as smaller spheres and M atoms are located at the centre of the octahedra. Oxygen positions indicated are partially occupied and represent the excess oxygen positions

of K_2NiF_4-type nickelates remains predominantly p-type electronic in the entire pO_2 range. For these compounds, the bulk ionic transport occurs *via* anisotropic diffusion of interstitial ions in the rock-salt-type layers and vacancies in the perovskite layers.[198] Oxygen permeation through the dense material is reported to be limited by the surface exchange rate.[199] The hole transport is lower than that in the perovskite analogues, but still sufficient for practical applications. Small amounts of A^{2+} doping leads to a higher conductivity and lower oxygen content, with a negative impact on the oxygen diffusivity, which is determined by the interstitial anion migration.[200] Evidence for the diffusion path of oxide ions in a K_2NiF_4-type mixed conductor was firstly reported by Yashima *et al.*[201] for the $(Pr_{0.9}La_{0.1})_2(Ni_{0.74}Cu_{0.21}Ga_{0.05})O_{4+\delta}$ compound through a high-temperature neutron powder diffraction study. The nuclear-density distribution shows the two-dimensional network of the O2-O3-O2 diffusion paths of oxide ions where O2 represents the oxygen at a $(0, 0, z)$ position, and O3 represents the interstitial oxygen located at a $16n$ site (see Figure 2.19). This result is also consistent with the anisotropic transport of oxide ions in $La_2NiO_{4+\delta}$.[202,203]

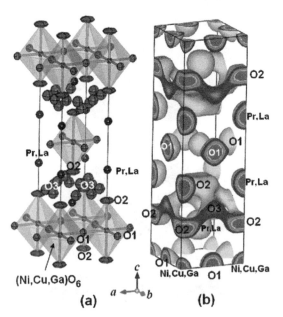

Figure 2.19 Representation of the (a) crystal structure and (b) isosurface of nuclear density at 0.05 fm $Å^{-3}$ for the $(Pr_{0.9}La_{0.1})_2(Ni_{0.74}Cu_{0.21}Ga_{0.05})O_{4+\delta}$ compound determined *in situ* at 1015.6 °C. Reprinted with permission from Yashima *et al.*, 2008[201]. Copyright (2008) American Chemical Society

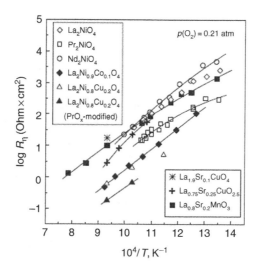

Figure 2.20 Illustration of the area-specific polarisation resistance as a function of the temperature for different porous oxide cathodes in air. All data correspond to electrochemical cells with YSZ electrolytes, except for $La_2Ni_{0.9}Co_{0.1}O_{4+\delta}$ and $La_2Ni_{0.8}Cu_{0.2}O_{4+\delta}$, where LSGM electrolytes were used. Reprinted with permission from Tsipis and Kharton, 2008[1]. Copyright (2008) Springer Science + Business Media

Figure 2.20 presents the electrochemical behaviour of different nickelate-based electrodes and it is clear that for the $Ln_2NiO_{4+\delta}$ (Ln = La, Nd, Pr) series, relatively low polarisation resistances were observed.[126] The high performance of the Pr analogue may be a result of the metastability of the praseodymium nickelate, which decomposes into PrO_x and $Pr_4Ni_3O_{10-\delta}$, having higher ionic and electronic conductivities.[194] Kim et al.[204] performed AC impedance measurements and suggested that the electrode reaction is limited by the surface exchange reaction for the $La_2NiO_{4+\delta}$. These results are in agreement with the enhancement of the electrochemical activity of $La_2Ni_{0.8}Cu_{0.2}O_{4+\delta}$[205] and $LaNi_{0.5}Fe_{0.5}O_{3-\delta}$[206] by surface modification with praseodymium oxide.

Much of the work on these materials has focused on the chemical and electrical characterisation. Compatibility of these materials with the electrolyte is also an important issue to be addressed. For this purpose, Munnings et al.[207] reported on the stability and reactivity of the $La_2NiO_{4+\delta}$ with the LSGM electrolyte. They found an effect on the surface stoichiometry and that this stoichiometry change adversely affected the cathode performance. They also observed that the best performance of the $La_2NiO_{4+\delta}$ cathode was on surfaces with elevated levels of strontium and magnesium. Thermodynamic modelling calculations made by Solak et al.[208] also show that $La_2NiO_{4+\delta}$ is not chemically compatible with the LSGM electrolyte not only at fabrication conditions but also at operation conditions. On the contrary, Sayers et al.[195] recently found that there is no evidence of secondary phase formation on $La_2NiO_{4+\delta}$-LSGM mixtures from the obtained diffraction data over a period of 72 h at temperatures of up to 1000 °C. They also found that there is significant reactivity between $La_2NiO_{4+\delta}$ and CGO after 24 h at 900 °C, with the formation of a higher order Ruddlesden–Popper $(La_{n+1}Ni_nO_{3n+1})$ phase as one of the reaction products.

Another interesting material presenting the K_2NiF_4-type structure is $Sr_{2-x}La_xMnO_{4+\delta}$. Munnings et al.[209] reported that this material presents similar TEC to the most common electrolyte materials and is also chemically stable over a wide range of oxygen partial pressures. Sun et al. recently studied the reactivity between $Sr_{1.4}La_{0.6}MnO_{4+\delta}$ and a CGO electrolyte and no reaction was found after heat treatment at 1000 °C for 12 h.[210] However, these materials have been shown to have relatively low electronic conductivities of the order of 2 S cm^{-1} and are therefore unlikely to be used as cathodes.

2.2.2.5 Barium Strontium Cobalt Ferrite (BSCF)

$Ba_{0.5}Sr_{0.5}Co_{0.8}Fe_{0.2}O_{3-\delta}$ is a cubic perovskite material in the $BaCoO_{3-\delta}$-$SrCoO_{3-\delta}$ system, which was first developed for high-temperature oxygen permeation membrane applications.[211] Shao and Haile[42] first proposed this material for intermediate temperature fuel cell applications and reported high power densities of about 1 W cm^{-2} at 600 °C and 0.4 W cm^{-2} at 500 °C on a BSCF-SDC-Ni/SDC cell when operated with humidified hydrogen and air, in the anode and cathode side, respectively. Wang et al.[212] calculated the TEC for the $Ba_{0.5}Sr_{0.5}Co_{0.8}Fe_{0.2}O_{3-\delta}$ composition from high temperature XRD data and they obtained a value of 11.5×10^{-6} K^{-1}, which is comparable with most of the electrolytes used in SOFCs. Higher values for the TEC ($\sim$20 $\times$ 10^{-6} K^{-1}) were obtained from neutron diffraction data by McIntosh et al. for the same composition,[213], and also measured by Zhu et al. (19.2–22.9 $\times$ 10^{-6} K^{-1}).[214] In this work they also reported on the chemical compatibility of BSCF with 8YSZ and 20 CGO. They found no reaction up to 800 °C but above 800 °C severe reactions were detected. In terms of chemical compatibility, similar results were also obtained by Duan et al.[49] BSCF cathodes have, however, also been tested with LSGM as the electrolyte with no apparent reaction between phases.[215] Nevertheless best results were found when using the BSCF cathode with CGO as the electrolyte. For example, Liu et al. obtained $\sim$1300 mW cm^{-2} at 600 °C on a BSCF-CGO-Ni/CGO cell.[216]

Li et al. showed by AC impedance spectroscopy that Nd-doped BSCF presents higher electrical conductivity than BSCF. It also presents better electrochemical performance than the BSCF cathode, probably determined by a combination of the high electrical conductivity and the amount of oxide ion vacancies in the cathode.[217]

2.2.2.6   Cerium Niobate

Cerium niobate ($CeNbO_{4+\delta}$) is of interest for SOFC applications due to its wide range of oxygen stoichiometries, varying from $CeNbO_4$ to $CeNbO_{4.33}$.[218] It also presents a large open structure, which is potentially beneficial for fast oxide ion conduction. The crystal structure of this material at room temperature is monoclinic fergusonite ($I2/a$), and on heating exhibits a phase transition at approximately 750 °C in air to a tetragonal scheelite ($I4_1/a$).[219] The scheelite material exhibits a reasonably high total conductivity (0.030 S cm^{-1} at 850 °C) and also presents mixed ionic/p-type electronic conduction (ionic transference number = 0.4),[220] due to the mixed valence of Ce and interstitial oxygen anions. In this work, it was found that the high temperature scheelite polymorph is unstable at low partial oxygen pressures. Packer et al.[221] have studied the oxide ion diffusion in cerium niobate and found a relatively high value for $D^* = 8.29 \times 10^{-8}$ cm^2 s^{-1} at 850 °C and a good surface exchange coefficient ($k^* = 6.85 \times 10^{-7}$ cm s^{-1} at 850 °C). Despite its high oxide ion diffusion, the electronic conductivity is currently too low for SOFC cathode applications. More work is therefore required in order to improve this aspect of the material, but improvements could be achieved through the use of composites (with an electronic component) and also via substitution on the Ce site,[222] with for instance La, or on the Nb site[223] with vanadium. The full range of possible substitutions is extensive and should be carefully considered to optimise all aspects of cathode performance.

Recent studies by in-situ powder XRD revealed that $CeNbO_4$ decomposed to CeO_2 and $CeNb_3O_9$[224] under a pure oxygen atmosphere. The decomposition temperature is shown to decrease with decreasing partial pressure of oxygen, and decomposition of $CeNbO_{4+\delta}$ at lower partial pressures of oxygen is kinetically hindered, as can be seen in Figure 2.21.

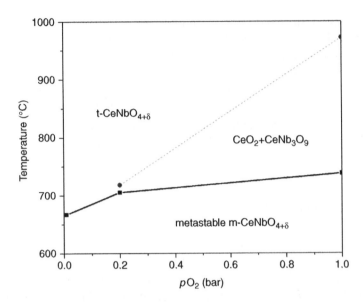

Figure 2.21   Illustration of the decomposition temperature of $CeNbO_{4+\delta}$ as a function of the partial pressure of oxygen. Data from Vullum and Grande[224]

2.2.2.7   Double Perovskites, $A_2B_2O_{6-\delta}$

$GdBaCo_2O_{5+\delta}$ (GBCO) adopts the double perovskite structure with orthorhombic symmetry, in which Co ions are coordinated in square pyramids (CoO_5) and octahedra (CoO_6), which alternate along the b-axis.[225] The Ba cations are ordered in alternating (001) layers with oxygen vacancies mainly located along (100), in the $[GdO]_x$ planes (see Figure 2.22).[226] This material was reported to have rapid oxygen transport kinetics at low temperatures,[227] and was first proposed for IT-SOFC cathode applications by Chang et al.[228] They found by AC impedance spectroscopy that GBCO cathode materials exhibit good performance at low temperatures on CGO electrolytes. Furthermore, high oxygen surface exchange ($k^* = 2.8 \times 10^{-7}$ cm s^{-1}) and reasonable oxide ionic diffusivity ($D^* = 4.8 \times 10^{-10}$ cm^2 s^{-1}) were obtained for GBCO at 575 °C.[229] These authors also reported an area specific resistance (ASR) of 0.25 Ω cm^2 at 625 °C for a GBCO/ CGO/GBCO cell. Similar results were obtained by Li et al.[230] for the GBCO cathode on the CSO electrolyte. Peña-Martìnez et al.[231] also reported no reactivity and good electrochemical performance of the GBCO cathode with the LSGM electrolyte, indicating that this is one of the most exciting alternatives to the ABO_3 cathodes discussed earlier.

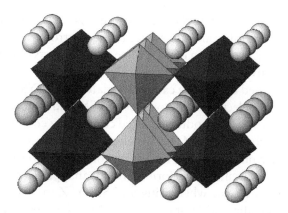

Figure 2.22 Representation of the double perovskite structure adopted by the $GdBaCo_2O_5$. Gd atoms are shown as light grey spheres, Ba atoms are shown as dark grey spheres and Co atoms are located at the centre of the polyhedra

The effect of Sr substitution for Ba in the GBCO material on the structural chemistry has been investigated.[232] The $GdBa_{1-x}Sr_xCo_2O_{5+\delta}$ system exhibits a structural change from an orthorhombic *Pmmm* ($x = 0$) to tetragonal *P4/mmm* ($x = 0.2-0.6$) to a further orthorhombic *Pnma* ($x = 1$) structure. The $x = 0.2$ and 0.6 samples exhibit higher power density when tested as an SOFC cathode than either the $x = 0$ and $x = 1$ samples, partly due to faster oxygen transport within the tetragonal structure. While the parent $GdBaCo_2O_{5+\delta}$ sample suffers from interfacial reaction with LSGM and CGO electrolytes at 1100 °C, Sr substitution for Ba greatly improves the chemical stability of $GdBa_{1-x}Sr_xCo_2O_{5+\delta}$.

The analogous material $PrBaCo_2O_{5+\delta}$ (PBCO) has also been suggested as a SOFC cathode due to its unusually rapid oxygen transport kinetics at low temperatures (300–500 °C).[233,234] Zhu et al.[235] demonstrated the suitability of PBCO for IT-SOFC applications. They performed electrochemical measurements on a PBCO/SDC/Ni-SDC single cell obtaining a maximum power density of about 850 mW cm^{-2} at 650 °C. Moreover, the samarium derivative ($SmBaCo_2O_{5+\delta}$) presents good electrochemical properties and has also been recently proposed as a cathode material for IT-SOFC.[236]

2.2.2.8 Mixed Conducting Composites

The use of composite cathodes generally includes an electronically conducting material and a solid electrolyte. By the introduction of the solid electrolyte, the electrochemical reaction zone is enlarged, and the microstructural stability and adherence to the electrolyte is therefore also enhanced. The incorporation of crystalline nanocatalysts onto the cathode could also improve the performance of the electrodes.[237–239]

The most studied composite cathode to date is LSM/YSZ.[240–243] The connectivity of the electronically conducting LSM and ionically conducting YSZ phases is crucial to produce a mixed conducting composite, because the ionic conductivity of LSM and the electronic conductivity of YSZ are negligible under normal SOFC operating conditions. Results from electrochemical measurements performed by Haanappel *et al.*[244] showed the highest performance when using an LSM/YSZ mass ratio of 50/50. Recently, Princivalle and Djurado reported on the fabrication of LSM/YSZ composite films prepared by electrostatic spray deposition (ESD) with various compositions and microstructures.[245] By this technique they prepared LSM/YSZ composites with simultaneous graded composition and porosity adjusting the YSZ content and nozzle-to-substrate distance at a constant substrate temperature. Electrical and electrochemical properties of these promising composites are still unknown. Huang *et al.* also reported an improvement on the performance of LSM-based cathodes through doping with Co-containing compounds, forming the $La_{0.8}Sr_{0.2}Mn_{0.75}Co_{0.25}O_3$/YSZ composite.[246]

Composites of LSM with other electrolyte materials such as scandia-stabilised zirconia (ScSZ),[247,248] CGO,[249] $Ce_{0.7}Bi_{0.3}O_2$ (CBO)[250] or yttria-stabilised bismuth oxide (YSB)[251] have also been shown to improve the cathode performance. Another interesting manganite-based composite is the $La_{0.8}Sr_{0.2}Sc_{0.1}Mn_{0.9}O_{3-\delta}$-ScSZ. Using a mass ratio of 80/20, Zheng *et al.*[252] showed high power densities of about 1200 mW cm^{-2} at 800 °C on an anode supported Ni-ScSZ/ScSZ/$La_{0.8}Sr_{0.2}Sc_{0.1}Mn_{0.9}O_{3-\delta}$-ScSZ cell.

MIECs used in combination with the electrolyte material (usually CSO or CGO) to form a composite also present superior performance in comparison with the pure MIEC, as shown, for example, in the $Sm_{0.5}Sr_{0.5}CoO_3$-$Sm_{0.2}Ce_{0.8}O_{1.9}$ (75/25 wt%),[253] $SrCo_{0.8}Fe_{0.2}O_{3-\delta}$-$La_{0.45}Ce_{0.55}O_{2-\delta}$ (50/50 wt%),[254] $LaNi_{0.6}Fe_{0.4}O_3$-CSO (50/50 wt%)[255] and $Pr_{0.7}Sr_{0.3}Co_{0.9}Cu_{0.1}O_{3-\delta}$-CSO (65/35 wt%)[256] composites. Adding an ionic conducting phase to the LSCF electrode also reduces the cathode polarisation, as reported for LSCF-CGO and LSCF-CSO,[257–260] LSCF-LSGM[261] and LSCF-YSZ[262] composites.

Camaratta and Wachsman[263] recently reported on the $Bi_2Ru_2O_7$-$Bi_{1.6}Er_{0.4}O_3$ composite applying pure $Bi_2Ru_2O_7$, a pyrochlore oxide, as a current collector to the electrode surfaces. The lowest value of ASR they obtained was 0.73 Ω cm^2 at 500 °C and 0.03 Ω cm^2 at 700 °C, one of the lowest SOFC electrode ASR values reported to date, making this composite a good candidate for IT-SOFCs.

Concerning the Ruddlesden–Popper series, Laberty et al.[264] reported on the performance of SOFC cathodes with lanthanum-nickelate-based composites. They showed that lanthanum nickelate performs poorly when used as a single-phase cathode in yttria-stabilised, zirconia-based air-H_2 button cells at 800 °C. However high power densities of up to 2.2 W cm^{-2} were measured using a $La_2NiO_{4+\delta}$-CSO composite bilayer cathode.

BSCF composites, such as BSCF-$LaCoO_3$,[265] BSCF-CSO[266] and BSCF-$Sm_{0.5}Sr_{0.5}CoO_{3-\delta}$[267] have also been proposed as good IT-SOFC cathodes. For example, the BSCF-$LaCoO_3$ composite presents an ASR as low as 0.21 Ω cm^2 at 600 °C for the 30 vol% of LSC-BSCF.[265]

2.2.3 Ceramic and Sulfur Tolerant Anodes

As discussed earlier in Section 2.1.2, conventional cermet anodes suffer from issues of redox stability, coking and sulfur poisoning. In an effort to develop high durability anodes that overcome these limitations the modification of typical cermet materials to offer greater resistance to coking and sulfur poisoning has been considered. Elsewhere novel ceramic anodes have been proposed to achieve the same objective. Each of these potential materials solutions are attractive and will be addressed in turn.

2.2.3.1 Redox Tolerant Cermets

A number of approaches have been adopted to modify the current Ni-based cermets including substitution of Ni with alternative metals. Replacement of Ni with Cu has been reported to reduce the affinity of Ni for hydrocarbon cracking that leads to coking of the anode. In developing Cu cermets many strategies have demonstrated successful anode performance. Park et al. pioneered the use of Cu-based cermets demonstrating that SOFCs can operate effectively with a Cu-ceria cermet, where the ceria acts as a catalyst for the oxidation of hydrocarbons and Cu is used as it is an excellent electronic conductor.[268,269] Synthesis of the cermets has generally focused on the use of impregnation methods to infiltrate the Cu phase into the YSZ parent phase. Jung et al. have investigated the use of both aqueous $Cu(NO_3)$ and aqueous $Cu(NO_3)$/urea[270] to introduce Cu and have studied the effect of the starting solution on the particle

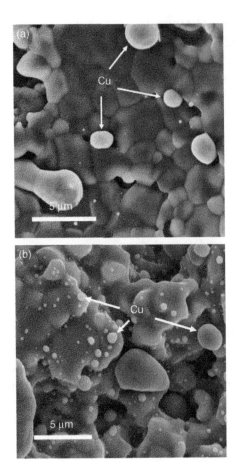

Figure 2.23   Scanning electron microscopy image showing the incorporation of Cu by infiltration into a novel anode. (a) Nitrate and (b) nitrate/urea infiltration. Reprinted with permission from Jung *et al.*, 2006[270]. Copyright (2006) Elsevier

morphology and ultimately fuel cell performance. A typical micrograph illustrating the distribution of Cu achieved is shown in Figure 2.23.

Cu is a poor catalyst with respect to the C-H bond and hence overcomes the issues of carbon deposition, allowing methane and alternative higher hydrocarbons to be used as the fuel. This is illustrated by the direct comparison of the performance of anodes both with and without the ceria content (Figure 2.24), as detailed by He *et al.*[271] Indeed Gorte *et al.* demonstrated that a cell with a Cu-ceria anode operating on n-butane, toluene, methane, ethane and 1-butene had reasonable power densities of 0.12 W cm^{-2} at 700 °C.[269] However, the power densities obtained are relatively modest, as acknowledged by the authors, and compare

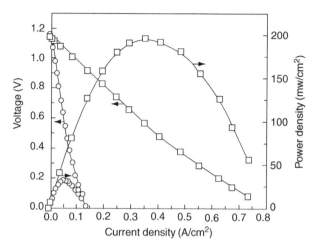

Figure 2.24 Fuel cell performance curves for a YSZ-based SOFC with a Cu impregnated anode. Open squares are data with a ceria layer, open circles are data obtained from a cell without the ceria layer. Reprinted with permission from He *et al.*, 2003[271]. Copyright (2003) *J. Electrochem. Soc.* – American Institute of Physics

unfavourably with cells using H_2 as a fuel (0.3 W cm^{-2})[269] and to those cells operating with Ni-YSZ anodes on H_2 at 800 °C (1.0 W cm^{-2}).[272] These comparisons are perhaps somewhat unfair as the Ni-YSZ-based cells have undergone decades of materials, composition and microstructural optimisation that could conceivably enhance the performance of the Cu-based system proposed. As illustration of this Costa-Nunes *et al.*[273] developed a Cu-CeO$_2$-YSZ anode system using CO as a fuel that achieved higher performance than with H_2 fuel, achieving up to 370 mW cm^{-2}. A further development has been demonstrated by Ye *et al.*[274] where a Cu-CeO$_2$-ScSZ cermet was tested with H_2 and C_2H_5OH/H_2O environments, but poor performance resulted at all temperatures, achieving a maximum power density of 372 mW cm^{-2} at 800 °C and only 130 mW cm^{-2} at 700 °C.

To illustrate the effect of Cu content on carbon formation a series of Cu-Ni alloys were incorporated into YSZ-based anodes and significant suppression of coking achieved. It was also noted that the use of alloys is a relatively unexplored domain and of interest as the catalytic activity of the alloy is not directly related to the properties of the parent metals, and hence the alloy activity is not simply a summation of the two metals individual properties.[275]

Of course the use of Cu-based cermets raises the issue of both chemical and thermal compatibility between components used in the cermet and recent work has indicated that YSZ and CuO will react, producing a monoclinic ZrO_2 phase.[276] Whilst CuO is not used in the anode itself many manufacturing processes are initiated with the metal oxide and subsequently reduced. Starving the fuel cell of fuel during thermal excursions could also cause oxidation of the anode and hence reactivity of the oxides is significant. Although Ruiz-Morales et al.[276] reported reactivity they also indicated that the formation of monoclinic ZrO_2 did not adversely affect the polarisation resistance of the cell which was attributed to the compensation of the low conductivity of the ZrO_2 phase by the high electronic conduction of Cu.

Further studies have been undertaken to characterise the potential of cermets based upon Cu-CSO.[270, 277–282] In most cases it was found that additional CeO_2 had to be incorporated into the composition as CSO is a relatively poor catalyst for hydrocarbon oxidation. There was speculation that despite Cu having no hydrocarbon oxidation activity, the catalysis of water gas shift reactions by Cu could have a beneficial effect and hence a comparison of Cu-CSO-CeO_2 and Au-CSO-CeO_2 was performed. Measurements in gas atmospheres of H_2 and n-butane at 650 °C, showed no significant difference in performance, indicating that Cu serves as a purely electronic conductor in these cermets[279] and that CeO_2 adopts a critical catalytic role in hydrocarbon oxidation.

Further modification of the composition of cermets by introducing LSGM to the CeO_2:Cu ratio was also investigated with the intention of using these anodes at intermediate temperatures. Three ratios of Cu:Ni in LSGM were studied (0.7:0.3, 0.5:0.5, 0.3:0.7) with the conclusion being that LSGM showed good stability towards the metallic components at temperatures below 1000 °C, however the poor catalytic activity of the Cu negated the use of these materials.[283] A proposal to introduce CeO_2 to this composition was considered to enhance the catalysis. In this case however only H_2 gas was considered and no hydrocarbon content included in the fuel feed. In contrast, An et al.[284] achieved a far superior performance on a thick (440 μm) LSGM electrolyte with a Cu-CeO_2-LSGM anode in both H_2 and C_4H_{10} fuels, attaining a maximum power density of 220 mW cm^{-2} at 700 °C. Analysis of the electrode data obtained from these cells suggested that the anode was limiting the cell performance as two-thirds of the cell polarisation resistance could be attributed to the anode.

Of course there is no reason to restrict the cermet to a single metal type and efforts have been expended to introduce Fe and Co to the Ni-YSZ cermet.[285] Through the use of a combustion synthesis technique it has been shown that nanometric scale powders with high surface areas,

Table 2.5 Electrochemical performance of metal-YSZ cermets produced by co-precipitation. After Grgicak et al.[286] Samples prepared at different pH values

Sample	R1 (Ω cm^{-2})	R2 (Ω cm^{-2})	R3 (Ω cm^{-2})
Ni-YSZ-2	9	11	36
Ni-YSZ-3	14	46	21
Cu-YSZ-3	31	55	95
Cu-YSZ-5	10	28	38
Co-YSZ-2	7	15	20
Co-YSZ-3	4	1	15

essential for catalytic activity, can be easily produced. In this work the Ni-Fe composition was determined to have the highest surface area by BET, with 33 m^2 g^{-1}. Whilst Ni and Cu cermets are the most prevalent there has also been some interest in Co-YSZ cermets[286] with comparison of the effect of synthesis conditions on the overall performance of the three types of anode. Significant differences in the impedance spectra of the three anodes was observed (Table 2.5) which the authors ascribe to differences in the particle size ratio between the YSZ phase and the metal. In each case three components were identified which are attributed to high, medium and low frequency responses corresponding to charge transfer, dual phase boundaries and mass transport, respectively.

One further materials solution that has been proposed is the use of substituted YSZ in the cermet. Commonly the use of Ti as a substituent has been attempted,[287,288] with a typical composition being $Y_{0.2}Ti_{0.18}Zr_{0.62}O_{1.9}$ (YZT) combined with CuO. These cermets were shown to have the requisite high electronic conductivity but unfortunately they underwent Cu segregation at temperature leading to a degradation of the microstructure of the electrode. However, electrochemically they were identified as favouring hydrogen oxidation reactions rather than H$_2$ evolution, making these materials good candidates for SOFCs. In this work, however poor electrode microstructure led to relatively poor performance. Further optimisation of the YZT concept was achieved through the further substitution of Ce giving $Y_{0.15}Zr_{0.57}Ti_{0.13}Ce_{0.15}O_{1.925}$ which was then tested with both NiO and CuO as cermets.[288] The best performance was identified for the YSTC sample with 60% CuO at 500 °C, with conductivity five times higher than that for the corresponding NiO cermet. Again, it was found that degradation of the microstructure occurred with the Cu-based cermets on increasing temperature, and hence is a concern when considering the Cu-based cermets for SOFCs. These authors also found that performance of all of the composites tested was inferior to the

Ni-YSZ standard and conclude that this may be due to the microstructure, again indicating the difficulty of designing a functional anode that competes with or improves upon Ni-YSZ.

2.2.3.2 Ceramic Anodes

Perhaps a logical step in the evolution of anode materials is the development of single phase ceramic or oxide anodes. Evidently this solution presents some significant materials challenges, perhaps mostly concerned with stability at low pO_2, but offers the potential of sulfur tolerance and thus higher durability. Given the large array of possible oxides that could be tested as anodes it is perhaps unsurprising that efforts were directed towards familiar structural chemistries, and hence early studies have considered perovskite-type anodes,[289–291] with excursions to tungsten bronzes[292,293] and double perovskites.[294] Relatively few studies of the tungsten bronze type, $A_{0.6}BO_3$ compositions have been undertaken, with the best performance found to be with the $Sr_{0.2}Ba_{0.4}Ti_{0.2}Nb_{0.8}O_3$ material that achieved a conductivity of 10 S cm^{-1} at 10^{-20} atm.[292,293] Despite this, overall cell performance was found to be inferior to current standard materials and hence further work has not been forthcoming with this class of materials. $Sr_2Mg_{1-x}Mn_xMoO_{6-\delta}$, a double perovskite, was developed to extend the lifetime of anodes and findings confirmed that if only the Sr_2MgMoO_6 composition was employed sulfur tolerance was much improved, indicating the negative effect of Mn in this case. However, despite this initial promise no further advances in double perovskite anodes have been presented.

One of the earliest reports of a novel perovskite oxide anode was concerned with the use of $La_{0.6}Sr_{0.4}Co_{0.2}Fe_{0.8}O_{3-\delta}$[295] as an anode rather than a cathode with the intention being to directly oxidise methane at the anode. These authors investigated the stability of LSCF in anodic conditions consisting of temperature programmed reaction of 3% CH_4 in 6% O_2 with an unspecified balance gas. From these measurements it was evident that the catalyst was active for methane decomposition but that there were concerns over the stability of the composition at elevated temperatures for timescales applicable to devices. Further developments have subsequently focused on the use of Mn- and Cr-based lanthanide perovskites with the aim being to stabilise the material at low pO_2 and high temperature.

Several authors have investigated the substitution of $La_{1-x}Sr_xCrO_{3-\delta}$, using Mn,[289,290] Ru,[296] Ni[297] and Fe[298,299] with studies of the Fe substituted chromates amongst the earliest ceramic anode studies. Ramos and Atkinson[298] investigated the oxygen transport characteristics of the $La_{1-x}Sr_xCr_{1-y}Fe_yO_{3-\delta}$ material where $x = 0.2$, 0.4 and 0.6 under both oxidising and reducing conditions, finding that significant oxygen transport was present under a $H_2/N_2/H_2O$ atmosphere. Indeed these authors found that under oxidising conditions at a temperature of 800 °C oxygen diffusion coefficients (D^*) of the order of 10^{-8} cm^2 s^{-1} were obtained for materials where $x = 0.4$, whilst somewhat surprisingly under reducing conditions the diffusion coefficient at 800 °C was of the order of 10^{-7} cm^2 s^{-1}. A further increase in D^* on proceeding from $x = 0.2$ to $x = 0.4$ of two orders of magnitude was also observed. Whilst these data are encouraging there remain concerns over this material regarding its redox stability.

To overcome these concerns the majority of subsequent studies focused on the use of Mn substitution in the chromate lattice. Tao and Irvine[289] for example developed the $La_{0.75}Sr_{0.25}Cr_{0.5}Mn_{0.5}O_{3-\delta}$ (LSCM) composition investigating the structural chemistry and redox stability, finally investigating the performance of the material as an anode in a single fuel cell. Through a variety of testing conditions including atmospheres containing CH_4, 5% H_2 and/or 3% H_2O the optimum performance was achieved at a temperature of 900 °C giving a power density of 0.5 W cm^{-2} in a wet H_2 gas stream. Tao and Irvine claim this gives an anode that is comparable in performance with Ni-YSZ, with the additional benefit of activity towards oxidation of methane in dry atmospheres. Further work has indicated that improved performance can be achieved in these materials with A-site cation deficiency and led to the work of Raj et al. where oxygen transport data were obtained[300] for the $(La_{0.75}Sr_{0.25})_{0.95}Cr_{0.5}Mn_{0.5}O_{3-\delta}$ material. In this case, as was observed earlier with similar studies of the ferrites, diffusion coefficients were found to be significantly higher under a reducing atmosphere than in dry O_2. Of concern however was the behaviour of the surface exchange coefficients as a function of temperature in that non-Arrhenius behaviour was observed, which may indicate an increasing concentration of oxygen vacancies on increasing temperature is involved with the charge transfer at the surface of these oxides.

Evidently the requirement for novel redox and sulfur tolerant anodes requires understanding of the electronic conduction properties of materials as well as the ionic conductivity. In LSCM the dependence of electronic conductivity on composition and pO_2 has been detailed and related to the defect chemistry of the bulk oxide.[301] Here the authors used X-ray absorption spectroscopy to probe the cation valence and determined that

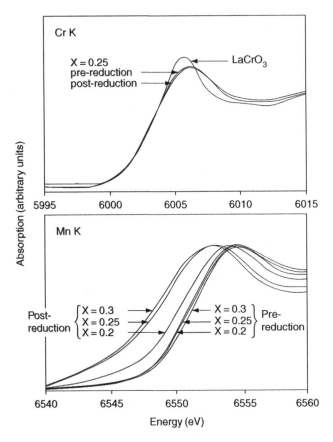

Figure 2.25 X-ray absorption spectra highlighting the redox behaviour of LSCM compositions identifying the Mn site as the most active. Reprinted with permission from Plint *et al.*, 2006[301]. Copyright (2006) Elsevier

the defect chemistry was controlled by Mn on the *B* site and there was no influence on the conductivity exhibited by the Cr on the *B* site, as is illustrated in Figure 2.25. Indeed the authors also suggest that the conductivity exhibited by this material can be attributed to the charge carriers associated with the concentration of Mn^{4+}, such that defects are generated and compensated electronically on the introduction of Sr. The origin of the Mn^{4+} is assumed to be from the disproportionation of Mn^{3+}.

Whilst LSCM has shown some promise as an anode alternative further materials developments have considered the adoption of composites containing this novel material. An initial attempt to improve anode performance involved the creation of a composite of LSCM with Cu.[302] In this work a cell performance of 250 mW cm^{-2} was achieved for an anode functioning on an LSGM electrolyte operating at 850 °C. Further development was considered by Ye et al.[274] who looked at a complex composite combining the LSCM with both an electrolyte component (ScSZ) and a metallic component, Cu. Evaluation of the performance of this anode layer indicated poor electrochemical performance, achieving only 140 mW cm^{-2}, lower than that obtained from a LSCM-Cu anode.[302] To overcome this deficiency in performance a layer of LSCM/Cu/YSZ was deposited on the anode and a significant increase in power density, to 584 mW cm^{-2}, was observed, however the value of this composite anode is questionable given the multicomponent nature of the anode with a total of six components including two different electrolyte materials.

Little further investigation of the LSCM system has been reported and work on perovskite type anodes appears to be directed towards the novel titanate type composition, $La_4Sr_8Ti_{11}Mn_{0.5}Ga_{0.5}O_{37.5-\delta}$[303,304] (LSTGM). Initial interest in this composition stemmed from the desire to investigate La substituted $SrTiO_3$ as a potential redox stable anode. In this case the material was further substituted with Ga and Mn and as a result extended point defects were formed, particularly disordered oxygen defects. Disorder was identified in these complex perovskites through the use of HRTEM,[305] and subsequent testing of the composition as an anode indicated performance comparable with Ni-YSZ with an LSM cathode on YSZ electrolyte.[303,305] Later improvements in performance were achieved through the use of composites of LSTGM with Cu,[303] leading to lower polarisation resistances, indicating that Cu impregnation of anodes is a likely route to enhanced performance in all anode materials.

Further oxide materials have been prepared and evaluated as potential anode materials, including Nb_2TiO_7[306] and $Gd_2Ti(Mo, Mn)O_7$[307,308] but have proved to be either extremely redox intolerant or possess very low ionic conductivities. Hence the most promising materials for oxide anodes remain materials structurally related to the ABO_3 perovskite.

2.3 MATERIALS DEVELOPMENTS THROUGH PROCESSING

As will be clear from the preceding discussion of materials developments for SOFCs many of the advances in performance of the individual components can be considered to be incremental in nature, with fractional improvements in redox stability, sulfur tolerance, ionic conductivity or thermomechanical stability. It is highly unusual for step changes in performance to be identified, but there have been recent reports and discussions surrounding the potential for nanostructured materials, particularly heterostructured composites, to massively enhance conduction properties within electrolyte layers. Garcia-Barriocanal *et al.*[309] produced a layered YSZ based on $SrTiO_3$ (YSZ/STO) substrates and report an eightfold increase in ionic conductivity, suggesting that devices based on YSZ could operate at close to room temperature, revolutionising SOFC technology. To achieve such significant increases, the authors suggest that the strain at the YSZ/STO interface is critical and allows facile oxygen diffusion pathways to open along this nanostructured interface. Of course these findings are the subject of scepticism as alternative models such as the incorporation of electronic conductivity would also explain these results. However, it is suggested that appropriate measurements were carefully performed to eliminate this possibility but until reproducible data are collected this advance is likely to be contested. A similar effect has been indentified, but to a lesser degree in heterostructured YSZ,[310] where the magnitude of the conductivity has been increased as a function of the number of layers of the electrolyte. Again these data were obtained from conductivity measurements and hence there is no direct measure of the ionic mobility within these structures, however these two sets of data, albeit from different materials and structures, indicate that a nanostructured layer will enhance SOFC performance. Whether these effects will be maintained over the lifetime of the device is an open question and, once the effects are fully explained, no doubt there will be increased interest in this field.

Closely related to these advances are the possibilities of exploiting anisotropic materials, such as the Ruddlesden–Popper structured materials detailed earlier. As the ionic mobility in these cathode materials is highly *ab* plane dependent, there is an inherent advantage to depositing high quality epitaxial thin films. Indeed preliminary data have been obtained from both single materials such as $La_2NiO_{4+\delta}$,[203] and also from single cell devices[311] illustrating the potential of these new materials. Deposition of epitaxial films, using either chemical vapour or pulsed laser deposition techniques, results in highly ordered films oriented closely to a substrate material, either $NdGaO_3$ or STO. This clearly raises the question of the effect of strain on the transport at the electrode/substrate interface and to date no significant enhancement has been determined, somewhat contradicting the studies on YSZ/STO films. These are complex systems and perhaps should not be directly compared, but it is clear that much further work is required to elucidate the details of interfacial transport, including use of modelling techniques.

2.4 PROTON CONDUCTING CERAMIC FUEL CELLS

Much of the preceding discussion has concerned the development of SOFCs based on oxide ion conducting ceramics. In these devices the electrolyte, cathode and anode all possess significant levels of oxygen ionic conduction and materials selection is then a compromise between the demands of each functional layer. As an alternative proton conducting ceramics have been suggested as the basis for a lower temperature SOFC. In this instance protons are incorporated into the oxide *via* the following reaction:

$$H_2O(g) + V_O^{\bullet\bullet} + O_O^x \leftrightarrow 2OH_O^{\bullet} \qquad (2.5)$$

where oxygen vacancies combine with H_2O to produce $[OH]^-$ defects in the lattice. This process consists of many potential steps but generally results in materials with significant proton conduction at temperatures below about 600 °C. As the temperature increases it is normally the case that oxide ion conduction increases and dominates, restricting the operating regime of ceramic proton conductors to relatively low temperatures. Clearly this could be considered an advantage in the search for new devices.

Proton conducting ceramics are commonly based upon the perovskite structure type with the cerate and zirconate types being most prevalent. Initial studies focused on the $SrCeO_3$- and $SrZrO_3$-based materials. Uchida *et al.* demonstrated sizeable proton conductivity in each of these materials[312–316] but due to concerns over the stability of these compositions attention shifted towards the Ba analogues. In these materials the key concern is their stability in CO_2 containing atmospheres.

2.4.1 Materials for Proton Conducting Solid Oxide Fuel Cells (PC-SOFCs)

Proton conducting ceramic fuel cells are currently considered as constituting devices with an electrolyte component from the $BaZrO_3$ and

$BaCeO_3$ families that demonstrates significant proton mobility. Barium cerate has been reported as the material with the highest proton conductivity over a wide temperature range but has also been shown to degrade rapidly in CO_2 containing atmospheres, resulting in the formation of surface carbonate species that block the ion conduction. In order to overcome this issue the substitution of Zr for Ce was proposed,[317,318] with the $BaZrO_3$ material also showing fast proton conduction. Unfortunately the ionic transport was identified as being lower than that of the analogous cerates and a further issue discovered: dense ceramics of the zirconates were difficult to produce, requiring high temperature sintering at >1700 °C. A solution to this aspect of materials development has been suggested – doping each of these compositions. Both $BaCe_{1-x}Y_xO_3$ (BCY) and $BaZr_{1-x}Y_xO_3$ (BZY) have been demonstrated as proton conducting electrolytes with later efforts directed towards solid solutions of these two materials giving electrolytes of composition $Ba(Zr, Ce, Y)O_3$. Several methods have been suggested to reduce the sintering temperature of these ceramics, including the use of spray pyrolysis[319] and addition of ZnO as a sintering aid.[320] The effect of the ZnO addition on the transport properties is not fully resolved.

Of course one of the advantages of the proton conductors is the relatively low temperature at which fast proton conduction is achieved, but it is also the case that many of the proton conductors have significant oxide ion conduction as well, and hence a good understanding of the temperature dependence of each of these properties is essential. Typically, the perovskites will have predominantly proton conduction until ~700 °C at which point oxide ion transport begins to dominate. This is of importance when considering fuel cell operation as the presence of both charge carriers can result in counter currents. A full review of the development of proton conducting ceramics for use in fuel cells is available in Kreuer.[321]

As the Ba- and Sr-based perovskites have some problems with processing and stability, attention has focused on the discovery of new materials. Haugsrud *et al.*[322–325] have suggested that the divalent substituted $LaNbO_4$ and $LaTaO_4$ series of materials are attractive candidates. Ca-doped $LaNbO_4$ was identified as having a conductivity of 10^{-3} S cm^{-1} at 800 °C[323] which was the highest reported for all of the Fergusonite type lanthanides. There has also been some interest in the La_3NbO_7 material with somewhat lower conductivity (10^{-4}–10^{-3} S cm^{-1}). Each of these materials indicates that considerable scope for development PC-SOFC devices exists, and could offer a viable low temperature alternative to the IT-SOFC.

2.5 SUMMARY

SOFCs consist of a complex multicomponent structure that relies heavily on inorganic materials chemistry to develop new compositions of enhanced functionality. In all aspects of both oxide ion and proton conducting based devices the structural and electrochemistry of ceramics is essential to future developments. In the preceding discussion we have highlighted a number of novel solutions for the electrolyte, anode and cathode functional components of SOFCs that overcome some of the materials limitations facing fuel cell developers. Much of this has concerned the reduction in operating temperature coupled with the durability of the materials in a wide range of operating conditions, including anode resistance to sulfur and coking. Enhanced properties may be achieved through the use of thin films of functional components, such as with yttria stabilised zirconia on $SrTiO_3$, but many questions as to the nature and origin of these enhancements remain. It is evident that there remains rich inorganic chemistry to be discovered and exploited in this vital technology area.

REFERENCES

[1] E. V. Tsipis and V. V. Kharton, *J. Solid State Electrochem.*, **12**, 1367 (2008).

[2] E. V. Tsipis and V. V. Kharton, *J. Solid State Electrochem.*, **12**, 1039 (2008).

[3] J. Fleig, *Annu. Rev. Mater. Res.*, **33**, 361 (2003).

[4] S. B. Adler, *Chem. Rev.*, **104**, 4791 (2004).

[5] N. P. Brandon, S. Skinner and B. C. H. Steele, *Annu. Rev. Mater. Res.*, **33**, 183 (2003).

[6] J. B. Goodenough, *Annu. Rev. Mater. Res.*, **33**, 91 (2003).

[7] J. W. Fergus, *J. Power Sources*, **162**, 30 (2006).

[8] L. J. Gauckler, D. Beckel, B. E. Buergler, E. Jud, U. R. Muecke, M. Prestat, J. L. M. Rupp and J. Richter, *Chimia*, **58**, 837 (2004).

[9] L. Gao and L. C. Guo, *J. Rare Earths*, **23**, 617 (2005).

[10] P. Holtappels, U. Vogt and T. Graule, *Adv. Eng. Mater.*, **7**, 292 (2005).

[11] J. P. P. Huijsmans, *Curr. Opin. Solid State Mater. Sci.*, **5**, 317 (2001).

[12] E. Baur and H. Preis, *Z. Elektrochem.*, **43**, 727 (1937).

[13] S. P. S. Badwal, F. T. Ciacchi and D. Milosevic, *Solid State Ionics*, **136**, 91 (2000).

[14] T. Horita, N. Sakai, H. Yokokawa, M. Dokiya, T. Kawada, J. Van Herle and K. Sasaki, *J. Electroceram.*, **1**, 155 (1997).

[15] S. Q. Hui, J. Roller, S. Yick, X. Zhang, C. Deces-Petit, Y. S. Xie, R. Maric and D. Ghosh, *J. Power Sources*, **172**, 493 (2007).

[16] T. Ishihara, H. Matsuda and Y. Takita, *J. Am. Chem. Soc.*, **116**, 3801 (1994).

[17] T. Ishihara, H. Matsuda and Y. Takita, *Solid State Ionics*, **79**, 147 (1995).

[18] T. Ishihara, H. Minami, H. Matsuda, H. Nishiguchi and Y. Takita, *Chem. Commun.*, 929 (1996).

[19] T. Ishihara, H. Matsuda, M. A. binBustam and Y. Takita, *Solid State Ionics*, 86–8, 197 (1996).

[20] T. Ishihara, J. A. Kilner, M. Honda and Y. Takita, *J. Am. Chem. Soc.*, 119, 2747 (1997).

[21] K. Q. Huang, M. Feng and J. B. Goodenough, *J. Am. Ceram. Soc.*, 79, 1100 (1996).

[22] K. Q. Huang, R. S. Tichy and J. B. Goodenough, *J. Am. Ceram. Soc.*, 81, 2565 (1996).

[23] K. Q. Huang, R. S. Tichy and J. B. Goodenough, *J. Am. Ceram. Soc.*, 81, 2576 (1998).

[24] K. Q. Huang, R. Tichy and J. B. Goodenough, *J. Am. Ceram. Soc.*, 81, 2581 (1998).

[25] J. W. Stevenson, K. Hasinska, N. L. Canfield and T. R. Armstrong, *J. Electrochem. Soc.*, 147, 3213 (2000).

[26] T. Ishihara, S. Ishikawa, C. Y. Yu, T. Akbay, K. Hosoi, H. Nishiguchi and Y. Takita, *Phys. Chem. Chem. Phys.*, 5, 2257 (2003).

[27] A. Atkinson, S. Barnett, R. J. Gorte, J. T. S. Irvine, A. J. McEvoy, M. Mogensen, S. C. Singhal and J. Vohs, *Nature Mater.*, 3, 17 (2004).

[28] S. P. Jiang and S. H. Chan, *J. Mater. Sci.*, 39, 4405 (2004).

[29] S. Primdahl and M. Mogensen, *J. Appl. Electrochem.*, 30, 247 (2000).

[30] D. Sarantaridis and A. Atkinson, *Fuel Cells*, 7, 246 (2007).

[31] V. Alzate-Restrepo and J. M. Hill, *Appl. Catal., A*, 342, 49 (2008).

[32] D. J. L. Brett, A. Atkinson, N. P. Brandon and S. J. Skinner, *Chem. Soc. Rev.*, 37, 1568 (2008).

[33] A. Lashtabeg and S. J. Skinner, *J. Mater. Chem.*, 16, 3161 (2006).

[34] S. J. Skinner, *Int. J. Inorg. Mater.*, 3, 113 (2001).

[35] B. C. H. Steele, *Solid State Ionics*, 134, 3 (2000).

[36] S. P. Jiang, *J. Mater. Sci.*, 43, 6799 (2008).

[37] A. Chen, G. Bourne, K. Siebein, R. DeHoff, E. Wachsman and K. Jones, *J. Am. Ceram. Soc.*, 91, 2670 (2008).

[38] A. Grosjean, O. Sanseau, V. Radmilovic and A. Thorel, *Solid State Ionics*, 177, 1977 (2006).

[39] P. Ried, P. Holtappels, A. Wichser, A. Ulrich and T. Graule, *J. Electrochem. Soc.*, 155, B1029 (2008).

[40] X. Y. Chen, J. S. Yu and S. B. Adler, *Chem. Mater.*, 17, 4537 (2005).

[41] V. Dusastre and J. A. Kilner, *Solid State Ionics*, 126, 163 (1999).

[42] Z. P. Shao and S. M. Haile, *Nature*, 431, 170 (2004).

[43] W. Zhou, R. Ran, Z. P. Shao, W. Zhuang, J. Jia, H. X. Gu, W. Q. Jin and N. P. Xu, *Acta Mater.*, 56, 2687 (2008).

[44] H. L. Zhao, W. Shen, Z. M. Zhu, X. Li and Z. F. Wang, *J. Power Sources*, 182, 503 (2008).

[45] Y. Zhang, J. Liu, X. Huang, Z. Lu and W. Su, *Solid State Ionics*, 179, 250 (2008).

[46] A. Y. Yan, M. Yang, Z. F. Hou, Y. L. Dong and M. J. Cheng, *J. Power Sources*, 185, 76 (2008).

[47] J. Pena-Martinez, D. Marrero-Lopez, J. C. Ruiz-Morales, B. E. Buergler, P. Nunez and L. J. Gauckler, *Solid State Ionics*, 177, 2143 (2006).

[48] Y. Lin, R. Ran, Y. Zheng, Z. P. Shao, W. Q. Jin, N. P. Xu and J. M. Ahn, *J. Power Sources*, 180, 15 (2008).

[49] Z. S. Duan, M. Yang, A. Yan, Z. F. Hou, Y. L. Dong, Y. Chong, M. J. Cheng and W. S. Yang, *J. Power Sources*, **160**, 57 (2006).

[50] M. Arnold, T. M. Gesing, J. Martynczuk and A. Feldhoff, *Chem. Mater.*, **20**, 5851 (2008).

[51] C. R. Xia, W. Rauch, F. L. Chen and M. L. Liu, *Solid State Ionics*, **149**, 11 (2002).

[52] Z. L. Tang, Y. S. Xie, H. Hawthorne and D. Ghosh, *J. Power Sources*, **157**, 385 (2006).

[53] H. Lv, Y. J. Wu, B. Huang, B. Y. Zhao and K. A. Hu, *Solid State Ionics*, **177**, 901 (2006).

[54] M. Koyama, C. J. Wen, T. Masuyama, J. Otomo, H. Fukunaga, K. Yamada, K. Eguchi and H. Takahashi, *J. Electrochem. Soc.*, **148**, A795 (2001).

[55] H. Fukunaga, M. Koyama, N. Takahashi, C. Wen and K. Yamada, *Solid State Ionics*, **132**, 279 (2000).

[56] T. Takahashi and H. Iwahara, *Energy Convers.*, **11**, 105 (1971).

[57] M. Kajitani, M. Matsuda and M. Miyake, *Solid State Ionics*, **177**, 1721 (2006).

[58] J. W. Stevenson, T. R. Armstrong, D. E. McCready, I. R. Pederson and W. J. Weber, *J. Electrochem. Soc.*, **144**, 3613 (1997).

[59] J. Bradley, P. R. Slater, T. Ishihara and J. T. S. Irvine, in *Solid Oxide Fuel Cells VIII (SOFC VIII)*, The Electrochemical Society, Pennington NJ, 315 (2003).

[60] V. Thangadurai and W. Weppner, *J. Electrochem. Soc.*, **148**, A1294 (2001).

[61] D. Lybye, F. W. Poulsen and M. Mogensen, *Solid State Ionics*, **128**, 91 (2000).

[62] T. Ishihara, H. Furutani, M. Honda, T. Yamada, T. Shibayama, T. Akbay, N. Sakai, H. Yokokawa and Y. Takita, *Chem. Mater.*, **11**, 2081 (1999).

[63] B. A. Khorkounov, H. Nafe and F. Aldinger, *J. Solid State Electrochem.*, **10**, 479 (2006).

[64] T. Ishihara, J. Tabuchi, S. Ishikawa, J. Yan, M. Enoki and H. Matsumoto, *Solid State Ionics*, **177**, 1949 (2006).

[65] M. Enoki, J. W. Yan, H. Matsumoto and T. Ishihara, *Solid State Ionics*, **177**, 2053 (2006).

[66] A. A. Yaremchenko, V. V. Kharton, E. N. Naumovich, D. I. Shestakov, V. F. Chukharev, A. V. Kovalevsky, A. L. Shaula, M. V. Patrakeev, J. R. Frade and F. M. B. Marques, *Solid State Ionics*, **177**, 549 (2006).

[67] N. S. Chae, K. S. Park, Y. S. Yoon, I. S. Yoo, J. S. Kim and H. H. Yoon, *Colloids Surf., A*, **313**, 154 (2008).

[68] R. Polini, A. Falsetti, E. Traversa, O. Schaf and P. Knauth, *J. Eur. Ceram. Soc.*, **27**, 4291 (2007).

[69] B. W. Liu and Y. Zhang, *J. Alloys Compd.*, **458**, 383 (2008).

[70] V. Thangadurai, P. S. Beurmann and W. Weppner, *Mater. Sci. Eng., B*, **100**, 18 (2003).

[71] S. Hashimoto, H. Kishimoto and H. Iwahara, *Solid State Ionics*, **139**, 179 (2001).

[72] P. Lacorre, F. Goutenoire, O. Bohnke, R. Retoux and Y. Laligant, *Nature*, **404**, 856 (2000).

[73] L. Malavasi, H. Kim, S. J. L. Billinge, T. Proffen, C. Tealdi and G. Flor, *J. Am. Chem. Soc.*, **129**, 6903 (2007).

[74] I. R. Evans, J. A. K. Howard and J. S. O. Evans, *Chem. Mater.*, **17**, 4074 (2005).

[75] F. Goutenoire, O. Isnard, R. Retoux and P. Lacorre, *Chem. Mater.*, **12**, 2575 (2000).

[76] P. Lacorre, *Solid State Sci.*, **2**, 755 (2000).

[77] S. Georges, F. Goutenoire, O. Bohnke, M. C. Steil, S. J. Skinner, H. D. Wiemhofer and P. Lacorre, *J. New Mater. Electrochem. Syst.*, **7**, 51 (2004).

[78] J. A. Collado, M. A. G. Aranda, A. Cabeza, P. Olivera-Pastor and S. Bruque, *J. Solid State Chem.*, **167**, 80 (2002).

[79] Z. S. Khadasheva, N. U. Venskovskii, M. G. Safronenko, A. V. Mosunov, E. D. Politova and S. Y. Stefanovich, *Inorg. Mater.*, **38**, 1168 (2002).

[80] S. Basu, P. S. Devi and H. S. Maiti, *J. Electrochem. Soc.*, **152**, A2143 (2005).

[81] S. Georges, F. Goutenoire, F. Altorfer, D. Sheptyakov, F. Fauth, E. Suard and P. Lacorre, *Solid State Ionics*, **161**, 231 (2003).

[82] S. Basu, P. S. Devi and H. S. Maiti, *Appl. Phys. Lett.*, **85**, 3486 (2004).

[83] D. Marrero-Lopez, D. Perez-Coll, J. C. Ruiz-Morales, J. Canales-Vazquez, M. C. Martin-Sedeno and P. Nunez, *Electrochim. Acta*, **52**, 5219 (2007).

[84] S. Georges, S. J. Skinner, P. Lacorre and M. C. Steil, *Dalton Trans.*, **3101** (2004).

[85] J. H. Yang, Z. Y. Wen, Z. H. Gu and D. S. Yan, *J. Eur. Ceram. Soc.*, **25**, 3315 (2005).

[86] J. X. Wang, X. P. Wang, F. J. Liang, Z. J. Cheng and Q. F. Fang, *Solid State Ionics*, **177**, 1437 (2006).

[87] A. Subramania, T. Saradha and S. Muzhumathi, *Mater. Res. Bull.*, **43**, 1153 (2008).

[88] J. B. Goodenough, J. E. Ruizdiaz and Y. S. Zhen, *Solid State Ionics*, **44**, 21(1990).

[89] A. Manthiram, J. F. Kuo and J. B. Goodenough, *Solid State Ionics*, **62**, 225 (1993).

[90] T. Schober, *Solid State Ionics*, **109**, 1 (1998).

[91] V. V. Kharton, F. M. B. Marques and A. Atkinson, *Solid State Ionics*, **174**, 135 (2004).

[92] T. Kobayashi, Y. Senoo, M. Hibino and T. Yao, *Solid State Ionics*, **177**, 1743 (2006).

[93] A. Rolle, R. N. Vannier, N. V. Giridharan and F. Abraham, *Solid State Ionics*, **176**, 2095 (2005).

[94] D. H. Gregory and M. T. Weller, *J. Solid State Chem.*, **107**, 134 (1993).

[95] T. Yao, Y. Uchimoto, M. Kinuhata, T. Inagaki and H. Yoshida, *Solid State Ionics*, **132**, 189 (2000).

[96] K. Kakinuma, H. Yamamura and T. Atake, *J. Therm. Anal. Calorim.*, **69**, 897 (2002).

[97] V. Jayaraman, A. Magrez, M. Caldes, O. Joubert, M. Ganne, Y. Piffard and L. Brohan, *Solid State Ionics*, **170**, 17 (2004).

[98] A. Yamaji, K. Kawakami, M. Arai and T. Adachi, *New Mater. Batteries Fuel Cells*, **575**, 343 (2000).

[99] K. Kakinuma, H. Yamamura, H. Haneda and T. Atake, *Solid State Ionics*, **154**, 571 (2002).

[100] J. B. Goodenough, *Solid State Ionics*, **94**, 17 (1997).

[101] G. Gattow and Schröder, Z. *Anorg. Allg. Chem.*, **318**, 176 (1962).

[102] M. Yashima and D. Ishimura, *Chem. Phys. Lett.*, **378**, 395 (2003).

[103] T. Takahashi, H. Iwahara and T. Arao, *J. Appl. Electrochem.*, **5**, 187 (1975).

[104] A. Watanabe and T. Kikuchi, *Solid State Ionics*, **21**, 287 (1986).

[105] H. Mizoguchi, K. Ueda, H. Kawazoe, H. Hosono, T. Omata and S. Fujitsu, *J. Mater. Chem.*, **7**, 943 (1997).

[106] M. J. Verkerk, K. Keizer and A. J. Burggraaf, *J. Appl. Electrochem.*, **10**, 81 (1980).

[107] J. R. Jurado, C. Moure, P. Duran and N. Valverde, *Solid State Ionics*, **28**, 518 (1988).

[108] M. J. Verkerk and A. J. Burggraaf, *J. Electrochem. Soc.*, **128**, 75(1981).

[109] A. Watanabe, *Solid State Ionics*, **79**, 84 (1995).

[110] H. Cahen, T. van den Belt, J. de Wit and G. Broers, *Solid State Ionics*, **1**, 411 (1980).

[111] T. Takahashi, H. Iwahara and T. Esaka, *J. Electrochem. Soc.*, **124**, 1563 (1977).

[112] A. Watanabe and A. Ono, *Solid State Ionics*, **174**, 15 (2004).

[113] N. M. Sammes, G. A. Tompsett, H. Nafe and F. Aldinger, *J. Eur. Ceram. Soc.*, **19**, 1801 (1999).

[114] H. Iwahara, T. Esaka, T. Sato and T. Takahashi, *J. Solid State Chem.*, **39**, 173 (1981).

[115] T. Takahashi, H. Iwahara and Y. Nagai, *J. Appl. Electrochem.*, **2**, 97 (1972).

[116] D. Mercurio, M. Elfarissi, J. C. Champarnaudmesjard, B. Frit, P. Conflant and G. Roult, *J. Solid State Chem.*, **80**, 133 (1989).

[117] M. Drache, P. Roussel and J. P. Wignacourt, *Chem. Rev.*, **107**, 80 (2007).

[118] A. Watanabe, *Solid State Ionics*, **40–1**, 889 (1990).

[119] A. Watanabe, *Solid State Ionics*, **86–8**, 1427 (1996).

[120] K. Z. Fung, J. Chen and A. V. Virkar, *J. Am. Ceram. Soc.*, **76**, 2403 (1993).

[121] N. X. Jiang and E. D. Wachsman, *J. Am. Ceram. Soc.*, **82**, 3057 (1999).

[122] A. Watanabe and M. Sekita, *Solid State Ionics*, **176**, 2429 (2005).

[123] F. Krok, I. Abrahams, W. Wrobel, S. C. M. Chan, A. Kozanecka, T. Ossowski and J. R. Dygas, *Solid State Ionics*, **175**, 335 (2004).

[124] A. A. Yaremchenko, V. V. Kharton, E. N. Naumovich, A. A. Tonoyan and V. V. Samokhval, *J. Solid State Electrochem.*, **2**, 308 (1998).

[125] K. Laurent, G. Y. Wang, S. Tusseau-Nenez and Y. Leprince-Wang, *Solid State Ionics*, **178**, 1735 (2008).

[126] F. Mauvy, J. C. Launay and J. Darriet, *J. Solid State Chem.*, **178**, 2015 (2005).

[127] R. Punn, A. M. Feteira, D. C. Sinclair and C. Greaves, *J. Am. Chem. Soc.*, **128**, 15386 (2006).

[128] N. X. Jiang, E. D. Wachsman and S. H. Jung, *Solid State Ionics*, **150**, 347 (2002).

[129] K. Reiselhuber, G. Dorner and M. W. Breiter, *Electrochim. Acta*, **38**, 969 (1993).

[130] X. P. Wang, G. Corbel, Q. F. Fang and P. Lacorre, *J. Mater. Chem.*, **16**, 1561 (2006).

[131] F. Abraham, M. F. Debreuillegresse, G. Mairesse and G. Nowogrocki, *Solid State Ionics*, **28**, 529 (1988).

[132] E. Pernot, M. Anne, M. Bacmann, P. Strobel, J. Fouletier, R. N. Vannier, G. Mairesse, F. Abraham and G. Nowogrocki, *Solid State Ionics*, **70–71**, 259 (1994).

[133] J. B. Goodenough, A. Manthiram, M. Paranthaman and Y. S. Zhen, in *Symposium on Solid State Ionics at the International Conference on Advanced Materials*, Strasbourg, France, 1991, p. 357.

[134] A. A. Yaremchenko, V. V. Kharton, E. N. Naumovich and V. V. Samokhval, *Solid State Ionics*, **111**, 227 (1998).

[135] E. Kendrick, M. S. Islam and P. R. Slater, *J. Mater. Chem.*, **17**, 3104 (2007).

[136] T. J. White and Z. L. Dong, *Acta Crystallogr., Sect. B*, **59**, 1 (2003).

[137] S. Nakayama, H. Aono and Y. Sadaoka, *Chem. Lett.*, 431 (1995).

[138] S. W. Tao and J. T. S. Irvine, *Mater. Res. Bull.*, **36**, 1245 (2001).

[139] J. E. H. Sansom, J. R. Tolchard, P. R. Slater and M. S. Islam, *Solid State Ionics*, **167**, 17 (2004).

[140] E. J. Abram, D. C. Sinclair and A. R. West, *J. Mater. Chem.*, **11**, 1978 (2001).

[141] L. Leon-Reina, M. C. Martin-Sedeno, E. R. Losilla, A. Cabeza, M. Martinez-Lara, S. Bruque, F. M. B. Marques, D. V. Sheptyakov and M. A. G. Aranda, *Chem. Mater.*, **15**, 2099 (2003).

[142] J. R. Tolchard, P. R. Slater and M. S. Islam, *Adv. Funct. Mater.*, **17**, 2564 (2007).

[143] M. S. Islam, J. R. Tolchard and P. R. Slater, *Chem. Commun.*, 1486 (2003).

[144] L. Leon-Reina, E. R. Losilla, M. Martinez-Lara, S. Bruque and M. A. G. Aranda, *J. Mater. Chem.*, 14, 1142 (2004).

[145] J. E. H. Sansom, J. R. Tolchard, M. S. Islam, D. Apperley and P. R. Slater, *J. Mater. Chem.*, 16, 1410 (2006).

[146] A. Orera, E. Kendrick, D. C. Apperley, V. M. Orera and P. R. Slater, *Dalton Trans.*, 5296 (2008).

[147] J. E. H. Sansom, D. Richings and P. R. Slater, *Solid State Ionics*, 139, 205 (2001).

[148] L. Leon-Reina, E. R. Losilla, M. Martinez-Lara, S. Bruque, A. Llobet, D. V. Sheptyakov and M. A. G. Aranda, *J. Mater. Chem.*, 15, 2489 (2005).

[149] L. Leon-Reina, J. M. Porras-Vazquez, E. R. Losilla and M. A. G. Aranda, *J. Solid State Chem.*, 180, 1250 (2007).

[150] A. Chesnaud, C. Bogicevic, F. Karolak, C. Estournes and G. Dezanneau, *Chem. Commun.*, 1550 (2007).

[151] H. Arikawa, H. Nishiguchi, T. Ishihara and Y. Takita, *Solid State Ionics*, 136, 31 (2000).

[152] E. J. Abram, C. A. Kirk, D. C. Sinclair and A. R. West, *Solid State Ionics*, 176, 1941 (2005).

[153] J. R. Tolchard, J. E. H. Sansom, P. R. Slater and M. S. Islam, in *OSSEP Workshop on Ionic and Mixed Conductors*, Aveiro, Portugal, 2003, p. 668.

[154] E. Kendrick, M. S. Islam and P. R. Slater, *Chem. Commun.*, 715 (2008).

[155] S. S. Pramana, W. T. Klooster and T. J. White, *Acta Crystallogr., Sect. B*, 63, 597 (2007).

[156] E. V. Tsipis, V. V. Kharton and J. R. Frade, *Electrochim. Acta*, 52, 4428 (2007).

[157] E. Kendrick, J. Kendrick, K. S. Knight, M. S. Islam and P. R. Slater, *Nat. Mater.*, 6, 871 (2007).

[158] F. Schönberger, E. Kendrick, M. S. Islam and P. R. Slater, *Solid State Ionics*, 176, 2951 (2005).

[159] J. M. S. Skakle and R. Herd, *Powder Diffraction*, 14, 195 (1999).

[160] M. Rozumek, P. Majewski, H. Schluckwerder, F. Aldinger, K. Kunstler and G. Tomandl, *J. Am. Ceram. Soc.*, 87, 1795 (2004).

[161] E. S. Raj, S. J. Skinner and J. A. Kilner, *Solid State Ionics*, 176, 1097 (2005).

[162] X. Kuang, M. A. Green, H. Niu, P. Zajdel, C. Dickinson, J. B. Claridge, L. Jantsky and M. J. Rosseinsky, *Nat. Mater.*, 7, 498 (2008).

[163] R. A. De Souza, M. S. Islam and E. Ivers-Tiffee, *J. Mater. Chem.*, 9, 1621 (1999).

[164] Y. Sakaki, Y. Takeda, A. Kato, N. Imanishi, O. Yamamoto, M. Hattori, M. Iio and Y. Esaki, *Solid State Ionics*, 118, 187 (1999).

[165] T. L. Wen, H. Tu, Z. Xu and O. Yamamoto, *Solid State Ionics*, 121, 25 (1999).

[166] Y. Takeda, H. Y. Tu, H. Sakaki, S. Watanabe, N. Imanishi, O. Yamamoto, M. B. Phillipps and N. M. Sammes, *J. Electrochem. Soc.*, 144, 2810 (1997).

[167] Y. Takeda, Y. Sakaki, T. Ichikawa, N. Imanishi, O. Yamamoto, M. Mori, N. Mori and T. Abe, *Solid State Ionics*, 72, 257 (1994).

[168] T. Ishihara, T. Kudo, H. Matsuda and Y. Takita, *J. Electrochem. Soc.*, 142, 1519 (1995).

[169] H. R. Rim, S. K. Jeung, E. Jung and J. S. Lee, *Mater. Chem. Phys.*, 52, 54 (1998).

[170] M. B. Phillipps, N. M. Sammes and O. Yamamoto, *Solid State Ionics*, 123, 131 (1999).

[171] M. A. Pena and J. L. G. Fierro, *Chem. Rev.*, 101, 1981 (2001).

[172] I. P. Marozau, V. V. Kharton, A. P. Viskup, J. R. Frade and V. V. Samakhval, *J. Eur. Ceram. Soc.*, **26**, 1371 (2006).

[173] V. V. Kharton, M. V. Patrakeev, J. C. Waerenborgh, A. V. Kovalevsky, Y. V. Pivak, P. Gaczynski, A. A. Markov and A. A. Yaremchenko, *J. Phys. Chem. Solids*, **68**, 355 (2007).

[174] A. Kovalevsky, V. V. Kharton, F. M. M. Snijkers, J. F. C. Cooymans, J. J. Luyten and F. M. B. Marques, *J. Membr. Sci.*, **301**, 238 (2007).

[175] H. L. Lein, K. Wiik and T. Grande, *Solid State Ionics*, **177**, 1795 (2006).

[176] M. Sogaard, P. V. Hendriksen and M. Mogensen, *J. Solid State Chem.*, **180**, 1489 (2007).

[177] J. Holc, D. Kuscer, M. Hrovat, S. Bernik and D. Kolar, *Solid State Ionics*, **95**, 259 (1997).

[178] L. Kindermann, D. Das, H. Nickel, K. Hilpert, C. C. Appel and F. W. Poulson, *J. Electrochem. Soc.*, **144**, 717 (1997).

[179] J. H. Wan, J. Q. Yan , and J. B. Goodenough, *J. Electrochem. Soc.*, **152**, A1511 (2005).

[180] M. Al Daroukh, V. V. Vashook, H. Ullmann, F. Tietz and I. A. Raj, *Solid State Ionics*, **158**, 141 (2003).

[181] H. Uchida, S. Arisaka and M. Watanabe, *J. Electrochem. Soc.*, **149**, A13 (2002).

[182] Y. Teraoka, T. Nobunaga, K. Okamoto, N. Miura and N. Yamazoe, *Solid State Ionics*, **48**, 207 (1991).

[183] M. Godickemeier, K. Sasaki, L. J. Gauckler and I. Riess, *Solid State Ionics*, **86–8**, 691 (1996).

[184] S. B. Adler, *Solid State Ionics*, **111**, 125 (1998).

[185] T. Kawada, K. Masuda, J. Suzuki, A. Kaimai, K. Kawamura, Y. Nigara, J. Mizusaki, H. Yugami, H. Arashi, N. Sakai and H. Yokokawa, *Solid State Ionics*, **121**, 271 (1999).

[186] R. H. E. van Doorn and A. J. Burggraaf, *Solid State Ionics*, **128**, 65 (2000).

[187] Y. Takeda, H. Ueno, N. Imanishi, O. Yamamoto, N. Sammes and M. B. Phillipps, *Solid State Ionics*, **86–8**, 1187 (1996).

[188] T. Hibino, A. Hashimoto, T. Inoue, J. Tokuno, S. Yoshida and M. Sano, *Science*, **288**, 2031 (2000).

[189] L. W. Tai, M. M. Nasrallah, H. U. Anderson, D. M. Sparlin and S. R. Sehlin, *Solid State Ionics*, **76**, 259 (1995).

[190] G. C. Kostogloudis and C. Ftikos, *Solid State Ionics*, **126**, 143 (1999).

[191] H. Y. Tu, Y. Takeda, N. Imanishi and O. Yamamoto, *Solid State Ionics*, **117**, 277 (1999).

[192] S. N. Ruddlesden and P. Popper, *Acta Cryst.*, **11**, 54 (1958).

[193] M. Zinkevich and F. Aldinger, *J. Alloys Compd.*, **375**, 147 (2004).

[194] A. V. Kovalevsky, V. V. Kharton, A. A. Yaremchenko, Y. V. Pivak, E. V. Tsipis, S. O. Yakovlev, A. A. Markov, E. N. Naumovich and J. R. Frade, *J. Electroceram.*, **18**, 205 (2007).

[195] R. Sayers, J. Liu, B. Rustumji and S. J. Skinner, *Fuel Cells*, **8**, 338 (2008).

[196] V. V. Kharton, A. P. Viskup, A. V. Kovalevsky, E. N. Naumovich and F. M. B. Marques, *Solid State Ionics*, **143**, 337 (2001).

[197] Q. Li, H. Zhao, L. H. Huo, L. P. Sun, X. L. Cheng and J. C. Grenier, *Electrochem. Commun.*, **9**, 1508 (2007).

[198] A. J. Jennings and S. J. Skinner, *Solid State Ionics*, **152**, 663 (2002).

[199] V. V. Kharton, A. P. Viskup, E. N. Naumovich and F. M. B. Marques, *J. Mater. Chem.*, **9**, 2623 (1999).

[200] S. J. Skinner and J. A. Kilner, *Solid State Ionics*, **135**, 709 (2000).

[201] M. Yashima, M. Enoki, T. Wakita, R. Ali, Y. Matsushita, F. Izumi and T. Ishihara, *J. Am. Chem. Soc.*, **130**, 2762 (2008).

[202] J. M. Bassat, P. Odier, A. Villesuzanne, C. Marin and M. Pouchard, *Solid State Ionics*, **167**, 341 (2004).

[203] M. Burriel, G. Garcia, J. Santiso, J. A. Kilner, J. C. C. Richard and S. J. Skinner, *J. Mater. Chem.*, **18**, 416 (2008).

[204] G. T. Kim, S. Y. Wang, A. J. Jacobson, Z. Yuan and C. L. Chen, *J. Mater. Chem.*, **17**, 1316 (2007).

[205] V. V. Kharton, E. V. Tsipis, A. A. Yaremchenko and J. R. Frade, *Solid State Ionics*, **166**, 327 (2004).

[206] V. V. Kharton, F. M. Figueiredo, L. Navarro, E. N. Naumovich, A. V. Kovalevsky, A. A. Yaremchenko, A. P. Viskup, A. Carneiro, F. M. B. Marques and J. R. Frade, *J. Mater. Sci.*, **36**, 1105 (2001).

[207] C. N. Munnings, S. J. Skinner, G. Amow, P. S. Whitfield and I. J. Davidson, *J. Fuel Cell Sci. Technol.*, **2**, 34 (2005).

[208] N. Solak, M. Zinkevich and F. Aldinger, *Solid State Ionics*, **177**, 2139 (2006).

[209] C. N. Munnings, S. J. Skinner, G. Amow, P. S. Whitfield and L. J. Davidson, *Solid State Ionics*, **177**, 1849 (2006).

[210] L. P. Sun, L. H. Huo, H. Zhao, Q. Li and C. Pijolat, *J. Power Sources*, **179**, 96 (2008).

[211] Z. P. Shao, W. S. Yang, Y. Cong, H. Dong, J. H. Tong and G. X. Xiong, *J. Membr. Sci.*, **172**, 177 (2000).

[212] H. H. Wang, C. Tablet, A. Feldhoff and H. Caro, *J. Membr. Sci.*, **262**, 20 (2005).

[213] S. McIntosh, J. F. Vente, W. G. Haije, D. H. A. Blank and H. J. M. Bouwmeester, *Chem. Mater.*, **18**, 2187 (2006).

[214] Q. S. Zhu, T. A. Jin and Y. Wang, *Solid State Ionics*, **177**, 1199 (2006).

[215] J. Pena-Martinez, D. Marrero-Lopez, J. C. Ruiz-Morales, B. E. Buergler, P. Nunez and L. J. Gauckler, *Solid State Ionics*, **177**, 2143 (2006).

[216] Q. L. Liu, K. A. Khor and S. H. Chan, *J. Power Sources*, **161**, 123 (2006).

[217] S. Y. Li, Z. Lu, X. Q. Huang and W. H. Su, *Solid State Ionics*, **178**, 1853 (2008).

[218] J. G. Thompson, R. L. Withers and F. J. Brink, *J. Solid State Chem.*, **143**, 122 (1999).

[219] S. J. Skinner, I. J. E. Brooks and C. N. Munnings, *Acta Crystallogr., Sect. C*, **60**, 137 (2004).

[220] E. V. Tsipis, C. N. Munnings, V. V. Kharton, S. J. Skinner and J. R. Frade, *Solid State Ionics*, **177**, 1015 (2006).

[221] R. J. Packer, E. V. Tsipis, C. N. Munnings, V. V. Kharton, S. J. Skinner and J. R. Frade, *Solid State Ionics*, **177**, 2059 (2006).

[222] R. J. Packer, S. J. Skinner, A. A. Yaremchenko, E. V. Tsipis, V. V. Kharton, M. V. Patrakeev and Y. A. Bakhteeva, *J. Mater. Chem.*, **16**, 3503 (2006).

[223] R. J. Packer, J. Barlow, A. Cott and S. J. Skinner, *Solid State Ionics*, **179**, 1094 (2008).

[224] F. Vullum and T. Grande, *Chem. Mater.*, **20**, 5434 (2008).

[225] M. Respaud, C. Frontera, J. L. Garcia-Munoz, M. A. G. Aranda, B. Raquet, J. M. Broto, H. Rakoto, M. Goiran, A. Llobet and J. Rodriguez-Carvajal, *Phys. Rev. B*, **6421**, 214401 (2001).

[226] A. Maignan, C. Martin, D. Pelloquin, N. Nguyen and B. Raveau, *J. Solid State Chem.*, **142**, 247 (1999).

[227] A. A. Taskin, A. N. Lavrov and Y. Ando, *Appl. Phys. Lett.*, **86**, 091910 (2005).

[228] A. M. Chang, S. J. Skinner and J. A. Kilner, *Solid State Ionics*, **177**, 2009 (2006).

[229] A. Tarancon, S. J. Skinner, R. J. Chater, F. Hernandez-Ramirez and J. A. Kilner, *J. Mater. Chem.*, **17**, 3175 (2007).

[230] N. Li, Z. Lu, B. O. Wei, X. Q. Huang, K. F. Chen, Y. Z. Zhang and W. H. Su, *J. Alloys Compd.*, **454**, 274 (2008).

[231] J. Peña-Martinez, A. Tarancon, D. Marrero-Lopez, J. C. Ruiz-Morales and P. Nunez, *Fuel Cells*, **8**, 351 (2008).

[232] J. H. Kim, F. Prado and A. Manthiram, *J. Electrochem. Soc.*, **155**, B1023 (2008).

[233] G. Kim, S. Wang, A. J. Jacobson, L. Reimus, P. Brodersen and C. A. Mims, *J. Mater. Chem.*, **17**, 2500 (2007).

[234] G. Kim, S. Wang, A. J. Jacobson, Z. Yuan, W. Donner, C. L. Chen, L. Reimus, P. Brodersen and C. A. Mims, *Appl. Phys. Lett.*, **88**, 024103 (2006).

[235] C. J. Zhu, X. M. Liu, C. S. Yi, D. Yan and W. H. Su, *J. Power Sources*, **185**, 193 (2008).

[236] Q. J. Zhou, T. M. He and Y. Ji, *J. Power Sources*, **185**, 754 (2008).

[237] T. Z. Sholklapper, V. Radmilovic, C. P. Jacobson, S. J. Visco and L. C. De Jonghe, *J. Power Sources*, **175**, 206 (2008).

[238] P. Datta, P. Majewski and F. Aldinger, *Mater. Chem. Phys.*, **107**, 370 (2008).

[239] H. Z. Zhang, H. Y. Liu, Y. Cong and W. S. Yang, *J. Power Sources*, **185**, 129 (2008).

[240] S. McIntosh, S. B. Adler, J. M. Vohs and R. J. Gorte, *Electrochem. Solid State Lett.*, **7**, A111 (2004).

[241] M. J. L. Ostergard, C. Clausen, C. Bagger and M. Mogensen, *Electrochim. Acta*, **40**, 1971 (1995).

[242] M. Juhl, S. Primdahl, C. Manon and M. Mogensen, *J. Power Sources*, **61**, 173 (1996).

[243] E. P. Murray, T. Tsai and S. A. Barnett, *Solid State Ionics*, **110**, 235 (1998).

[244] V. A. C. Haanappel, J. Mertens, D. Rutenbeck, C. Tropartz, W. Herzhof, D. Sebold and F. Tietz, *J. Power Sources*, **141**, 216 (2005).

[245] A. Princivalle and E. Djurado, *Solid State Ionics*, **179**, 1921 (2008).

[246] Y. Y. Huang, J. M. Vohs and R. J. Gorte, *J. Electrochem. Soc.*, **153**, A951 (2006).

[247] Z. W. Wang, M. J. Cheng, Y. L. Dong, M. Zhang and H. M. Zhang, *Solid State Ionics*, **176**, 2555 (2005).

[248] A. Hagiwara, N. Hobara, K. Takizawa, K. Sato, H. Abe and M. Naito, *Solid State Ionics*, **177**, 2967 (2006).

[249] N. T. Hart, N. P. Brandon, M. J. Day and N. Lapena-Rey, *J. Power Sources*, **106**, 42 (2002).

[250] H. Zhao, L. H. Huo and S. Gao, *J. Power Sources*, **125**, 149 (2004).

[251] Z. Y. Jiang, L. Zhang, K. Feng and C. R. Xia, *J. Power Sources*, **185**, 40 (2008).

[252] Y. Zheng, R. Ran, H. X. Gu, R. Cai and Z. P. Shao, *J. Power Sources*, **185**, 641 (2008).

[253] X. G. Zhang, M. Robertson, S. Yick, C. Deces-Petit, E. Styles, W. Qu, Y. S. Xie, R. Hui, J. Roller, O. Kesler, R. Maric and D. Ghosh, *J. Power Sources*, **160**, 1211 (2006).

[254] X. D. Zhu, K. N. Sun, N. Q. Zhang, X. B. Chen, L. J. Wu and D. C. Jia, *Electrochem. Commun.*, 9, 431 (2007).

[255] M. Bevilacqua, T. Montini, C. Tavagnacco, E. Fonda, P. Fornasiero and M. Graziani, *Chem. Mater.*, 19, 5926 (2007).

[256] C. J. Zhu, X. M. Liu, D. Xu, D. J. Wang, D. Yan, L. Pei, T. Q. Lu and W. H. Su, *J. Power Sources*, 185, 212 (2008).

[257] A. Esquirol, J. Kilner and N. Brandon, *Solid State Ionics*, 175, 63 (2004).

[258] F. Qiang, K. N. Sun, N. Q. Zhang, X. D. Zhu, S. R. Le and D. R. Zhou, *J. Power Sources*, 168, 338 (2007).

[259] Y. J. Leng, S. H. Chan and Q. L. Liu, *Int. J. Hydrogen Energy*, 33, 3808 (2008).

[260] J. D. Zhang, Y. Ji, H. B. Gao, T. M. He and J. Liu, *J. Alloys Compd.*, 395, 322 (2005).

[261] Y. Lin and S. A. Barnett, *Solid State Ionics*, 179, 420 (2008).

[262] J. Chen, F. L. Liang, L. N. Liu, S. P. Jiang, B. Chi, J. Pu and J. Li, *J. Power Sources*, 183, 586 (2008).

[263] M. Camaratta and E. Wachsman, *J. Electrochem. Soc.*, 155, B135 (2008).

[264] C. Laberty, F. Zhao, K. E. Swider-Lyons and A. V. Virkar, *Electrochem. Solid State Lett.*, 10, B170 (2007).

[265] W. Zhou, Z. P. Shao, R. Ran, P. Y. Zeng, H. X. Gu, W. Q. Jin and N. P. Xu, *J. Power Sources*, 168, 330 (2007).

[266] K. Wang, R. Ran, W. Zhou, H. X. Gu, Z. P. Shao and J. M. Ahn, *J. Power Sources*, 179, 60 (2008).

[267] W. X. Zhu, Z. Lu, S. Y. Li, B. Wei, J. P. Miao, X. Q. Huang, K. F. Chen, N. Ai and W. H. Su, *J. Alloys Compd.*, 465, 274 (2008).

[268] S. D. Park, J. M. Vohs and R. J. Gorte, *Nature*, 404, 265 (2000).

[269] R. J. Gorte, S. Park, J. M. Vohs and C. H. Wang, *Adv. Mater.*, 12, 1465 (2000).

[270] S. W. Jung, C. Lu, H. P. He, K. Y. Ahn, R. J. Gorte and J. M. Vohs, *J. Power Sources*, 154, 42 (2006).

[271] H. P. He, J. M. Vohs and R. J. Gorte, *J. Electrochem. Soc.*, 150, A1470 (2003).

[272] J. R. Wilson and S. A. Barnett, *Electrochem. Solid State Lett.*, 11, B181 (2008).

[273] O. Costa-Nunes, R. J. Gorte and J. M. Vohs, *J. Power Sources*, 141, 241 (2005).

[274] X. F. Ye, B. Huang, S. R. Wang, Z. R. Wang, L. Xiong and T. L. Wen, *J. Power Sources*, 164, 203 (2007).

[275] H. Kim, C. Lu, W. L. Worrell, J. M. Vohs and R. J. Gorte, *J. Electrochem. Soc.*, 149, A247 (2002).

[276] J. C. Ruiz-Morales, J. Canales-Vazquez, D. Marrero-Lopez, J. Pena-Martinez, A. Tarancon, J. T. S. Irvine and P. Nunez, *J. Mater. Chem.*, 18, 5072 (2008).

[277] C. Lu, W. L. Worrell, R. J. Gorte and J. M. Vohs, *J. Electrochem. Soc.*, 150, A354 (2003).

[278] C. Lu, S. An, W. L. Worrell, J. M. Vohs and R. J. Gorte, *Solid State Ionics*, 175, 47 (2004).

[279] C. Lu, W. L. Worrell, J. M. Vohs and R. J. Gorte, in *Solid Oxide Fuel Cells VIII (SOFC VIII)*, The Electrochemical Society, Pennington NJ, 773 (2003).

[280] C. Lu, W. L. Worrell, C. Wang, S. Park, H. Kim, J. M. Vohs and R. J. Gorte, *Solid State Ionics*, 152, 393 (2002).

[281] M. D. Gross, J. M. Vohs and R. J. Gorte, *J. Electrochem. Soc.*, 153, A1386 (2006).

[282] A. C. Tavares, B. L. Kuzin, S. M. Beresnev, N. M. Bogdanovich, E. K. Urumchin, Y. A. Dubitsky and A. Zaopo, *J. Power Sources*, 183, 20 (2008).

[283] T. M. He, P. F. Guan, L. G. Cong, Y. Ji, H. Sun, J. X. Wang and J. Liu, *J. Alloys Compd.*, **393**, 292 (2005).

[284] S. An, C. Lu, W. L. Worrell, R. J. Gorte and J. M. Vohs, *Solid State Ionics*, **175**, 135 (2004).

[285] A. Ringuede, J. A. Labrincha and J. R. Frade, *Solid State Ionics*, **141–142**, 549 (2001).

[286] C. M. Grgicak, R. G. Green and J. B. Giorgi, *J. Mater. Chem.*, **16**, 885 (2006).

[287] N. Kiratzis, P. Holtappels, C. E. Hatchwell, M. Mogensen and J. T. S. Irvine, *Fuel Cells*, **1**, 211 (2001).

[288] J. C. Ruiz-Morales, P. Nunez, R. Buchanan and J. T. S. Irvine, *J. Electrochem. Soc.*, **150**, A1030 (2003).

[289] S. W. Tao and J. T. S. Irvine, *J. Electrochem. Soc.*, **151**, A252 (2004).

[290] S. W. Tao and J. T. S. Irvine, *Nat. Mater.*, **2**, 320 (2003).

[291] A. Ovalle, J. C. Ruiz-Morales, J. Canales-Vazquez, D. Marrero-Lopez and J. T. S. Irvine, *Solid State Ionics*, **177**, 1997 (2006).

[292] P. R. Slater and J. T. S. Irvine, *Solid State Ionics*, **120**, 125 (1999).

[293] A. Kaiser, J. L. Bradley, P. R. Slater and J. T. S. Irvine, *Solid State Ionics*, **135**, 519 (2000).

[294] Y. H. Huang, R. I. Dass, Z. L. Xing and J. B. Goodenough, *Science*, **312**, 254 (2006).

[295] M. Weston and I. S. Metcalfe, *Solid State Ionics*, **113**, 247 (1998).

[296] A. L. Sauvet and J. Fouletier, *J. Power Sources*, **101**, 259 (2001).

[297] A. L. Sauvet and J. T. S. Irvine, *Solid State Ionics*, **167**, 1 (2004).

[298] T. Ramos and A. Atkinson, *Solid State Ionics*, **170**, 275 (2004).

[299] T. Ramos and A. Atkinson, *Ionic and Mixed Conducting Ceramics IV*, **2001**, 352 (2002).

[300] E. S. Raj, J. A. Kilner and J. T. S. Irvine, *Solid State Ionics*, **177**, 1747 (2006).

[301] S. M. Plint, P. A. Connor, S. Tao and J. T. S. Irvine, *Solid State Ionics*, **177**, 2005 (2006).

[302] J. Wan, J. H. Zhu and J. B. Goodenough, *Solid State Ionics*, **177**, 1211 (2006).

[303] J. C. Ruiz-Morales, J. Canales-Vazquez, C. Savaniu, D. Marrero-Lopez, P. Nunez, W. Z. Zhou and J. T. S. Irvine, *Phys. Chem. Chem. Phys.*, **9**, 1821 (2007).

[304] J. C. Ruiz-Morales, J. Canales-Vazquez, C. Savaniu, D. Marrero-Lopez, W. Z. Zhou and J. T. S. Irvine, *Nature*, **439**, 568 (2006).

[305] J. Canales-Vazquez, M. J. Smith, J. T. S. Irvine and W. Z. Zhou, *Adv. Funct. Mater.*, **15**, 1000 (2005).

[306] C. Reich, A. Kaiser and J. T. S. Irvine, *Fuel Cells: from Fundamentals to Systems*, **1**, 249 (2001).

[307] P. Holtappels, F. W. Poulsen and M. Mogensen, *Solid State Ionics*, **135**, 675 (2000).

[308] J. J. Sprague and H. L. Tuller, *J. Eur. Ceram. Soc.*, **19**, 803 (1999).

[309] J. Garcia-Barriocanal, A. Rivera-Calzada, M. Varela, Z. Sefrioui, E. Iborra, C. Leon, S. J. Pennycook and J. Santamaria, *Science*, **321**, 676 (2008).

[310] I. Kosacki, C. M. Rouleau, P. F. Becher, J. Bentley and D. H. Lowndes, *Solid State Ionics*, **176**, 1319 (2005).

[311] A. Yamada, Y. Suzuki, K. Saka, M. Uehara, D. Mori, R. Kanno, T. Kiguchi, F. Mauvy and J. C. Grenier, *Adv. Mater.*, **20**, 4124 (2008).

[312] H. Uchida, A. Yasuda and H. Iwahara, *Denki Kagaku*, **57**, 153 (1989).

[313] H. Uchida, N. Maeda and H. Iwahara, *Solid State Ionics*, **11**, 117 (1983).

[314] T. Yajima, H. Suzuki, T. Yogo and H. Iwahara, *Solid State Ionics*, **51**, 101 (1992).

[315] H. Iwahara, T. Yajima, T. Hibino, K. Ozaki and H. Suzuki, *Solid State Ionics*, **61**, 65 (1993).

[316] T. Hibino, K. Mizutani, T. Yajima and H. Iwahara, *Solid State Ionics*, **57**, 303 (1992).

[317] T. Pagnier, I. Charrier-Cougoulic, C. Ritter and G. Lucazeau, *Eur. Phys.l J.*, **9**, 1 (2000).

[318] I. Charrier-Cougoulic, T. Pagnier and G. Lucazeau, *J. Solid State Chem.*, **142**, 220 (1999).

[319] P. A. Stuart, T. Unno, R. Ayres-Rocha, E. Djurado and S. J. Skinner, *J. Eur. Ceram. Soc.*, **29**, 697 (2009).

[320] P. Babilo and S. M. Haile, *J. Am. Ceram. Soc.*, **88**, 2362 (2005).

[321] K. D. Kreuer, *Annu. Rev. Mater. Res.*, **33**, 333 (2003).

[322] R. Haugsrud and T. Norby, *J. Am. Ceram. Soc.*, **90**, 1116 (2007).

[323] R. Haugsrud and T. Norby, *Nat. Mater.*, **5**, 193 (2006).

[324] R. Haugsrud, B. Ballesteros, M. Lira-Cantu and T. Norby, *J. Electrochem. Soc.*, **153**, J87 (2006).

[325] R. Haugsrud and T. Norby, *Solid State Ionics*, **177**, 1129 (2006).

[326] S. C. Singhal and K. Kendal (Eds), *High Temperature Solid Oxide Fuel Cells: Fundamentals, Design and Applications*, Elsevier, Oxford, 2003.

[327] T. Ishihara, H. Matsuda and Y. Takita, *J. Electrochem. Soc.*, **141**, 3444 (1994).

[328] V. Thangadurai, G. N. Subbanna, A. K. Shukla and J. Gopalakrishnan, *Chem. Mater.*, **8**, 1302 (1996).

[329] P. N. Huang and A. Petric, *J. Electrochem. Soc.*, **143**, 1644 (1996).

[330] N. Liu, M. Shi, C. Wang, Y. P. Yuan, P. Majewski and F. Aldinger, *J. Mater. Sci.*, **41**, 4205 (2006).

[331] A. Petric, P. Huang and F. Tietz, *Solid State Ionics*, **135**, 719 (2000).

3

Solar Energy Materials

Elizabeth A. Gibson[a] and Anders Hagfeldt[b]

[a] *School of Chemistry, University of Nottingham, University Park, Nottingham NG7 2RD, UK*
[b] *Centre for Molecular Devices, Department for Physical and Analytical Chemistry, Uppsala University, Uppsala, Sweden*

3.1 INTRODUCTION

Solar energy conversion is a broad and rapidly expanding research field. The increasing global demand for energy in combination with concerns over the use of fossil fuels is encouraging the demand for renewable sources of fuel.[1] The umbrella of 'solar energy conversion' encompasses solar thermal, solar fuels, solar-to-electricity (photovoltaic, PV) technology and the great many subcategories below those.[2–5] In the following pages, some of the key aspects of PV technology and the ongoing effort to exploit the planet's largest energy source are illustrated. This multidisciplinary research area brings together fundamental and applied aspects of, amongst others, materials science, synthetic inorganic and organic chemistry, photochemistry, electrochemistry, semiconductor physics and engineering. The main focus of this chapter will be the inorganic materials: the core of current and developing PV technology and a research area crucial to the future of solar energy conversion.

3.1.1 The Solar Spectrum

In 2009 the average global primary energy consumption reached 14.9 TW.[6] This value is set to increase, according to predictions, doubling by

---

Energy Materials Edited by Duncan W. Bruce, Dermot O'Hare and Richard I. Walton
© 2011 John Wiley & Sons, Ltd.

2050 and tripling by the end of the century.[5] To meet this demand alone, additional sources of fuel will be required if inflated fuel prices and political conflicts are to be prevented, since most of the current demand is met by fossil fuels, which are finite and are located in only a few locations in the world. In addition, pressure is being put on governments to reduce the amount of CO_2 and other greenhouse gases emitted in an effort to prevent global warming. Sunlight is clean, abundant [more energy from sunlight strikes the Earth in 1 h (4.3×10^{20} J) than all the energy consumed on the planet in a year (4.1×10^{20} J)], readily available, secure so far from geopolitical tension, and, in use, poses no threat to our environment through pollution or to our climate through greenhouse gases.[7]

The sun emits electromagnetic radiation as a black body at 5800 K. This means an energy distribution increasing in intensity from the ultraviolet (UV) to the visible, with a maximum around 500 nm, tailing into the infrared (IR) and some in the radiowave, microwave, X-ray and γ-ray regions. By the time this radiation has passed through space and the Earth's atmosphere, the intensity is around 120 000 TW. The X-rays, γ-rays and UV radiation below 200 nm in wavelength are absorbed selectively by nitrogen and oxygen in the atmosphere, UV radiation between 200 nm and 300 nm is absorbed by O_3 in the stratosphere, IR radiation above 700 nm is partially absorbed by CO_2, O_3 and water, and 30% of the visible radiation 400–700 nm is reflected back by the atmosphere or the Earth's surface. Figure 3.1 shows the resulting spectrum; the peak power of the sun occurs in the yellow region of the visible region, at about 2.5 eV. At high noon on a cloudless day, the surface of the Earth receives around 1000 W of solar power per square metre (1000 W m^{-2}).

As compounds have discrete molecular orbitals, semiconductor crystals have bands where electrons are located and gaps in between where they are not. The highest energy band where electrons are located is the valence band and, in the dark, the lowest lying unoccupied band is called the conduction band. The gap in between them, where there are no energy states, is called the band gap. When a photon with sufficient energy (larger than the band gap) is absorbed, the energy is transferred to an electron in the valence band and it is promoted to a higher energy state in the conduction band, similar to an excited state in an organic dye. In the valence band, where the electron was, a positively charged vacancy (or hole) is created. In PV cells, an inorganic semiconductor e.g. Si (or an organic dye) absorbs photons above a certain threshold energy [the band gap of the material or the difference between the highest occupied

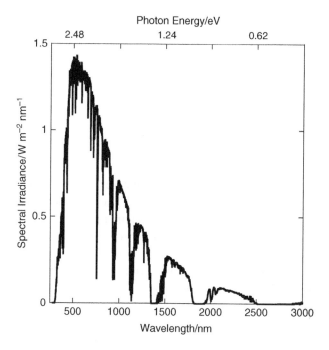

Figure 3.1 Solar energy spectrum (AM 1.5) in terms of radiation energy *vs* photon wavelength. Reprinted from reference[7] with permission from NREL

molecular orbital (HOMO) and lowest unoccupied molecular orbital (LUMO) energy levels of the dye]. The electron–hole pairs are separated and move in opposite directions until they are collected and the resulting current passes through wires to an external circuit to perform work. The difference in potential energy of the electrons and holes results in the cell voltage.

The band gap energy (E_g) of semiconductors can be determined from UV–visible absorption spectroscopy since

$$\lambda_g(\text{nm}) = 1240/E_g(\text{eV}) \tag{3.1}$$

where λ_g is the fundamental absorption edge.

For a direct transition (where the momentum of the electron in the conduction band is the same as the hole in the valence band):

$$\alpha = A\left(h\upsilon - E_g\right)^{1/2} \tag{3.2}$$

where α is the the absorption coefficient, A is a constant (Equation 3.3), h is Planck's constant, υ is the frequency of absorbed light at the onset of absorption ($h\nu$ is the energy of the photon at frequency ν) and E_g is the band gap.

$$A = \frac{q^2 x_{vc}^2 (2m_r)^{3/2}}{\lambda_0 \varepsilon_0 h^3 n} \tag{3.3}$$

where m_r is the reduced mass, defined as

$$m_r = \frac{m_h^* m_e^*}{m_h^* + m_e^*} \tag{3.4}$$

and where m_e^* and m_h^* are the effective masses of the electron and hole, respectively, q is the elementary charge, n is the refractive index, ε_0 is the vacuum permittivity and x_{vc} is a matrix element.

For an indirect transition (where the momentum of the electron in the conduction band is not the same as the hole in the valence band and a phonon as well as a photon is required; this lowers the probability of the transition and therefore the value of α):

$$\alpha \propto \frac{(h\nu - E_g + E_p)^2}{\exp\left(\dfrac{E_p}{kT}\right) - 1} + \frac{(h\nu - E_g - E_p)^2}{1 - \exp\left(-\dfrac{E_p}{kT}\right)} \tag{3.5}$$

where E_p is the phonon energy.

Therefore, whether a transition is direct or indirect can be determined from the absorption spectrum of the material. If a plot of $h\nu$ vs α^2 forms a straight line, the band gap is direct, and the band gap energy can be determined by extrapolating the straight line to $\alpha = 0$. If a plot of $h\nu$ vs $\alpha^{1/2}$ forms a straight line, there is an indirect band gap, the energy of which can be determined by extrapolating the straight line to $\alpha = 0$.[8]

3.1.2   The Photovoltaics Industry

The PVs industry is expanding rapidly, spurred on by government incentives and lower production costs. The number of installations and the amount of investment follow an exponential trend. Between 2004 and 2007 the total global investment in solar energy increased 20-fold to $12.4 billion in 2007.[9] Applications of PVs fall into four main categories.[10] Off-grid, non-domestic systems were the first commercial applications for terrestrial PVs, including telecommunications, water pumping, navigational aids, vaccine refrigeration and other applications where small amounts of electricity have a high value. Off-grid domestic systems are used to power houses not connected to the utility electricity network. In this category PVs are used for low-power applications (1 kW), lighting and refrigeration. PVs are the most cost-competitive means of supplying electricity in remote areas (at around $13 W^{-1}) and a huge potential market exists in the developing world where the majority of the population has no access to electricity. However, off-grid systems total only 4% of the total installed PV systems. Instead, most of the recent growth has been in (largely government subsidised) grid-connected systems. Grid-connected centralised systems include centralised power stations where the electricity generated contributes to bulk power rather than being supplied to one customer. The year 2007 saw a threefold increase in growth of large-scale systems. Grid-connected distributed systems are integrated to a customer premises grid, such as on a roof top, which is connected to the electrical network. The estimated total installed capacity in 2009 was over 20 GW,[a][10] most of this was grid-connected with an installed price of around $4–6 W^{-1}; more than 6 GW was installed in 2009 alone and over 90% of that was installed in Germany, Italy, USA, Japan, and France.

3.1.3 Terminology

PV devices are compared under standard conditions such as 100 mW cm^{-2} (1 sun) solar radiation of (air mass) AM1.5 spectrum[7] at 25 °C. The most common way devices are compared is by their current–voltage (J–V) characteristics (Figure 3.2).

Most solar cells work on the basis of p-n junctions (see below) where an n-type material is in contact with a p-type material or an n- or a p-type material is in contact with a redox electrolyte. Ideally, the J–V behaviour follows the single exponential relationship known as the diode equation:

$$\text{Dark current} = J_{\text{dark}}(V) = J_0 \left(e^{qV/kT} - 1 \right) \qquad (3.6)$$

where J_0 is a constant, q is electronic charge, k is the Boltzmann constant, and T is temperature (K).

The photocurrent at short-circuit (J_{SC}) is given by

$$J_{\text{SC}} = q \int I_s(E) \Phi(E) dE \qquad (3.7)$$

---

[a] 1 GW is approximately the equivalent of a nuclear or medium-sized coal fired power plant or two off-shore wind farms. Note, 7.8 GW is double the capacity of the combined Drax power stations in the UK.

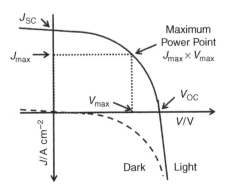

Figure 3.2 Typical J–V characteristics for a PV cell in the dark and under illumination

where I_s is solar photon flux, E is photon energy, and Φ is quantum yield.[b] The broader the absorption, the greater the fraction of incident light converted, and the higher the photocurrent. The higher the charge transfer and/or collection efficiency, the higher the quantum yield and the higher the current. The net current of the solar cell device is the difference between the dark current that flows as a result of the applied potential difference, and the photocurrent in the opposite (positive) direction:

$$\text{Net current} = J(V) = J_{SC} - J_{dark}(V) = J_{SC} - J_0\left(e^{qV/kT} - 1\right) \qquad (3.8)$$

However, in real devices ideal behaviour is rarely observed since losses occur and so a non-ideality factor, γ, which accounts for the transport mechanism of the dark current, is introduced:

$$J(V) = J_{SC} - J_{dark}(V) = J_{SC} - J_0\left(e^{qV/\gamma kT} - 1\right) \qquad (3.9)$$

The photovoltage at open circuit [*i.e.* $J(V) = 0$] is:

$$V_{OC} = \frac{kT}{q}\ln\left(\frac{J_{SC}}{J_0} + 1\right) \qquad (3.10)$$

The maximum power point of the cell is the values of J and V (J_{max} and V_{max}) at which the maximum rectangle in the figure meets the J–V curve. The 'ideal' rectangle would be drawn from J_{SC} and V_{OC} and deviation

---

[b] Φ = External Quantum Efficiency (EQE) = Incident Photon-to-current Conversion Efficiency (IPCE).

from this defines a term called the 'fill factor' (FF), which is the ratio of the maximum experimentally obtainable power to the product of the short-circuit current and the open-circuit voltage:

$$FF = \frac{J_{max} \times V_{max}}{J_{SC} \times V_{OC}} \tag{3.11}$$

$FF = 1$ when $J_{max} = J_{SC}$ and $V_{max} = V_{OC}$.

This 'squareness' of the J–V plot depends on the I_0, γ, the shunt resistance and the series resistance of the device. I_0 and γ in the diode equation limit the maximum FF and typical values for commercial devices are 0.7–0.8.[12] The series resistance of the cell, including the resistance of the substrate, the charge-transfer resistance at the counter electrode or charge collectors, charge-transport resistance through the semiconductor, ion (mass) transport resistance in the case of the photoelectrochemical devices, adversely affects the FF.[13] The shunt resistance is due to leakage of current in loss reactions in the device and also lowers the FF.

Solar cell performance is usually compared in terms of the overall solar conversion efficiency, η, defined in Equation 3.12, which includes the parameters described above.

$$\eta = \frac{P_{out}}{P_{in}} = \frac{P_{max}}{I \times A} = \frac{J_{SC} \times V_{OC} \times FF}{I \times A} \tag{3.12}$$

where P_{max} is maximum power, I is irradiance ($= 100$ mW cm^{-2} for AM 1.5) and A is cell area (cm^2).

Another useful measurement is the spectral response of the cell, the Incident Photon-to-current Conversion Efficiency[c] (IPCE),[14] given in Equation 3.13, or the Absorbed Photon-to-current Conversion Efficiency (APCE),[d] given in Equation 3.14.

$$\text{IPCE} = 1240 \times \frac{J_{SC}}{P_{in} \times \lambda} \tag{3.13}$$

$$\text{APCE} = \text{IPCE}/\text{LHE} \tag{3.14}$$

where LHE is light harvesting efficiency $= 1 - 10^{-A}$, where A is the absorbance.

---

[c] For DSSCs: IPCE = LHE $\times \eta_{inj} \times \eta_{cc}$, where η_{inj} is charge injection efficiency and η_{cc} is charge collection efficiency.
[d] APCE = Internal Quantum Efficiency (IQE).

In addition, PV modules are compared by cost per rated power (*e.g.* $ W^{-1}) at standard conditions:

$$\$ \, W^{-1} = \frac{\text{Module cost} + \text{BOS and installation cost} + \text{tracker cost}}{\text{Peak Power Generation}} \quad (3.15)$$

$$\text{Peak Power Generation} = \text{efficiency} \times 1000 \; W_p \; m^{-2} \quad (3.16)$$

where BOS is the balance of systems costs (in the range of $250 m^{-2} for c-Si cells).[15]

Because of day/night and time-of-day variations in insolation and cloud cover, the average electrical power produced by a solar cell over a year is about 20% of its W_p rating.[5]

Alternatively, there is the Levelised Cost of Electricity (LCOE), $ kWh^{-1}, over the lifetime of the system, defined by:[15]

$$\$kWh^{-1} = \frac{\text{Module cost} + \text{BOS and installation cost} + \text{tracker cost} + \text{O and M cost}}{\text{Total Energy Generated}}$$

$$(3.17)$$

where O and M is operation and maintenance.

The $ W_p^{-1} cost figure of merit can be converted to $ kW h^{-1} by scaling by a factor of 20: $1 W_p^{-1} \sim $0.05 kWh^{-1}.[5]

Another useful comparison can be kWh yr^{-1} under real operating conditions (Equation 3.18), or the energy produced per rated power, kWh/kW$_{STC}$ under standard conditions.

$$\text{kWh yr}^{-1} = \eta \times \text{irradiance yr}^{-1} \quad (3.18)$$

where irradiance yr$^{-1} \sim 1700$ kWh m^{-2} depending on the location and angle of the system.

PV systems are often compared in terms of their energy payback time:

$$\text{Energy payback time (yr)} = \frac{\text{energy used to make the system} \left(\text{kWh m}^{-2} \right)}{\text{energy produced by the system} (\text{kWh m}^{-2} \text{yr}^{-1})}$$

$$(3.19)$$

Economic payback time (yr)

$$= \frac{\text{cost} \left(\$ \, m^{-2} \right)}{\eta \times \text{irradiance yr}^{-1} (\text{kWh m}^{-2}) \times \text{electricity price} (\$ \, kWh^{-1})}$$

$$(3.20)$$

Some values of energy payback times for different PV technologies will be given in the following sections.

3.2 DEVELOPMENT OF PV TECHNOLOGY

Solar cell technologies can be divided into three generations.[16] The first is the established technology such as crystalline silicon. The second includes the emerging thin-film technologies that have just entered the market, while the third generation covers future technologies, which are not yet commercialised.

3.2.1 First Generation: Crystalline Silicon (c-Si)

The first generation includes established technologies.[17] Silicon-based systems make up around 90% of the current PV market and most are manufactured in Europe and Asia. The raw materials (silicon wafers) are expensive, shortages have had a massive impact on price, and high purity is required, therefore over half the cost is that of the silicon wafers. Nonetheless the cost per watt has decreased exponentially over the last few decades, leading to module retail prices of around $4.45 W^{-1} in the USA and €4.34 W^{-1} in Europe, due to increases in efficiency, which have now reached 23% for modules and 25% in the laboratory.[18]

Although silicon is the second most abundant element on the Earth's surface, the high purity required and the energy-intensive processing required means the cost of the silicon feedstock is high. Solar grade silicon is supplied at a reasonable cost as a by-product of (and is therefore reliant on the strength of) the high-technology electronics industry.[19] However, the PVs industry has now out-grown and overtaken the electronics industry and the demand for silicon is greater than the supply of rejected material. Monocrystalline cells are cut from a silicon boule that is grown from a single large crystal, which has in turn grown in only one plane, whereas polycrystalline cells are cut from a silicon boule that is grown from a multifaceted crystalline material that grows in multiple directions.[20] Polycrystalline solar cells have a lower efficiency due to the relative impurity of the silicon and therefore require a larger area. Sharp and Q-Cells are giants of the c-Si PVs industry, but China is now the world leader in producing c-Si-based PV cells and modules, with a capacity of over 2300 MW yr^{-1}.

The International Energy Agency (IEA) has estimated the energy payback time of c-Si PV modules, incorporated as a grid-connected, rooftop installation, as 1.5–2 years, based on 2006 figures for the cost of a 13–14% efficient c-Si-based system, irradiated at 1700 kWh m^{-2} yr^{-1} and a performance ratio (which accounts for losses in the system, due to

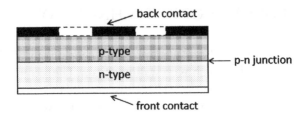

Figure 3.3 A schematic diagram of a c-Si solar cell[23,24]

e.g. shading, heat loss, DC–AC conversion, compared with the module) of 0.75.[21] In 2009, as a result of the global economic crisis, the spot price fell as low as \$67 kg^{-1} (from above \$300 kg^{-1} the previous year) corresponding to \$0.5 W^{-1} (assuming *ca* 8.5 g W^{-1}).[22] Such a price would make c-Si-based PV devices profitable at a selling price of \$2 W^{-1} and competitive with electricity from fossil fuels.

A schematic diagram of the basic principle of the c-Si solar cell is shown in Figure 3.3.[23,24] The n-type Si anode is doped with a Group 15 electron-donor element, *e.g.* phosphorous. The p-type Si cathode is doped with Group 13 electron-acceptor element, *e.g.* boron. An electric field exists at the junction between the two materials. Absorption of a photon at the p-n junction generates an electron–hole pair (the electron is promoted to the conduction band, leaving a hole in the valence band), which are separated with help from the electric field. The difference in potential in the anode and cathode drives a current (J):

$$J = \sigma E \qquad (3.21)$$

where E is the electric field strength and σ is the conductivity

$$\sigma = e(\mu_n n + \mu_h p) \qquad (3.22)$$

where e is the elementary charge, μ_n is the mobility of the electrons, μ_h is the mobility of the holes, n is the density of electrons and p is the density of holes.

The quantities μ_n and μ_h are properties of the material and are temperature dependent. For c-Si typical values for μ_h and μ_n are 500 cm^2 V^{-1} s^{-1} in n-type Si and 1200 in cm^2 V^{-1} s^{-1} in p-type Si, respectively.[25,26]

The efficiency of the device is the ratio of power output to input power. The thermodynamic limit to solar energy conversion is 93%.[27] The loss processes in a simple p-n junction solar cell are shown in Figure 3.4. Photons with energy lower than the band gap do not have sufficient energy to promote an electron to the conduction band and therefore do not contribute to the power output. Process 1 occurs when a photon with

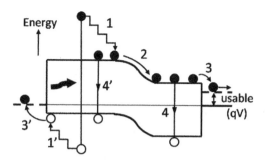

Figure 3.4 Loss processes in a standard solar cell: 1, thermalisation loss; 2 and 3, junction and contact voltage loss; 4, recombination loss. Reprinted from Green, 2001[28] with permission from John Wiley & Sons, Ltd

energy larger than the band gap is absorbed. The resulting photoexcited electron–hole pair quickly loses any energy it may have in excess of the band gap by thermalisation. This means higher-energy photons do not contribute to the output more than lower-energy photons. The maximum efficiency is, therefore, limited by the band gap of the absorber; 44% for an optimum band gap of 1.1 eV.[29] Another loss process is recombination between the electron and hole pair (4). The probability of this process is a property of the material (defects, grain boundaries, *etc.*) and determines the carrier lifetime. The process can be radiative and therefore measured spectroscopically. To account for radiative recombination, and the voltage losses (*e.g.* processes 2 and 3) in the device, Shockley and Queisser treated the solar cell as a black body, absorbing photons from the sun at 6000 K and re-emitting them at 300 K, and estimated the maximum efficiency to be much lower than the estimate above; 30% for a material with a band gap of 1.1 eV.[29] These, in addition to reflection losses, mean the actual experimental efficiencies for single-junction solar cells are much lower than the predicted limits, typically between 15 and 20%, and the record efficiency is 25% for a c-Si cell.[17]

Figure 3.5 shows the 20% learning curve for PV modules.[30] A log–log plot of price *vs* cumulative volume for the period 1979–2005 is a straight line from which the experience factor can be taken. In the case of PVs the experience factor is 80%, meaning that module prices reduce 20% for every doubling of cumulative volume. A number of forecasts have been made using this learning curve along with the annual growth rate of the industry to predict when PV technologies will become competitive with fossil fuels. A number of forecasts have been made based on the PV learning curve, ranging from optimistic assumptions that the market

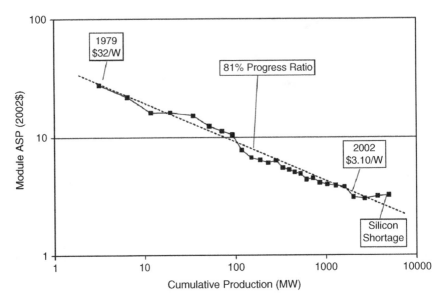

Figure 3.5 Historical plot of module price showing classic experience curve behaviour. Reprinted from Swanson, 2006[30] with permission from John Wiley & Sons, Ltd

growth will continue at a rate of 30% a year to less optimistic growth rates of 10–15% per year. While the exact values vary, the general agreement is that, following the 20% learning curve, module prices will not fall below the $1 W_p^{-1} threshold until the early 2020s, corresponding to a cumulative production capacity of about 120 GW_p.[31–34] If solar energy is to fill the 'energy gap' between current and future usage or indeed replace current nonrenewable fuels, a substantial reduction in cost and vast improvement in efficiency are required much faster than would be accomplished by following the present experience curve. To change the slope of this trend drastically, and escape the dependency on silicon supply, new technologies which improve the cost-to-efficiency ratio are needed.

3.2.2 Second Generation: Thin-Film Technologies

Thin-film technologies are becoming a competitive class of PV, doubling production from 2006 to 2007. The majority is produced in the USA where investment in alternative thin-film technologies is highest, but China should soon overtake the USA in terms of production capacity. The market share of thin-film technologies increased from only a few per cent in 2000, to 16% in 2009, compared with 37% for monocrystalline Si.[35] It is expected that the growth in these alternative technologies should exceed that of c-Si over the coming years. The advantages of thin-film solar cells include the ease of manufacture of large areas at lower cost, a wider range of applications including building integration, higher kWh/kW$_{STC}$ output, higher battery charging current, attractive appearance and a range of possible deposition techniques resulting in the ability to assemble devices using flexible substrates.[36] Materials that are suitable for thin-film PVs must have direct band gaps (high extinction coefficients) and maintain good electronic properties in polycrystalline form (*i.e.* minimum recombination at grain boundaries).[17]

3.2.2.1 Amorphous Silicon

The most established thin-film technology is amorphous silicon (a-Si).[37] The silicon is deposited by plasma-enhanced vapour deposition at low temperatures (200–500 °C), enabling a range of low-cost, lightweight and flexible substrates to be used, *e.g.* glass, steel or plastic. It can be made n-or p-type by introducing dopants such as gaseous B_2H_6 or PH_3 during the deposition. The 5–20 atom% hydrogen from the silane source improves the electronic quality by passivating the broken Si-Si bonds, enabling the material to be suitable for solar cell applications, such as watches and calculators. a-Si:H also no longer behaves like an indirect band gap material like c-Si and has a high extinction coefficient, more like direct band gap materials, and so thin films can be used. The as-deposited a-Si has a band gap of 1.7 eV which can be tuned between 1.3 eV and 2.0 eV by co-deposition with GeH_4 or CH_4. The efficiencies are lower than crystalline solar cells due to internal energy losses, resulting from the disordered material (Figure 3.6). Free carrier mobilities for electron and holes are at least two orders of magnitude lower than c-Si, only 10 and 1 cm^2 V^{-1} s^{-1}, respectively. However, a-Si outperforms c-Si by 10–20% in terms of kWh/kW$_{STC}$ because a-Si has a lower temperature coefficient for power loss (0–0.2% per °C) compared with c-Si (0.4–0.6% per °C). The cost is only slightly lower than for c-Si as a consequence of expensive manufacturing equipment and the cost to produce high-quality transparent conducting oxide (TCO) layers. Often the most expensive component of thin-film devices is the transparent metal oxide[e] coated glass substrate.

---

[e] *E.g.* fluorine-doped tin oxide (FTO) or indium tin oxide (ITO = Sn-doped In_2O_3).

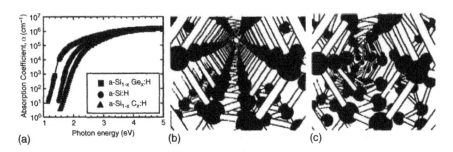

Figure 3.6 The optical absorption of a-Si:Ge:H and a-Si:C:H films (a); the tetrahedrally bonded crystal structure of c-Si (b); and the absence of long-range order in a-Si (c). Reprinted from Wronski, 2000[37]. Copyright (2000) IEEE

Since the most active region is the p-n junction, the aim of the a-Si device is to extend this junction and keep the n- and p-doped regions as thin as possible, with an undoped (intrinsic) region in between. An electric field is established in the undoped region, the strength of which depends on the width. Poorer quality or degraded silicon (see below) requires a stronger electric field and therefore thinner films. However, this may mean that not all of the light is harvested and therefore cells 100–200 nm thick are stacked above each other. Device structures include triple junction cells on flexible, stainless steel substrates, micromorph tandem cells on glass, double junction a-Si/a-Si or a-Si/a-SiGe on glass and a-Si single junctions on glass.[36]

a-Si devices suffer from light-induced degradation. The Staebler–Wronski effect causes a 10–15% relative efficiency loss in performance during the first 10 h of light exposure before stabilising.[38] After stabilisation, losses of 1% yr^{-1} match c-Si.[36] Stable laboratory efficiencies of 13.5% and module efficiencies around 5.3–6.4% have been achieved corresponding to a cost of $3.5–8.6 W_p^{-1}. The expanding industry now accounts for around 6% of the PV market with a number of 10–30 MW_p yr^{-1} capacity plants based in the USA, China and Japan, with Sharp being the market leader with a 1 GW_p yr^{-1} capacity facility.

3.2.2.2   Cadmium Telluride

Heterojunction solar cells are constructed from p-type absorber layers and n-type, wide-band gap layers (windows) to form a p-n junction.[39] Compound semiconductors such as CdTe are used which, unlike Si and Ge, have direct band gaps and, therefore, high extinction coefficients, making them suitable for thin-film applications.

CdTe is a crystalline compound with a cubic zinc blende (sphalerite) crystal structure (lattice constant of 6.481 Å), a direct band gap of 1.5 eV, an ideal match to the solar spectrum, and an extinction coefficient around 5×10^4 cm^{-1}.[40] The intrinsic defects include cadmium interstitials and cadmium vacancies, and extrinsic doping can be achieved using In (donor) substitution or Cu, Ag, Au (acceptor) substitution for Cd. The mobilities have been measured to be up to 1100 cm^2 V^{-1} s^{-1} for electrons and up to 8 cm^2 V^{-1} s^{-1} for holes. Dopant densities up to 10^{17} cm^{-3} have been achieved for n- and p-type materials, but controlling the stoichiometry is difficult and causes problems when forming low-resistance, ohmic contacts to the material. When in contact with CdS, a p-n junction is formed with the depletion field mostly in the CdTe layer. The grain boundaries can be doped to be more p-type than the bulk to reduce minority carrier recombination in the polycrystalline material.

The p-n heterojunction device is shown in Figure 3.7.[39] The highest efficiency is achieved in the superstrate configuration although CdTe solar cells can be grown in substrate configurations on metal foils or polymers. A conducting layer of ITO (for low temperatures) or SnO$_2$ (for high temperatures) is deposited on the glass substrate which forms an ohmic contact with the CdS layer. Deposition of CdS is followed by post treatment in a reducing atmosphere or in the presence of CdCl$_2$ to increase the grain size and reduce defect density. A thin layer of SnO$_x$ may be deposited between the ITO and the CdS to prevent shunting. The CdS n-type window and the CdTe layer can both be deposited by electrodeposition, physical vapour deposition, close-space sublimation (CSS), screen printing or spray pyrolysis; the orientation and grain size are dependent on the deposition conditions. High-temperature methods such as CSS give rise to larger grain sizes up to 10 μm. Post treatment such as dipping in CdCl$_2$, followed by heating in an oxygen-containing atmosphere at 400 °C, then rinsing to remove excess CdCl$_2$, is carried out

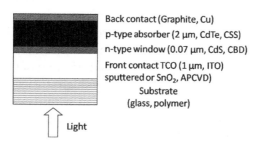

Back contact (Graphite, Cu)
p-type absorber (2 μm, CdTe, CSS)
n-type window (0.07 μm, CdS, CBD)
Front contact TCO (1 μm, ITO)
sputtered or SnO$_2$, APCVD)
Substrate
(glass, polymer)

Light

Figure 3.7   Superstrate configuration of a CdTe solar cell. Adapted from Romeo *et al.*, 2004[39] with permission from John Wiley & Sons, Ltd

to improve the performance of the device by causing some structural changes that result in a higher effective acceptor concentration.[41] Lastly an ohmic contact with a transparent charge collector is formed. There is not a metal with a large enough work function (5.7 eV) to give a direct ohmic contact to p-type CdTe, so a heavily doped degenerate layer is formed on the surface of the material by etching with bromine-methanol to give a Te-rich surface. p-Type dopants such as Cu, Hg, P or Au are then deposited, followed by heat treatment, before depositing the secondary conductor such as graphite or gold. However, diffusion of these metals into the CdS/CdTe or CdS/ transparent conducting oxide (TCO) interface causes degradation of the cell and so higher efficiencies are achieved when a buffer layer is incorporated.[39,42]

Laboratory efficiencies of 16.7% and module efficiencies of 10.9% have been reported.[43] The theoretical conversion efficiency is considered to be 28%, however, more recently a more conservative estimate of 17.5% has been proposed.[44] The drawbacks to CdTe systems include the toxicity and low abundance of the materials, temperature-dependent efficiencies and only an average light tolerance. Nonetheless, CdTe solar cells have entered the market. The leading manufacturers of CdTe modules are First Solar in the USA, with a capacity of over 1 GW_p yr^{-1}, and Calyxo GmbH, a Q-cells subsidiary, in Germany, who have a 25 MW_p yr^{-1} and a 6 MW_p yr^{-1} capacity facility and boast an energy-pay-back time of approximately 1.5 years for their systems.

3.2.2.3   Copper Indium (Gallium) Diselenide

Another class of heterojunction solar cells are $CuInSe_2$-based devices, formed from p-n junctions with CdS thin films.[39] $CuInSe_2$ is a ternary compound with a band gap of 2.4 eV and is stable as a chalcopyrite or sphalerite structure.[45] Chalcopyrite (lattice constant $a = 0.5789$ and $c = 1.162$ Å) is stable at room temperature up to 810 °C. The band gap is direct and approximately 1.02 eV at room temperature, with an absorption coefficient above 5×10^4 cm^{-1}. On substitution of Ga for In or S for Se, the band gap increases up to 1.68 eV. For high efficiency devices, band gaps between 1.20 eV and 1.25 eV are used with [Ga]/[In + Ga] ratios between 25% and 30%.[39] Different stoichiometries give rise to different intrinsic defects and hence electronic properties, e.g. Cu and In (acceptor) vacancies (excess Se) give rise to p-type character and Se vacancies lead to n-type material (for solar cells a slightly Cu-deficient material is used).

For p-type materials, mobilities of 15–150 $cm^2 V^{-1} s^{-1}$ and carrier densities of 0.15–2 $\times 10^{17} cm^{-3}$ at 300 K have been measured.

Cu(In, Ga)Se$_2$ (CIGS) systems are capable of high efficiencies (19.9% in the laboratory,[46] 16.7% in sub-modules[43] and 13.5% in modules), suffer from little degradation and the high extinction coefficient of the CIGS absorber means materials costs can be kept low (films 2 μm thick are sufficient for light capture) even though indium is rare. The material can be deposited on a range of substrates and since CIGS has a tunable band gap (1.0–2.7 eV) they are suitable for tandem cells.[36] The manufacturing costs result from complicated fabrication methods involving vacuum processes but energy consumption is low compared with c-Si and other thin-film PVs.

Polycrystalline compound semiconductor solar cells can be arranged in either a superstrate configuration or a substrate configuration (Figure 3.8). The superstrate configuration facilitates low cost encapsulation of solar modules and can be incorporated into tandem solar cells as a means of harvesting the complete solar spectrum.[39] However, CIGS cells grown in the substrate configuration give higher efficiencies since it is better suited to the assembly conditions (see below) but requires an encapsulation layer and/or glass to protect the cell surface.

The structure of a CIGS cell includes a CdS layer deposited by chemical bath deposition (CBD) where the sample is dipped in ammonia solution containing CdSO$_4$ and thiourea. The CuInSe$_2$ layer can be deposited using a range of methods such as physical vapour deposition (co-evaporation)

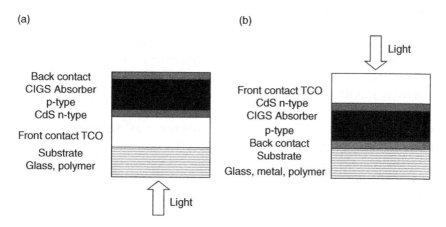

Figure 3.8 CIGS solar cell configurations; (a) superstrate configuration; (b) substrate configuration. Adapted from Romeo *et al.*, 2004[39] with permission from John Wiley & Sons, Ltd

of the three or four components onto a heated substrate, allowing variation of the component ratios and graded band gap absorber layers.[47]
Alternatively, the metallic elements can be deposited successively, followed
by reactive annealing of the resultant multi-layer structure in a selenium
or selenium/sulfur atmosphere, with partial substitution of In with
Ga and of S with Se to match the band gap with the solar spectrum.[36,39]
The molybdenum back contact allows diffusion of Na from the soda lime
glass to the CIGS layer where it is incorporated (0.1 atom%), improving
efficiency, V_{OC} and fill factor.[48]

When the device is assembled in a superstrate configuration, the transparent, current-collecting Al- or B-doped ZnO layer acts as a barrier
between the glass and the CIGS. This layer prevents diffusion of Na
from the glass substrate. Therefore, in order to obtain V_{OC} and FFs
comparable with devices assembled in the substrate configuration,
co-evaporation of Na_2Se or Na_2S during CIGS deposition is required.[39]
This is also the case for cells built on metal or polymer substrates. Whilst
higher Ga content also improves the V_{OC}, this is offset by a reduction in
short-circuit current density due to reduced absorption and collection
efficiency. This is advantageous, however, in a module where high
current densities give rise to higher resistance losses (see below).

As for CdTe systems, the carrier density in CdS is larger than in $CuInSe_2$
so the depletion layer, and therefore the region where electron–hole
pairs are generated, is in the $CuInSe_2$, close to the interface because of
the high extinction coefficient of the $CuIn_{1-x}Ga_xSe_2$. This minimises
minority carrier recombination at the metallurgical interface. As for
CdTe, $CuInSe_2$ grain boundaries, which are parallel to the current-flow
direction, can be passivated by dopants such as oxygen and by low-
temperature, post-processing heat treatments to make them more p-type
and reduce recombination of the minority carriers (electrons).

A champion CIGS-based thin film solar cell was prepared and characterised at NREL by Repins et al.[46] A record efficiency of 19.9% was
measured for a 0.41 cm^2 device in which, during the latter stages of
vacuum deposition (100 Å), the chalcopyrite was deposited without
gallium, improving the quality of the material near the surface by reducing the number of surface defect states. An improved (5.5 ns) charge
carrier lifetime was measured and a maximum carrier density of 2×10^{16}
cm^{-3} was measured.

CIS modules are produced by Würth Solar who have a 3 MW_p yr^{-1}
production capacity, Avancis who have a 20 MW_p yr^{-1} plant producing
11% efficient modules in Germany and Showa Shell in Japan who
opened a 60 MW_p yr^{-1} capacity module factory in 2008 to add to their

20 MW$_p$ yr^{-1} facility. Global Solar, who operate a 4 MW$_p$ yr^{-1} capacity plant in the USA and a 35 M$_p$ yr^{-1} capacity plant in Germany, Nanosolar, Miasol and Honda Soltec are some of the leading producers of CIGS modules. In Europe, a joint venture between Q-cells and Solibro AB, who hold the current world record for a CIGS mini-module with 16.6% efficiency,[43] operate a 25 MW$_p$ yr^{-1} capacity factory.

3.2.2.4   Scale Up

For commercialisation of thin-film PVs, manufacturing costs should fall below $1 W^{-1}, which requires large-scale (>100 MW) manufacture at high throughput rates and high yield (93% is possible). High efficiency is often traded for high yield, speed and materials utilisation since uniformity and cost are most important for commercial use. Large-scale modules are typically built from a number of cells defined by laser scribing connected in series on one substrate (m^2 glass plates) to form a monolithic device (Figure 3.9);[41] The device is encapsulated in e.g. ethyl vinyl acetate. A second piece of glass can be used to cover the cells and the ethyl vinyl acetate cured to seal the device, with current leads attached to the contacts. Aside from glass, thin-film devices can be manufactured on kilometre-long rolls of flexible plastic or metal foil. Since the thin-film devices rely on several layers of different material ranging between 10 nm and 2 µm in thickness, strict control over the deposition is critical in large-scale manufacture of thin-film PV modules.

Module efficiencies are usually much lower than cell efficiencies on account of the losses that increase with size. The cell width (w) is determined by the resistivity of the current collector, e.g. ZnO, and must be optimised to balance resistive losses and area losses. The maximum sheet

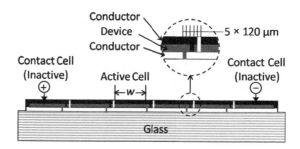

Figure 3.9   Schematic cross-sectional view of a polycrystalline thin film PV module. Adapted from Birkmire and Eser, 1997[41] with permission from Annual Reviews

resistance to avoid current loss from absorption of light in ZnO is $10 \, \Omega \, \text{sq}^{-1}$[f]. Since resistive power losses are proportional to the ratio between J_{SC} and V_{OC}, it is better to have a higher V_{OC} and low current than *vice versa*, although for CIGS devices the best efficiencies as yet have high currents and low voltages.[41] Other problems that escalate when increasing the size of the devices include the adhesion of the Mo charge collector on the glass substrate, the quantity of Na doping in CIGS devices, the need for $CdCl_2/O_2$ post treatment and CdTe contacting. Many of the tricks that are used to overcome these problems and achieve record efficiencies, such as graded or profiled interfaces, are not worthwhile economically when the process is scaled up. Also, processes that are not fully understood can become more important when the system is scaled up (such as Na doping). It may also be impractical or costly to use high purity raw materials or substrates, which are used in the research laboratory for large scale manufacture.

Many of the processes that are standard in the laboratory become less viable economically on a larger scale. For example, on a laboratory scale, four-source, physical vapour deposition is used to deposit CIGS layers, which requires a vacuum, a slow deposition rate of $0.05 \, \mu\text{m} \, \text{min}^{-1}$, a moderate substrate temperature of 550 °C and high temperatures for the sources (1300 °C for Cu, 1100 °C for In and Ga). To keep the manufacturing costs low, a high throughput is necessary and large modules need to be built at a rate of one every 1–2 min, at a high yield.[36,41] This means an increase in deposition rate or a reduction in thickness. Fortunately the absorbers used in thin-film technologies have such high extinction coefficients that 1 µm material is thick enough for absorption of all or most of the light. Back reflectors can also be incorporated to reduce the amount of material needed. Alternatives to vacuum deposition techniques with lower equipment costs and faster deposition rates exist for multicrystalline technologies like CIGS and CdTe, *e.g.* screen printing followed by rapid thermal processing,[36] or close-spaced sublimation where the cost is reduced because the deposition rate is higher ($1 \, \mu\text{m} \, \text{min}^{-1}$). The same technique can be used for the CdS window. However, a-Si deposition requires a high vacuum to prevent Si-O bond formation which lowers the performance.

Many of the materials used in second generation solar cells can be deposited onto flexible substrates, enabling roll-to-roll processes and therefore higher throughput rates. However, most laboratory processes

---

[f] Ohms-per-square is a conventional unit of sheet resistance, commonly measured using a four-point probe technique, and is independent of sample dimensions. It is used for films of negligible (uniform) thickness where the resistivity is equal to the electrical resistance (Ω) multiplied by the width and divided by the length (*i.e.* the resistance of a square sample).

suit a batch manufacturing process, especially post treatments like $CdCl_2$ treatment and selenisation. In cases where post-treatment is required or heating up and cooling down steps are involved, the deposition of materials is not rate limiting in a batch process. Parallel batch process where *e.g.* sixty modules are manufactured at the same time would produce one module per minute.

For scale-up, monolithic configurations are often used where laser scribing is used to separate the cells to increase the manufacturing efficiency (automation) and reliability. This is an advantage of thin-film technologies over c-Si which requires mechanical connection and soldering of each cell to form a module and is often a source of failure of the device, as well as introducing extra steps in the manufacturing process. Nonetheless, 10–20% losses are imposed when going to monolithic configurations and integrated modules on account of additional resistances from connections, a loss in active area from scribing and optical absorption from the encapsulent.

In commercial manufacturing, environmental issues become more important and the toxicity of the materials becomes more of a problem. Cd is toxic and carcinogenic. CdS and CdTe are more stable than Cd itself (which is a by-product of Zn refining) so are less hazardous than the starting material; however some of the processes used in the laboratory become unsuitable for use on a large scale. For example, chemical bath deposition is hazardous in the workplace because of the large quantities of basic aqueous solutions of Cd and S ions. Deposition of excess material on the container surface, the formation of particles of CdS, the requirement for fresh solution needed for each substrate and the need to remove Cd ions before disposal of solution each reduces the process efficiency and has consequences for health and safety. Likewise, during sputtering and CSS processes not all the material goes on the substrate and fine powders that are left in the reaction vessel need to be removed. Also, since many of the raw materials are expensive, optimisation of the equipment to reduce waste and additional recycling steps are needed.

Whilst it has been demonstrated that thin-film PV modules can be produced at a reasonable cost, the limitation to the technology comes from the availability of the materials. Table 3.1 compares the abundance of the components and the current (2004) production. Whilst a-Si-based devices are not reliant on the rate of production of the raw materials, CdTe PVs are severely limited by the abundance and rate of mining of Te. Similarly CIGS technology is limited by the rate of production of In, Se and Ga. The scarcity and toxicity of the many of the components means that careful disposal or, more likely, recycling of the materials is required. Since manufacturing costs are so dependent on scale, reductions in the

Table 3.1 Availability of materials for thin-film active layers. Adapted from Keshner and Arya, 2004[49] with permission from NREL

Device material	Amount required for 30 M m² yr⁻¹ (t p.a.)	Abundance in Earth's crust (ppb)	Current amount mined worldwide (t p.a.)	Potential production from % abundance (t p.a.)	Limit by current production (GW p.a.)	Limit by potential production (GW p.a.)
Amorphous silicon single junction						
a-Si:H	37	282 000 000	4 100 000	73 723 260 000	222 826	4 006 698 913
ZnO	45	70 000	8 700 000	18 300 100	388 393	816 969
Al	32	83 200 000	127 000 000	21 750 976 000	7 839 506	1 342 652 840
Additional or alternate materials						
Ag	125	70	18 300	18 300ᵃ	293	293
SnO₂	109	2000	283 000	522 860	5 192	9 592
Ge	7	5400	44	1 411 722	13	403 349
CdTe single junction						
CdS	19					
Cd	15	200	18 700	52 286	4 121	11 521
S	4	2 600 000	58 000 000	679 718 000	45 312 500	531 029 688
CdTe	446					
Cd	219	200	18 700	52 286	282	789
Teᵇ	228	1	130	2000	2	29
Back contact						
Al	32					
Al	32	83 200 000	127 000 000	21 750 976 000	13 096 875	2 243 069 400

Cu (In .85 Ga .15) Se$_2$

Mo	412	1500	112 000	392 145	979	3427
Cu(In,Ga)Se$_2$	464					
Cu	88	55 000	13 200 000	14 378 650	539 020	587 150
In[c]	139	100	335	26 143	9	676
Ga[d]	14	15 000	70	3 921 450	18	1 014 168
Se[b]	223	50	1500	4000	24	65
ZnO/CdS	2					
ZnO/CdS	112	70 000	8 700 000	18 300 100	279 643	588 218
Al grid	8	83 200 000	127 000 000	21 750 976 000	60 476 190	10 357 607 619

[a] 'Potential production' is an order-of-magnitude estimate based on the ratio of silver production to silver abundance in the Earth's crust. Abundance in the Earth's crust is an order of magnitude estimate and only an indication of possible mining production.

[b] Se and Te are both by-products of Cu refining and are not economical to mine directly. Cu anode slimes are 41 000 metric tons annually worldwide with an average of 5% Te and 10% Se. Thus, Se production could be increased to 4 000 metric tons p.a. and Te production to 2 000 metric tons pa.

[c] In is currently a by-product of Zn refining, which is declining from year to year. Recycling of In is important to meet current demand. Although the abundance numbers suggest In production could be increased, it tends to be found only in low concentrations with base metals such as Zn and Cu. Therefore, increasing production may be difficult without a substantial increase in price.

[d] Ga is obtained mostly from Al production. It is also a by-product of Zn production. In addition, there are a small number of other mineral deposits (100 ppm)

The availability of In, Ga, Se and Te could be greatly extended if the films could be made 5–10 times thinner, 0.2–0.4 μm, with light trapping to enhance light absorption in the thinner layer. Limitations to the light trapping effectiveness might be the low reflectivity (50%) of the Mo back contact and seed layer for Cu(In, Ga)Se$_2$.

Source: USGS mineral commodity summaries (available from http://minerals.usgs.gov. Accessed November 2010). Reprinted from M. S. Keshner and R. Arya, Study of potential cost reductions resulting from super-large-scale manufacturing of PV modules, October 2004, NREL/SR-520-36846 (available from www.nrel.gov Accessed 2009-10-07).

amount of material required, as well as very high efficiencies, are essential for second generation PVs to become cost competitive and meet the energy demand.

3.2.3   Third Generation: Nanotechnology/ Electrochemical PVs

Despite improvements in efficiency and the increase in demand for renewable sources of electricity, solar power currently supplies only 0.1% of the primary energy demand. This can be attributed to the high installation cost of PV systems, 50% of which is the cost of the module. Solar electricity tariffs cost around three to five times that of conventional residential electricity tariffs, yet grid-connected systems are the highest growing applications. To compete with c-Si in this market, the next generation of solar cells should perform close to maximum efficiency over a range of light conditions and temperatures, have good long-term stability and, in order to lower installation costs, need to be durable, flexible and attractive so that they can be applied to building integrated PVs *e.g.* on flexible steel and architectural glass. To be able to supply electricity to the potential market in the developing world, the installation costs must be lowered considerably. For solar electricity to penetrate the electricity segment in the developed world, installed solar system costs need to drop from around $8–10 W_p^{-1} to $3 W_p^{-1} (corresponding to $0.15 kWh^{-1}). For these reasons, the third generation solar cells are designed significantly to reduce the manufacturing costs of PV systems (Figure 3.10).[16]

The thermodynamic limit on the conversion of sunlight into electricity is around 93%. Standard first and second generation, single junction solar cells are quantum converters, ideally converting one photon to one electron in the load. This limits the efficiency to about 31%. However, Würfel has estimated a maximum efficiency for an infinite stack of semiconductors, each absorbing a small photon energy interval, of 67% for nonconcentrated sunlight (86% for concentrated light).[27] To achieve energy conversion efficiencies above the Shockley–Queisser limit,[g] double or triple the efficiencies of first and second generation technology, but with low manufacturing costs, new materials and configurations, will need to be developed.[30] The third generation approaches being investigated include dye-sensitised titania solar cells,

---

[g] Photons of energy higher than the band gap lose their excess energy as heat, and photons of energy lower than the band gap are not absorbed.

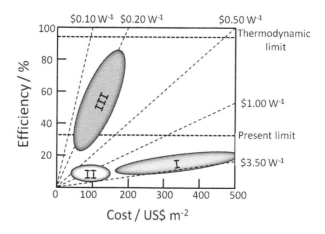

Figure 3.10 Efficiency and cost projections for first, second and third generation PV technology. Adapted from Green, 2001[28] with permission from John Wiley & Sons, Ltd

organic PVs,[50,51] tandem cells, 'hot electron' utilisation and materials that generate multiple electron–hole pairs.[17,52]

3.2.3.1 Multijunction Cells

The most straightforward route to overcoming the Shockley–Queisser limit, by avoiding the key loss process in single junction solar cells, is the construction of tandem cells.[28,53] Multiple cells with different band gaps are assembled on top of each other (Figure 3.11) and each cell converts a narrow range of photon energies close to its band gap. By arranging the cells with the highest band gap at the top (illumination side), the energy of the absorbed photons are only a little higher than the band gap. For an infinite stack of independently operated cells, an efficiency of 86.8% is possible. Current matching of the cells enables them to be connected in series so that losses at the contacts are minimised. However, the efficiency is limited by the lowest performing cell.

3.2.3.1.1 Silicon-Based

Because it is relatively straightforward to tune the band gap of a-Si over a wide range of band gaps (1.3–2.0 eV), multijunction a-Si-based solar cells are being developed.[54,55] However, a-Si:Ge is inferior to a-Si:H and so limits the efficiency. This is accounted for in the design of the device, and tricks such as textured reflectors are used to boost the efficiency of the

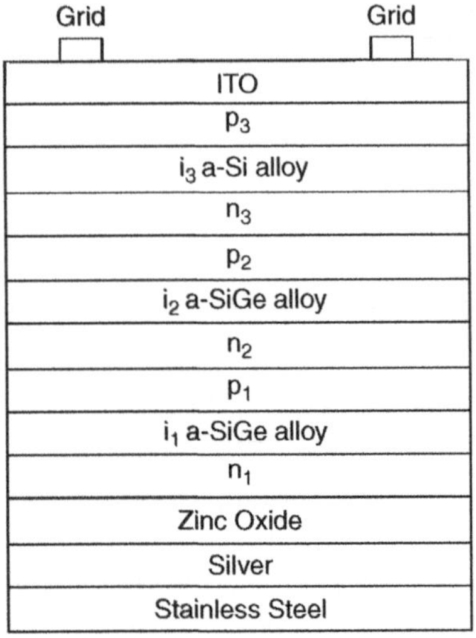

Grid   Grid

| ITO |
| P₃ |
| i₃ a-Si alloy |
| n₃ |
| P₂ |
| i₂ a-SiGe alloy |
| n₂ |
| P₁ |
| i₁ a-SiGe alloy |
| n₁ |
| Zinc Oxide |
| Silver |
| Stainless Steel |

Figure 3.11 Schematic diagram of a triple-junction cell. Reprinted from Guha, 1997[53]. Copyright (1997) Elsevier

lower layer (Figures 3.11 and 3.12). Stabilised efficiencies up to 12.5% have been confirmed for these systems.[43]

An alternative approach is to use microcrystalline Si as the lower cell instead of a-Si:Ge (Figure 3.13).[56] Two silicon layers – one amorphous and one crystalline – are deposited on a glass substrate. The amorphous layer absorbs the visible part of the solar spectrum, while the crystalline layer absorbs the near-infrared light. This lowers the amount of Si used in comparison with conventional technologies, reducing material and

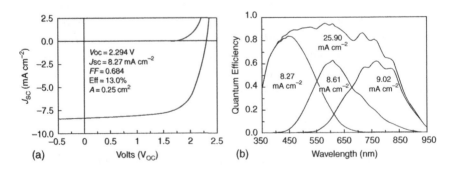

Figure 3.12 (a) Current–voltage and (b) quantum efficiency characteristics of an a-SiGe:H/a-SiGe:H/a-Si:H n-i-p stacked cell. Reprinted from Wronski, 2000[37]. Copyright (2000) IEEE

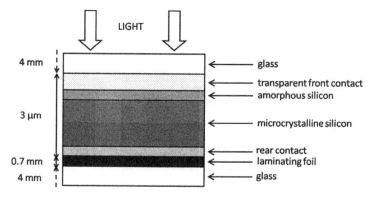

Figure 3.13   Structure of a c-Si/a-Si device[16]

manufacturing costs but with reasonably high efficiencies in excess of
10% and an energy payback time of approximately 2 years.[56]

Triple-cell configurations are preferred over double-cell structures, since
the top layer needs to be thick to match the current of the thick multicrystalline layer. This gives rise to poorer stability and long deposition
times as the multicrystalline layer is several micrometres thick and the
deposition rate is slow (5 Å s^{-1}) in order to control the crystal growth.
The small band gap (1.1 eV) means the open circuit voltage is low.

3.2.3.1.2   III–V

The conversion efficiency of InGaP/GaAs-based multijunction solar cells
has been improved in the same way.[57,58] A schematic illustration of the
InGaP/(In)GaAs/Ge triple-junction solar cell reported by Takamoto *et al.*
is shown in Figure 3.14. The device pictured gave an efficiency of 31.5%
under standard conditions.[59]

3.2.3.2   Dye-Sensitised Solar Cells

3.2.3.2.1   Introduction

With the development of molecular solar cells, or so-called dye-sensitised
solar cells (DSSCs),[h] conventional solid-state PV technologies are now
challenged by devices functioning at a molecular and nano level.[60–62]
Record efficiencies of up to 12%, promising stability data, passing for

---

[h] Other names of the technology are dye-sensitised nanostructured solar cells, mesoscopic
injection solar cells, nanosolar cells, the artificial leaves, or Grätzel cells.

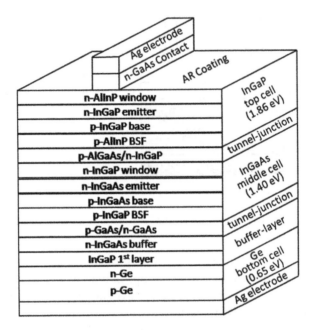

Figure 3.14   A schematic illustration of the InGaP/(In)GaAs/Ge triple-junction solar cell and key technologies for improving conversion efficiency. Reprinted from Takamoto *et al.*, 2005[59] with permission from John Wiley & Sons, Ltd

example the critical 1000 h stability test at 80 °C with a durable efficiency of 8–9%, and means of energy-efficient production methods have been accomplished.[63] The prospect of low-cost investments and fabrication are key features. DSSCs are also relatively better compared with other solar cell technologies under diffuse light conditions and at higher temperatures. The possibilities to design solar cells with respect to shape, colour and transparency and to integrate them into different products open up new commercial opportunities.

DSSC research groups are established around the world with the biggest activity in Europe, Japan, Korea, China, and Australia. The field is growing fast and can be illustrated by the fact that about two to three research articles are being published every day. The industrial interest in DSSCs is strong, with large multinational companies such as BASF, Bosch and Tata Steel in Europe, Toyota, Sharp, Panasonic, Sony, Nippon Oil and Samsung in Asia developing the technology. A pilot processing line is set up in the company G24i, Wales, making prototypes to recharge mobile phones. Research companies such as Dyesol, Australia, Solaronix, Switzerland, and Peccell, Japan, are expanding, focusing on selling material components and equipment.

The principle of the DSSC has also become a part of the core-chemistry and energy science teaching and research. Text books have sections or chapters dealing with DSSCs;[8,64] laboratory kits have been developed for educational purposes under the slogan 'make your own solar cell'. Not only energy science, but also photochemistry, photoelectrochemistry, materials science and transition metal co-ordination chemistry have significantly benefited from DSSC research.

3.2.3.2.2 The State of the Art
In the seminal *Nature* paper of 1991, O'Regan and Grätzel reported an order-of-magnitude increase for DSSC type solar cell efficiency to 7–8%.[60] With regards to stability, a turnover number of 5×10^6 was measured for the sensitiser (a trimeric ruthenium complex).[65,66] This was followed up by the introduction of the ruthenium bipyridyl-based N3 dye (Figure 3.15), giving efficiencies around 10%.[14]

Since the initial work in the beginning of the 1990s, a wealth of DSSC components and configurations have been developed. Perhaps a key concept for the future success of DSSCs in this regard is 'diversity'. DSSC technology is a technology to be explored and is full of nuances including versatile materials and devices. At present several thousands of dyes have been investigated. Not as many, but certainly hundreds of electrolyte systems and mesoporous films with different morphologies and compositions have been studied and optimised. For DSSCs at present, in the official table of world record efficiencies for solar cells, the record is held by the Sharp company in Japan at 10.4 ± 0.3%.[43] A criterion to qualify for these tables is that the solar cell area is at least 1 cm². For smaller cells a certified conversion efficiency of 11.1% has been reached using the black dye (Figure 3.15) as the sensitiser.[67]

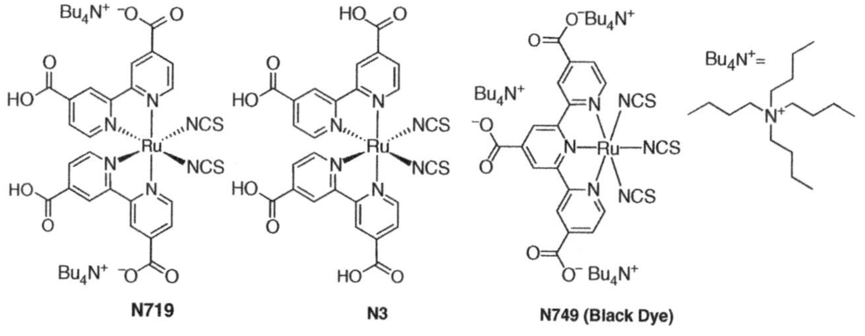

Figure 3.15 Structures of the standard ruthenium-based dyes used in DSSCs

Early on in DSSC research the classical dyes N3, with its salt analogue N719 and the black dye (N749) were developed (Figure 3.15).[63] The IPCE values are close to 80% for both sensitisers across the visible part of the solar spectrum. The N3 dye starts to absorb light at around 800 nm whereas the photocurrent onset is red-shifted for the black dye to 900 nm. However, the IPCE increases only gradually from the absorption onset to shorter wavelengths due to relatively low extinction coefficients of these sensitisers.

There is a lot to be gained by developing sensitisers with high extinction coefficients particularly in the near-infrared region of the solar spectrum. The highest photocurrent density[67,68] of the black dye under AM1.5 conditions is close to 21 mA cm^{-2} which can be compared with the theoretical maximum photocurrent of 33 mA cm^{-2} from the AM1.5 spectrum with an absorption threshold of 900 nm. Improving the solar light absorption in the 650–950 nm domain (there is a dip in the solar irradiance spectrum at 950 nm) would then be one of the main directions to take aiming for DSSC efficiencies above 15%. The other direction is to increase the photovoltage by replacing the conventionally used I^-/I_3^- redox couple with a system having a more positive redox potential. These are key challenges facing present DSSC research and will be further discussed below.

During the last two to three years the advent of heteroleptic ruthenium complexes furnished with an antenna function has raised the performance of the DSSC to a new level. Two examples of these dyes are C101 and Z991 (Figure 3.16).

Compared with the classical DSSC Ru dyes, their extinction coefficients are higher and the spectral response is shifted to the red. The solar cell efficiency of these types of sensitisers has increased continuously over the last 2 years and efficiencies above 12% have been reported.[63] There are sometimes arguments that DSSCs have not developed much since the

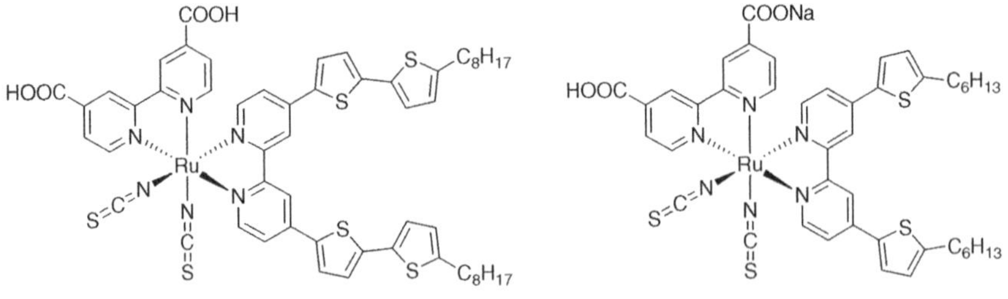

Figure 3.16 Two examples of heteroleptic ruthenium sensitisers, C101 and Z991, giving DSSC efficiencies above 12%

breakthrough in the early 1990s. The record efficiencies plateaued at 10–11% and the most efficient devices have remained essentially unchanged from the original concepts. It is, however, a complex system. The realisation of the need to handle such complexities, with the plethora of material components to explore, make the 12% efficiency, now reached with one of the new classes of dyes, a strong indicator for further progress in improving DSSC performance. Another essential performance indicator is long-term stability. Here, the progress during the last 15 years has been steady and accelerated durability tests have been passed with higher and higher efficiencies. Most of the earlier work has been reviewed.[69,70] As an example of recent results, the C101 sensitiser maintains outstanding stability at efficiency levels over 9% under light soaking at 60°C for more than 1000 h. This is achieved by the molecular engineering of the sensitiser, but also, very importantly, by the use of robust and nonvolatile electrolytes, such as ionic liquids and adequate sealing materials. Good results on overall system endurance have been reported for several years and these results are presently being confirmed under real outdoor conditions. Because of the direct relevance to the manufacturing of commercial products, little work is published on these subjects and it is, for example, difficult to find information on processing issues as well as sealing materials and methods. The industrial development and commercialisation of DSSCs was the topic of the DSSC-IC 3 conference in Nara, Japan, April 2009. Many encouraging results were presented, giving confidence that the DSSCs can match the stability requirements needed to sustain outdoor operation for many years. Still, a lot of research and development activities need to be performed and in particular more data on outdoor field tests is required. One important research topic is to develop protocols for accelerated long-term stability tests relevant to DSSC technologies.

A schematic of the interior of a DSSC showing the principle of how the device operates is shown in Figure 3.17.

The typical configuration is as follows. At the heart of the device is the mesoporous oxide layer composed of a network of TiO_2 nanoparticles which have been sintered together to establish electronic conduction. Typically, the film thickness is ca 10 μm and the nanoparticle size 10–30 nm in diameter; the porosity is 50–60%. The mesoporous layer is deposited on a transparent conducting oxide on a glass or plastic substrate. A typical scanning electron microscopy (SEM) image of a mesoporous TiO_2 film is shown in Figure 3.18.

Attached to the surface of the nanocrystalline film is a monolayer of the charge-transfer dye. Photoexcitation of the latter results in the injection of

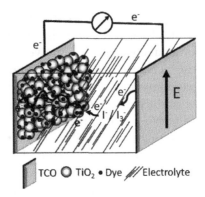

TCO ◯ TiO₂ • Dye ///Electrolyte

Figure 3.17 A schematic illustration of the interior of a DSSC

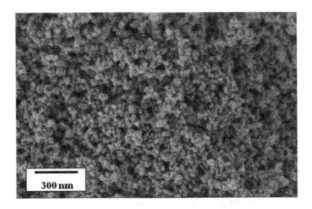

300 nm

Figure 3.18 SEM image of a mesoporous TiO₂ film. Reprinted from Lindström *et al.*, 2002[161]. Copyright (2002) Elsevier

an electron into the conduction band of the oxide leaving the dye in its oxidised state. The dye is restored to its ground state by electron transfer from the electrolyte, usually an organic solvent containing the iodide/triiodide redox system. The regeneration of the sensitiser by iodide intercepts the recapture of the conduction band electron by the oxidised dye. The I_3^- ions formed by oxidation of I^- diffuse a short distance (<50 μm) through the electrolyte to the cathode, which is coated with a thin layer of platinum catalyst where the regenerative cycle is completed by electron transfer to reduce I_3^- to I^-. For a DSSC to be durable for more than 15 years outdoors the required turnover number is 10^8, which is satisfied by the ruthenium complexes mentioned above. The voltage generated under illumination corresponds to the difference between the electrochemical potential of the electron at the two contacts, which generally for DSSCs is the difference between the Fermi level of the mesoporous TiO_2 layer and the redox potential of the electrolyte. Overall, electric power is generated without permanent chemical transformation.

The basic electron transfer processes in a DSSC, as well as the potentials for a state-of-the-art device based on the N3 dye adsorbed on TiO_2 and I^-/I_3^- as redox couple in the electrolyte, are shown in Figure 3.19.

Besides the desired pathway of the electron transfer processes (processes 2, 3, 4 and 7) described in Figure 3.19, the loss reactions 1, 5 and 6 are indicated. Reaction 1 is direct recombination of the excited dye reflected by the excited state lifetime. Recombination reactions of injected electrons in the TiO_2 with either oxidised dyes or with acceptors in the electrolyte are numbered 5 and 6, respectively. In principle, electron transfer to I_3^- can occur either at the interface between the nanocrystalline oxide and the electrolyte or at areas of the anode contact (usually a fluorine-doped tin oxide layer on glass) that are exposed to the electrolyte. In practice, the second route can be suppressed by using a compact blocking layer of oxide deposited on the anode by spray pyrolysis.[71,72] Blocking layers are mandatory for DSSCs that utilise one-electron redox systems such as cobalt complexes[73–75] or for cells using solid organic hole-conducting media.[76,77]

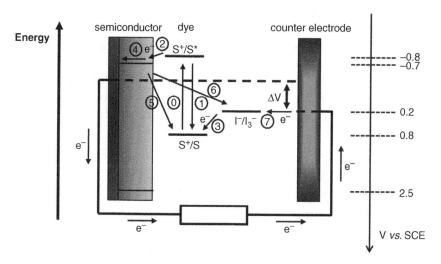

Figure 3.19 The basic electron transfer processes in a DSSC. The reactions are indicated by numbers and are described in the text

As mentioned above, hundreds of alternatives to the components used in conventional DSSCs have been investigated. As for the sensitisers, ruthenium complexes have given the best results since the early days; these dyes have recently been reviewed by Robertson.[78] Osmium and iron complexes and other classes of organometallic compounds, such as phthalocyanines and porphyrins, have also been developed. Metal-free organic dyes are catching up showing efficiencies close to 10% with the use of indoline dyes.[79,80] Moreover, chemically robust organic dyes showing promising stability results have recently been developed by several groups.[81–84] Organic dyes for DSSCs have been reviewed by Mishra et al.[85]

With regards to the hole-conducting medium, electrolytes based on the I^-/I_3^- redox couple have been the preferred choice. Organic nitrile-based solvents give the highest efficiencies, whereas gelification of the solvent or ionic liquids are used for the best stability, but somewhat compromise the efficiency. However, the I^-/I_3^- couple is not necessarily unique. Successful results have also been achieved with other redox systems such as cobalt-based systems, $SCN^-/(SCN)_3^-$ and $SeCN^-/(SeCN)_3^-$,[86–88] and organic systems based on TEMPO[89] and TDP[90] (Figure 3.20). Interesting developments have been made for solid-state DSSCs, using organic hole conductors[91] and inorganic p-type semiconductors, such as CuI[92] and CuSCN.[93] Recent overviews for photoanode materials can be found in references[94–96]. The preferred oxides are so far TiO_2, ZnO, SnO_2 and Nb_2O_5.

Different morphologies based on nanoparticles, nanofibres and tubes, and core-shell structures have been developed. In this chapter, works on TiO_2 and ZnO will be summarised. Different counter electrode (cathode) materials will also be reviewed. The most common counter electrode is a platinised conducting glass but also carbon materials and conducting polymers have been developed. For third generation solar cells, the goal is to deliver electricity at a large-scale, competitive price with all types of energy sources, i.e. less than $0.5 W_p^{-1}. This means very effective solar cells that are produced by techniques that permit facile mass production. In particular, the

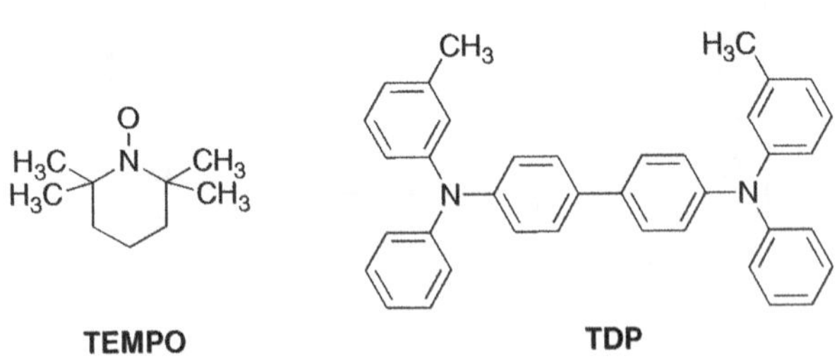

TEMPO   **TDP**

Figure 3.20   Structures of some organic redox mediators used in DSSCs

main direction for research on third generation solar cells is to find ways of achieving energy conversion efficiencies above the so called Shockley–Queisser limit of 31%.[16] For most of the development of third generation PV systems nanotechnology will be essential, utilising for example, properties in the quantum-size domains. Hence, DSSC technology provides an interesting launching pad. As a specific example work towards the possible construction of tandem DSSCs will be considered.

DSSC devices can be designed by appropriate choice of the sensitiser to absorb and convert incident photons quantitatively to electric current in selective spectral regions of the solar emission, while maintaining high transparency in the remaining wavelength range. A further advantage of sensitised nanostructured electrodes is that their short-circuit photocurrent output can readily be varied by changing the film thickness and effective pore size. This, along with the ease of formation of multilayer structures by simple techniques such as screen printing, constitutes a great advantage with regard to conventional solar cells, facilitating the fabrication and optimisation of tandem cells.

3.2.3.2.3   Operational Principles

The much simplified picture of the energetics and kinetics for a working DSSC device, which emerged in the early research,[61] is still useful as an introduction of working principles. The chemical complexity of the device must, however, be understood and mastered to improve our ability to identify predictive materials and optimised structure/function relationships.[97] The present understanding of these processes is now covered with specific emphasis on the electron transport through the mesoporous TiO_2 electrode. The reader is also directed to recent review articles[97–99] on these topics.

In Figure 3.21 the same schematic picture as in Figure 3.19 is used but includes the kinetic data for the different electron transfer processes taking place at the oxide/dye/electrolyte interface, for a state-of-the-art DSSC , based on TiO_2, including a ruthenium complex sensitiser and a liquid I^-/I_3^- electrolyte.

Electron Injection

One of the most astounding findings in DSSC research is the ultrafast injection from the excited ruthenium complex in the TiO_2 conduction band. Although the detailed mechanism of this injection process is still under debate, it is generally accepted that a fast, femtosecond (fs) component is observed for these types of sensitisers, directly attached to an oxide surface.[100–103] This would then be one of the fastest chemical processes

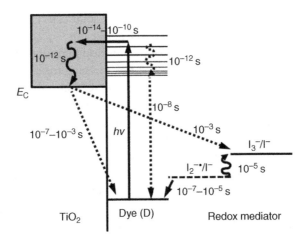

Figure 3.21 Overview of processes and typical time constants in a ruthenium-based DSSC with iodide/triiodide electrolyte. Recombination processes are indicated by broken arrows

known to date. There is a debate for a second, slower, injection component on the picosecond (ps) timescale. The reason for this could, on one hand, be due to an intersystem crossing of the excited dye from a singlet to a triplet state. The singlet state injects on the fs timescale whereas the slower component arises from the relaxation time of the singlet to triplet transition and from a lower driving force in energy between the triplet state and the conduction band of the TiO$_2$.[103] Another view of the slower component is that it is very sensitive to sample condition and originates from dye aggregates on the TiO$_2$ surface.[102] For DSSC device performance, the timescales of the injection process should be compared with direct recombination from the excited state of the dye to the ground state. This is given by the excited state lifetime of the dye which, for typical ruthenium complexes used in DSSC, is 30–60 ns.[104] Thus, the injection process itself has not generally been considered to be a key factor limiting device performance. Interestingly, Koops *et al.* observed a much slower electron injection in a *complete* DSSC device with a time constant of around 150 ps. This would then be slow enough for kinetic competition between electron injection and excited state decay of the dye, with potential implications for the overall DSSC performance.[105] A slower injection in the sub-ps time regime could be even more severe for organic dyes, for which the excited state decay time could be less than nanoseconds. Research on the electron injection process will thus continue to be an important topic in DSSC research and needs to be extended to other classes of sensitisers beside the ruthenium complexes.

Recombination of Electrons in the Semiconductor with Oxidised Dyes or Electrolyte Species

During their relatively slow transport through the mesoporous TiO_2 film, electrons are always within only a few nanometres distance of the oxide/electrolyte interface. Recombination of electrons with either oxidised dye molecules or with acceptors in the electrolyte is therefore a possibility. The recombination of electrons with the oxidised dye molecules competes with the regeneration process, which usually occurs on a timescale of sub-microseconds to microseconds. The kinetics of the back electron-transfer reaction from the conduction band to the oxidised sensitiser follow a multiexponential time law, occurring on a microsecond to millisecond timescale depending on electron concentration in the semiconductor and, thus, the light intensity. The reasons suggested for the relatively slow rate of this recombination reaction are: (i) weak electronic coupling between the electron in the solid and the Ru^{III} centre of the oxidised dye; (ii) trapping of the injected electron in the TiO_2; and (iii) the kinetic impediment due to the inverted Marcus region.[106] Application of a potential to the mesoporous TiO_2 electrode has a strong effect.[107–110] When the electron concentration in the TiO_2 particles is increased, a strong increase in the recombination kinetics is found. Under actual working conditions, the electron concentration in the TiO_2 particles is rather high and recombination kinetics may compete with dye regeneration.

Recombination of electrons in TiO_2 with acceptors in the electrolyte is, for the I^-/I_3^- redox system, generally considered to be an important loss reaction, in particular under working conditions of the DSSC device when the electron concentration in the TiO_2 is high. The kinetics of this reaction are determined from voltage decay measurements and normally referred to as the electron lifetime. Lifetimes observed with the I^-/I_3^- system are very long (1–20 ms under one sun light intensity) compared with other redox systems used in DSSCs, explaining the success of this redox couple. The mechanism for this recombination reaction remains unsettled, but appears to be dominated by the electron trapping–detrapping mechanism in the TiO_2.[111] Recently, a lot of attention has been drawn to the effects of the adsorbed dye on the recombination of TiO_2 electrons with electrolyte species. There are several reasons: first, adsorption of the dye can lead to changes in the conduction band edge of TiO_2 because of changes in surface charge. This will lead to a larger driving force for recombination. Secondly, dyes can either block or promote reduction of acceptor species in the electrolyte.[97] The size of the oxide particle, and thus the surface-to-volume ratio, is also expected to have a significant effect on electron lifetime.[112,113]

Regeneration of the Oxidised Dyes

The interception of the oxidised dye by the electron donor, normally I^-, is crucial for obtaining good collection yields and high cycle life of the sensitiser. The ground-state redox potential of the N3 dye, as well as that of several other ruthenium complexes and organic sensitisers measured when the dye is adsorbed to the surface of the mesoporous TiO_2 film, is around 1 V against the normal hydrogen electrode (NHE). The redox potential of I^-/I_3^- is close to 0.4 V and the potential of the conduction band edge of TiO_2 is located at -0.5 V (Figure 3.19). Using a transition energy of 1.65 eV for the excitation of the dye to the LUMO level,[63] the excited state potential of N3 is at -0.65 V. Hence, the driving force for electron injection is 0.15 eV. In contrast, the regeneration consumes 0.6 eV, which is clearly excessive. How much driving force is actually required for efficient regeneration is, at present, debated.[114] Grätzel reports that a minimum driving force of 0.2–0.3 eV is required due to the two-electron character of the iodide oxidation reaction, which passes through I_2^- radicals as intermediates.[63] The regeneration time depends naturally on the dye and electrolyte used. For the state-of-the art ruthenium complexes and the I^-/I_3^- redox couple, regeneration times are in the microsecond time domain. For a turnover number (cycle life of the sensitiser in the DSSC device) above 10^8 (which is required for a DSSC lifetime of 20 years in outdoor conditions) the lifetime of the oxidised dye must be longer than 100-s. This is achieved by the best-performing ruthenium complexes.

3.2.3.2.4   Electron Transport in Mesoporous TiO_2 Electrodes

The 'driving force' for transport of electrons and ions in the DSSC is the gradient of free energy, or *electrochemical potential*, which is equivalent to the gradient of the Fermi energy. This is the thermodynamic picture of charge transport in general and the reader interested in the parallels between solid-state solar cells, such as silicon and thin-film solar cells, and DSSCs should consult the book by Würfel.[64] For DSSCs, electrons injected into the TiO_2 from the photoexcited dye must move through the network of interconnected oxide nanoparticles to reach the anode. The excess electronic charge in the oxide arising from injection of electrons is balanced by a net positive ionic charge in the electrolyte arising from the regeneration of the dye by electron transfer from iodide ions. The density of ionic charges in the electrolyte is high ($\sim 10^{20}$ cm^{-3}), so that excess charges in the mesoporous oxide are effectively shielded and electric fields are screened out over short distances. As a consequence, it appears that electrons are collected at the anode by diffusion.[115–117] The strong coupling between ionic and electronic charges means that the diffusion

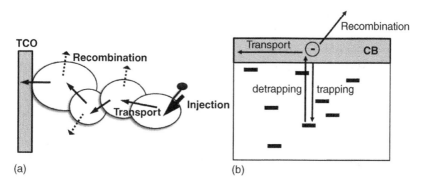

(a)   (b)

Figure 3.22 (a) The main processes determining the collection efficiency of electron transport in DSSCs. (b) Schematic drawing of the trapping/detrapping process. Electrons in trap states are thermally excited to the conduction band (CB) where they are mobile and can undergo transport towards the conducting substrate or recombination to the electrolyte or oxidised dyes

process is ambipolar.[118,119] However, the diffusion of electrons is also complicated by trapping and detrapping processes that influence the time response of the current to external perturbations.

A schematic figure of electron transport and the trapping/detrapping processes is shown in Figure 3.22. In contrast to the notion that electron transport occurs by diffusion, it is observed that the electron transport depends on the incident light intensity, becoming more rapid at higher light intensities.[120,121] This can be explained by a diffusion coefficient that is light intensity dependent, or, more correctly, dependent on the electron concentration and Fermi level in the TiO_2. The measured value of the diffusion coefficient is orders of magnitude lower than that determined for single crystalline TiO_2 anatase. These observations are usually explained using a multiple trapping (MT) model.[121–125] Electrons are considered to be mostly trapped in localised states below the conduction band, from which they can be thermally 'detrapped' to the conduction band where they can move before they are trapped again. Experiments suggest that the density and energetic location of such traps is described by an exponentially decreasing tail of states below the conduction band.[122,124]

Experimentally measured diffusion coefficients are thus effective values of the electron diffusion coefficient (D_n) rather than the corresponding free electron values D_0. The lifetime of the electrons in the mesoporous TiO_2 electrode is also light intensity dependent and the effective lifetime, τ_n, is treated in relation to the free electron lifetime in an analogous way as the diffusion coefficient. It has been observed that D_n increases with intensity, whereas τ_n decreases with intensity in such a way that the product, $D_n \times \tau_n$, is almost independent of intensity.[124] An important figure of merit for DSSCs is the diffusion length, L_0, since it is a measurement of how far an electron diffuses toward the electrical contact before it is lost by electron transfer to acceptor species at the oxide/dye/ electrolyte interface. The diffusion length is given by:

$$L_0 = \sqrt{D_0 \tau_0} \qquad (3.23)$$

To get an expression for the measured effective diffusion length, L_n, we need to consider models which consider trapping and detrapping of electrons. Emphasis has been placed on the quasi-static treatment of Bisquert and Vikhrenko[126] since it has the merit of being testable and suitable for incorporation into detailed models of the response of DSSCs to external perturbations. Their treatment shows that D_n and τ_n depend on trap occupancy and hence on the Fermi level. The quasi-static approximation predicts that effective τ_n is given by

$$\tau_n = \left(1 + \frac{\partial n_t}{\partial n_c}\right) \tau_0 \qquad (3.24)$$

whereas D_n is given by

$$D_n = \left(1 + \frac{\partial n_t}{\partial n_c}\right)^{-1} D_0 \qquad (3.25)$$

The $\partial n_t / \partial n_c$ term and its inverse reflect the way that the densities of trapped and free electrons, n_t and n_c, respectively, vary with changes in the quasi-Fermi level. When D_n and τ_n are measured at the same trap occupancy, according to the quasi-static approximation,[126] the effective diffusion length becomes independent of light intensity and one obtains

$$L_n = \sqrt{D_n \tau_n} = \sqrt{D_0 \tau_0} \qquad (3.26)$$

This approach has been used in a number of studies.[127,128]

Measurements of charge transport properties for DSSC can be performed with steady-state methods or by time-dependent techniques. The latter are generally based on two types of perturbation. The first type uses a time-dependent illumination signal (photomodulation) to change the rate of electron injection onto the conduction band of the semiconductor oxide. The preferred perturbation is a small amplitude illumination, superimposed on steady background light. The amplitude should be small enough that the system response is linear. The perturbation can be a short laser pulse, a rectangular pulse, or a sinusoidal intensity profile.

The measured response is the photocurrent as in intensity modulated photocurrent spectroscopy (IMPS) or photovoltage as in intensity modulated photovoltage spectroscopy (IMVS). The response can also be followed in the time domain, *i.e.* photocurrent and photovoltage decay measurements where the transients are followed after the light has been switched off. IMPS and photocurrent decay measurements give information of the electron transport and estimates of D_n. Voltage measurements give on the other hand estimates of the electron lifetime, τ_n. The second type of perturbation involves modulation of the voltage applied to the DSSC, either in the dark or under illumination. This is the basis for impedance measurements, which have become standard techniques to characterise DSSCs.[129,130] Impedance techniques measure the effective values D_n and τ_n.

Often, D_n is obtained under short-circuit conditions and τ_n under open-circuit conditions. Since the trap occupancy in the mesoporous oxide is very different under these two extreme conditions, Equation 3.26 is not valid. A very interesting method has been proposed by O'Regan *et al.*[131] based on small-amplitude photovoltage rise and decay time measurements, in which D_n and τ_n can be measured at the same trap occupancy. Another technique in which this is possible is impedance measurements under illumination at open circuit.[132]

By our own and others recent investigations, it turns out that the quasi-static approximation remains valid under a wide range of experimental conditions. Recent results indicate, however, that there are circumstances when the trapping/detrapping does not hold.[99] The two main assumptions of the model are: electron transfer *via* surface states will be neglected and the reaction orders with respect to electrons and triiodide will be taken to be 1. From this it follows that the photovoltage should increase by 59 mV for every decade of light intensity at 298 K. In practice, DSSCs are always non-ideal to some extent, and generally the photovoltage varies by more than 59 mV per decade of intensity (values as high as 110 mV per decade are not uncommon). The origin of this non-ideality is not understood but it has important consequences for the interpretation of many of the experimental methods used to characterise charge-transfer and transport processes in complete DSSC devices. Other transport models such as charge (polaron) hopping between trap states[133,134] have also been suggested and the question is still open which transport model is most valid and under which conditions.

A direct measurement of the effective diffusion length, without the need of estimating D_n and τ_n, is to use a steady-state technique based on IPCE measurements. The IPCE values are measured for illumination from the anode substrate side (SE) and from the electrolyte side (EE).[135] The ratio of the collection efficiency, obtained from the IPCE measurements, for SE and EE illumination is sensitive to the ratio of L_n/d, where d is the film thickness of the mesoporous film. This method to determine L_n was originally described by Södergren et al.[135] for DSSCs, adapting the work by Lindquist et al.[136] Recently, it has been discussed by Halme et al.[137] and by Barnes et al.[138] Good DSSCs are generally characterised by L_n values that are two to three times the film thickness.

Interestingly, in the most recent studies of charge transport processes in DSSCs it has been observed that the electron diffusion lengths obtained by non-steady-state methods are generally higher than those found by the steady-state method. The origin of the electron traps remains obscure at present: they could correspond to trapping of electrons at defects in the bulk or surface regions of the mesoporous oxide or to Coulombic trapping due to interaction of electrons with the cations of the electrolyte.[133,134,139] Open questions like these, together with the often observed nonideal behaviour of the multiple trapping processes, emphasise the need of further fundamental investigations in DSSC research.

3.2.3.2.5   Anode Materials
Typically the class of materials chosen for applications in DSSCs is metal oxides including TiO_2, ZnO, Nb_2O_5, WO_3 and SnO_2 which can easily adsorb dyes functionalised with anchoring groups such as carboxylic or phosphonic acids. Figure 3.23 shows a selection of metal oxides that were applied as anode materials in DSSCs using the N3 dye.[140] In the next

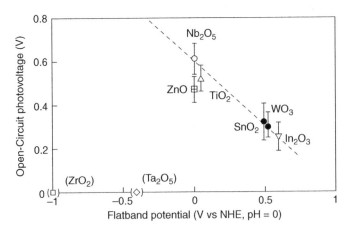

Figure 3.23 Semiconductor flat-band potentials and the open circuit photovoltage (V_{OC}) of dye-sensitised semiconductor cells measured under 520 nm monochromatic light. Reprinted from Sayama et al., (1998)[141]. Copyright (1998) American Chemical Society

section we will describe some of the recent work regarding optimisation of the geometry and optical and electrochemical properties of the metal oxide semiconductors applied in DSSCs.

Nanostructured TiO_2 Anodes

TiO_2 is a stable oxide, which has a high refractive index ($n = 2.4–2.5$)[141] and is, therefore, used widely as a white pigment in paint, toothpaste, sunscreen and food (E171). Other energy and environmental applications include photocatalysis (including water splitting) electrochromic devices, hydrogen storage, sensors and photovoltaics.[142–144] Because of its high band gap ($E_g = 3.2$ eV), TiO_2 is transparent in the visible region of the electromagnetic spectrum but can be modified to absorb visible light by doping, tuning the crystal size[145] or adsorption of a dye. TiO_2 is intrinsically doped *via* oxygen vacancies.[146]

Titanium dioxide occurs in three crystal modifications, namely rutile, anatase and brookite. The most thermodynamically stable form is rutile but anatase, formed in low-temperature preparations, has been shown to give much higher efficiencies in the DSSC because of its higher band gap [E_g(anatase) $= 3.2$ eV, E_g(rutile) $= 3.0$ eV].[147] The effective mass[i] of an electron in anatase ($m^* \approx m_0$) is also much smaller than that in rutile ($m^* = 20m_0$).[148] The effective mass of a hole in anatase is similar to that of the effective mass of an electron in anatase, $m_h^* = 0.01–0.1$ m_0.[149] Values for the electron mobilities vary with the preparation conditions and are strongly temperature dependent, but as an indication, Forro *et al.* reported 20 cm^2 V^{-1} s^{-1} for a single crystal of anatase.[150] The static dielectric constant is 31 for anatase[151] and 170 for rutile.[151–153]

Over recent years, substantial progress has been made in the development of methods to synthesise new anatase nanostructures such as nanoparticles, nanorods, nanowires, nanobowls, nanosheets and nanotubes, and mesoporous materials such as aerogels, opals, and photonic materials.[150,154–157] These methods include sol and sol-gel, micelle and inverse micelle, hydrothermal, solvothermal, sonochemical, microwave deposition techniques, direct oxidation, chemical vapour deposition, physical vapour deposition, and electrodeposition. For DSSC, the most common deposition technique is the sol-gel method from hydrolysis

---

[i] The effective mass is a term used in solid-state physics describing the properties of an electron in the presence of an electric or magnetic field, in the conduction band (or hole in the valence band) of a semiconductor crystal. It has different values in different materials and is used in charge transport calculations or to calculate the density of states. The value is usually smaller than the real electron mass, $m_0 = 9.11 \times 10^{-31}$ kg or 0.511 MeV. For a more detailed description see Memming[8] and Brus.[232]

of a titanium precursor such as a titanium(IV) alkoxide. Acid or basic hydrolysis gives materials of different shapes and properties, and the rate of hydrolysis, temperature and water content can be tuned to produce particles and chains of different sizes. A detailed review by Chen and Mao compares numerous techniques and structures.[143]

Nanoparticles   The breakthrough in efficiency of DSSCs came about as a result of the application of a nanocrystalline, mesoporous electrode, which resulted in a massive increase in surface area over flat substrates, allowing significantly more dye molecules to be adsorbed (Figure 3.24).[60,62] Typically, TiO_2 working electrodes comprise of a three-dimensional network of interconnected, 15–20 nm-sized nanoparticles on a conducting glass (FTO- or ITO-coated) transparent substrate with a porosity of 50–60%. Typically small anatase crystals (*ca* 10 nm) are formed from hydrolysis of a precursor such as $TiCl_4$ or a titanium alkoxide and larger particles are then obtained by hydrothermal treatment in an autoclave that induces Ostwald ripening of the particles. The drawback of the disordered nanoparticle arrays is the large number of interparticle interfaces, which impede the transfer of electrons through the network.

Figure 3.25 shows an example where 30% higher current was achieved when anatase was used compared with rutile.[158] The SEM studies of the particle shape and size showed that the surface area of the rutile film was about 25% less than that of anatase film. The absorption spectra showed

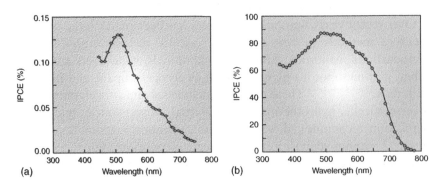

(a)   Wavelength (nm)   (b)   Wavelength (nm)

Figure 3.24 The incident-photon-to-current conversion efficiency plotted as a function of the excitation wavelength for TiO_2 electrodes, sensitised with N3: (a) single-crystal anatase cut in the (101) plane; (b) nanocrystalline anatase film. The electrolyte consisted of a solution of 0.3 M LiI and 0.03 M I_2 in acetonitrile. Reprinted with permission from Grätzel, 2001[62]. Copyright (2001) Macmillan Publishers Ltd

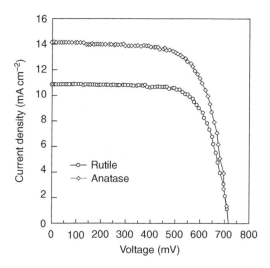

Figure 3.25 Comparison of the *J–V* characteristics of dye-sensitised rutile and anatase films of the same thickness (11.5 μm) at one-sun illumination. Reprinted from Deb, 2005[159]. Copyright (2005) Elsevier

that the rutile film contained about 35% less adsorbed dyes than the anatase film. Nonetheless, rutile has a higher refractive index and larger rutile crystals have been used to form scattering layers to increase the efficiency of DSSCs.

A method of preparing nanoparticles of defined size can be achieved by using polymer templates such as poly(ethylene glycol)-block-poly(propylene glycol)-block-poly(ethylene glycol) (PEO-PPO-PEO, Pluronic®).[159,160] Zukalová *et al.* prepared templated films using Pluronic P123 as template (Figure 3.26).[159] The *J–V* characteristics for the cells assembled from the sensitised electrodes are shown in Figure 3.27. The templated film gave a photoconversion efficiency of 4.04% (J_{SC} = 7.05 mA cm^{-2}, V_{OC} = 0.799 V and FF = 0.72). A randomly orientated 0.95 μm film, tested under the same conditions, gave a lower efficiency, 2.74%.

Pressed TiO$_2$ At pressures of *ca* 1000 kg cm^{-2}, TiO$_2$ can be compressed into a film that is porous and mechanically stable, and suitable for electrodes for DSSCs.[161] The advantage of preparing films in this way is the avoidance of thermal treatment making it a suitable means of preparing TiO$_2$ nanostructured films on plastic substrates. It is also fast, only 1 s is required to press the film, making it favourable for continuous, roll-to-roll production. In this way, Boschloo *et al.* prepared

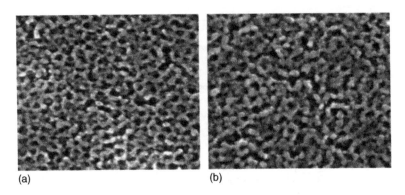

(a)   (b)

Figure 3.26   SEM images of PEO-PPO-PEO templated TiO_2 films (a) one-layer film; (b) three-layer film. The actual size of each image is 300×300 nm^2. Reprinted from Zukalová *et al.*, 2005[160]. Copyright (2005) American Chemical Society

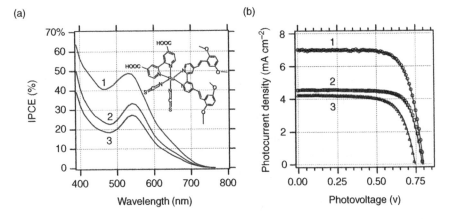

Figure 3.27   (a) IPCE and (b) photocurrent–voltage characteristics (AM 1.5) of a solar cell, based on TiO_2 films sensitised by N945 (inset). Pluronic-templated three-layer film; 1.0 μm thick (1), non-organised anatase treated by $TiCl_4$; 0.95 μm thick (2), non-organised anatase not treated by $TiCl_4$; 0.95 μm thick (3). Reprinted from Zukalová *et al.*, (2005)[160]. Copyright (2005) American Chemical Society

TiO_2 films, sensitised with N719, on ITO-PET [poly(ethylene terephthalate)] substrates which were assembled in a sandwich configuration with platinised SnO_2 on ITO-PET.[162] On illumination with 100 W m^{-2} simulated sunlight, an average efficiency of 4.5% was reported and an IPCE of 76% at the maximum, 530 nm. The solar cells were flexible and could be bent without loss in efficiency. The stability was poor due to permeability of the plastic substrate to small molecules, *e.g.* water.

Nanotubes A popular morphology attracting attention currently, is vertically orientated nanotube arrays, grown by potentiostatic anodisation of Ti metal (Figure 3.28).[163,164] The nanotubes are designed to have large internal surface area but more geometrical control and order than nanoparticle films. This makes them useful for gas sensing, photocatalysis, organic electronics, microfluidics, molecular filtration, drug delivery, and tissue engineering.[165]

In DSSC applications, the aim is to control the alignment (perpendicular to the charge collector) in order to provide a pathway for vectorial charge transfer to the charge collector. This improves the charge collection efficiency by promoting faster charge transport and slower recombination losses resulting from charge hopping across nanoparticle grain boundaries.[166]

Early work on TiO_2 nanotube arrays was carried out by Zwilling *et al.*, followed by Gong *et al.* shortly after, by electrochemical oxidation of titanium in a aqueous HF electrolyte.[167,168] TiO_2 nanotubes are typically grown from Ti thick films or foil in fluoride-based baths, but growth has also been achieved from sputtered Ti thin films on FTO coated conducting glass substrates for applications in DSSCs. The preparation conditions *e.g.* anodisation potential, time and temperature, electrolyte composition (water content, cation size, conductivity and viscosity) affect the features of the arrays,[169–171] *e.g.* length (<1000 μm), wall thickness (5–34 nm) and roughness, pore diameter (12–242 nm) and tube-to-tube

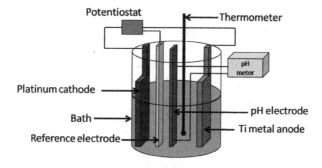

Figure 3.28 Schematic diagram of a three-electrode electrochemical cell in which the Ti samples are anodised. Fabrication variables include temperature, voltage, pH and electrolyte composition. Reprinted from Mor *et al.*, 2006[165]. Copyright (2006) Elsevier

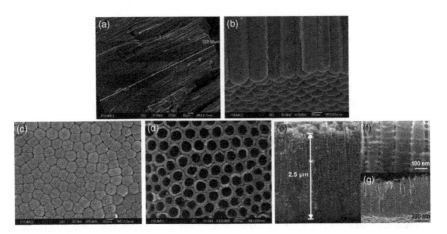

Figure 3.29 SEM images of nanotubes prepared under different conditions. (a) and (b) Cross-sectional, (c) bottom, and (d) top images of a smooth nanotube array grown at 60 V in an ethylene glycol electrolyte containing 0.25 wt% NH_4F. Reprinted from Grimes, 2007[167] by permission of the Royal Society of Chemistry. (e) Conventional anodic TiO_2 nanotubes, formed in 1 M H_2SO_4 with 0.15 wt% HF, with marked inhomogeneities; (f) a magnification of the structure; (g) an example of ripples on short tubes. Reprinted with permission from Macák et al., 2005[170]. Copyright (2005) Wiley-VCH Verlag GmbH & Co. KGaA

spacing (zero to tens of nanometres), which in turn affect the properties of the material (Figure 3.29).[172]

After deposition, the material is annealed at ca 480 °C in oxygen or air to give a crystalline, transparent film with an underlying barrier layer. Figure 3.30 shows examples of the differences between the as-deposited, amorphous films and the sintered, crystalline structures from X-ray diffraction measurements and Raman spectroscopy. In the Raman spectra, the tetragonal structure of anatase gives rise to six Raman-active transitions: three E_g modes centred at 144, 197, and 639 cm^{-1}, two B_{1g} modes at 399 and 519 cm^{-1}, and one mode of A_{1g} symmetry at 513 cm^{-1} overlapping with the B_{1g} mode at 519 cm^{-1}; two rather weak and broad bands are also observed at about 315 and 796 cm^{-1}, which can be attributed to disorder-induced or two-phonon scattering and the first overtone of B_{1g} at about 398 cm^{-1}, respectively.[176]

Three processes are believed to compete during the anodisation process: field-assisted oxidation of Ti to TiO_2, field assisted dissolution of Ti metal ions in the electrolyte and the chemical dissolution of Ti and TiO_2 due to etching by fluoride ions, enhanced by the presence of H^+ (Equations 3.27–3.29).[170] The chemical steps are not thought to

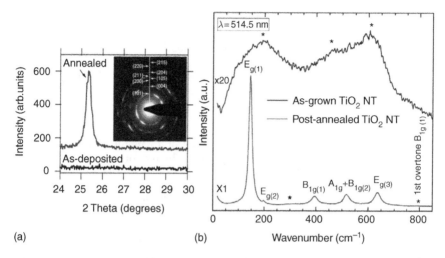

Figure 3.30 (a) X-ray diffraction patterns of as-deposited and annealed TiO_2 nanotubes. Inset shows the diffraction pattern of an annealed sample. Reprinted from Stergiopoulos et al., 2008[174] with permission from IOP Publishing Ltd. (b) Raman spectrum of the annealed TiO_2 tubes (bottom) in comparison with the as-grown tubes (top). Reprinted from Zhu et al., 2007[175]. Copyright (2007) American Chemical Society

contribute to the current. Figure 3.31 illustrates the key steps in the growth process. During the initial stages of deposition electronic conduction occurs until an insulating metal oxide film is formed over the substrate surface. After this, ionic conduction dominates. When dissolution and deposition reactions reach steady-state conditions, long tubes grow. A tubular structure forms because the rate is enhanced at the pore bottom because of localised acidification resulting from the oxidation and hydrolysis of elemental titanium.

$$2H_2O \rightarrow O_2 + 4e^- + 4H^+ \tag{3.27}$$

$$Ti + O_2 \rightarrow TiO_2 \tag{3.28}$$

$$TiO_2 + 6F^- + 4H^+ \rightarrow [TiF_6]^{2-} + 2H_2O \tag{3.29}$$

The thickness of the tubular structure stops increasing when the chemical dissolution rate of the oxide at the mouth of the tube (top surface) becomes equal to the rate of inward movement of the metal/oxide boundary at the base of the tube. Higher anodisation voltages promote the oxidation and field-assisted dissolution and increase the nanotube layer thickness before equilibrium is reached. A slightly different mechanism occurs in aqueous solvents compared with organic solvents: in aqueous solvents the process is subtractive (the nanotube array will be less than the

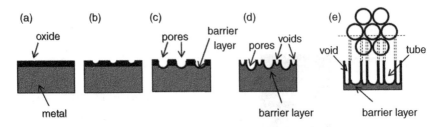

Figure 3.31 Schematic diagram of the evolution of a nanotube array at constant anodisation voltage: (a) oxide layer formation; (b) pit formation on the oxide layer; (c) growth of the pit into scallop-shaped pores; (d) metallic part between the pores undergoes oxidation and field assisted dissolution; and (e) fully developed nanotube array with a corresponding top view. Reprinted from Mor et al., 2006[164]. Copyright (2006) Elsevier

thickness of the titanium film); in organic solvents such as ethylene glycol, the process is additive (the nanotubes can be longer than the thickness of the titanium film) as material removed to form the pores goes in to build the walls. Extremely fast growth rates up to 15 μm h^{-1} have been achieved when ethylene glycol is used as the solvent.[172]

Anodisation is known to generate a thin TiO$_2$ barrier layer (Figure 3.31), shielding the Ti substrate from direct contact with the electrolyte. An insulating oxide layer forms between the nanotubes and the substrate during anodisation process, which causes a lower FF because of the increased series resistance and poor contact between the barrier layer and FTO substrate.[174] The barrier-layer thickness can be reduced using a stepwise reduction in the anodisation voltage, followed by an acid rinse to further thin the barrier layer.[175] Because the tubes are sealed at the substrate end, wetting of the surface is important for infiltration of the tube by dye solution or electrolyte.[170]

A number of groups have fabricated vertically aligned nanotube array-based DSSCs on flexible Ti substrates including foils,[173,174,176] grids, meshes[177] and wires.[178] It is hoped that the morphology will improve transport properties of the semiconductor and increase the efficiency of solid-state devices.[179] However, problems with surface wetting and infiltration of the narrow pores will need to be overcome. Two configurations for nanotube-based DSSCs are considered below.

Whilst a 6.89% photoconversion efficiency (N719 dye, $J_{SC} = 12.72$ mA cm^{-2}, $V_{OC} = 0.82$ V, FF = 0.663) has been achieved, for a nanotube-based DSSC on a Ti substrate illuminated through the counter electrode [Figure 3.32(a)], losses from reflections by the counter electrode and absorption of light from the electrolyte occur when using this geometry.[180] Instead, the front-side illuminated configuration [Figure 3.32(b)] is preferred

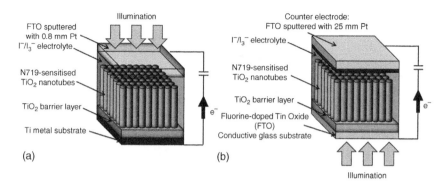

Figure 3.32 (a) Schematic diagram of back-side illuminated nanotube array (on foil) DSSC; (b) front-side illuminated nanotube array DSSCs. Reprinted from Mor *et al.*, 2006[165]. Copyright (2006) Elsevier

to obtain maximum efficiency.[181] Therefore, it is important that nanotubes can be grown on Ti-sputtered FTO-coated glass. The films are anodised until they become so thin they are discontinuous, at which point they are rinsed to remove fluoride and annealed at 525 °C (Figure 3.33). Despite the limitations to film thickness (less than 400 nm), an AM 1.5 photoconversion efficiency of 2.9% was obtained with N719 as a sensitiser. The efficiency was improved to 4.1% ($J_{SC} = 10.1$ mA cm^{-2}, $V_{OC} = 743$ mV, $FF = 0.55$) by use of a dye with a much higher extinction coefficient (54 500 M^{-1}).[182]

An important consequence for DSSCs of the use of the highly ordered nanotube arrays in comparison with nanoparticulate systems, is the improvement in photogenerated charge carrier lifetimes by well over an order of magnitude.[174,183,184] Zhu *et al.*[174] used frequency-resolved, modulated photocurrent/photovoltage spectroscopies[185,186] to study the electron transport and recombination rates in oriented TiO$_2$ arrays such as these, and compared them with nanoparticle-based films with similiar crystallite size and dye loading. DSSCs assembled from electrodes with the different morphologies were found to display comparable transport times (owing to a similar crystallite size and similar mechanistic properties governing the charge transport) but recombination was much slower in the nanotube films, suggesting fewer recombination sites are present in the nanotubes compared with nanoparticles (Figure 3.34).

These results suggested that the nanotube-based DSSCs have significantly higher (25%) charge-collection efficiencies (η_{cc}) than their nanoparticle-based counterparts.

$$\eta_{cc} = 1 - (\tau_c/\tau_r) \qquad (3.30)$$

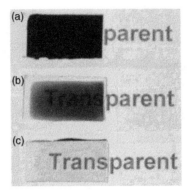

Figure 3.33 Appearance of the TiO$_2$ nanotube-array film at key stages in the fabrication process: (a) as-deposited Ti film; (b) film after anodisation; and (c) transparent film after heat treatment to crystallise the tubes and oxidise the remaining metallic islands. Reprinted from Mor et al., 2006[176]. Copyright (2006) American Chemical Society

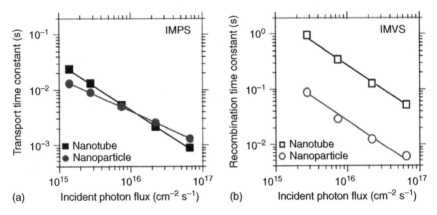

Figure 3.34 Comparison of (a) transport and (b) recombination time constants for nanotube- and nanoparticle-based DSSCs as a function of the incident photon flux for 680 nm laser illumination. Reprinted from Zhu et al., 2007[175]. Copyright (2007) American Chemical Society

The light-harvesting efficiencies of nanotube-based DSSCs were higher than those found for DSSCs incorporating nanoparticles owing to stronger internal light-scattering effects. The combination of these effects (Equation 3.31) gave rise to higher currents for nanotubes with the same film thickness as the nanoparticle-based electrodes (Figure 3.35).

$$J_{sc} = q\eta_{lh}\eta_{inj}\eta_{cc}I_0 \qquad (3.31)$$

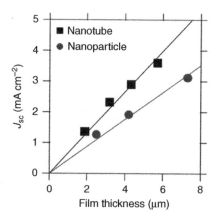

Figure 3.35 Comparison of the short-circuit photocurrent densities of dye-sensitised nanotube- and nanoparticle-based cells as a function of film thickness. The cells were illuminated by a 680 nm laser with an incident photon flux of 6.6×10^{16} cm^{-2} s^{-1}. Lines represent linear fits of data. Reprinted from Zhu *et al.*, 2007[175]. Copyright (2007) American Chemical Society

where q is the elementary charge, η_{lh} is the light-harvesting efficiency of a cell, η_{cc} is the charge-injection efficiency and η_{inj} is the charge injection efficiency.

Since the recombination time is relatively independent of film thickness for nanotube morphologies, thicker films can be prepared, increasing the optical density and therefore light harvesting without losing charge carriers to recombination before they reach the charge collector. Usually one expects that less recombination would mean a higher V_{OC} in the device, however the authors attribute the absence of this effect as due to the higher density of states in the nanotube array compared with the nanoparticle-based film. This observation was in contrast to that by Paulose *et al.* who observed a substantial increase in V_{OC} (860 mV) for nanotube-based DSSCs compared with nanoparticles (700–750 mV).[184] Jennings *et al.* also investigated the transport, trapping, and transfer of electrons in nanotube-based DSSCs.[187] They too found that TiO$_2$ nanotube cells exhibit high collection efficiencies for photoinjected electrons and estimated an electron diffusion length of the order of 100 μm.

Nanowires or Nanorods TiO$_2$ nanorods have been prepared using alumina nanotube templates.[154] The templates are prepared by anodisation of aluminium films on FTO. The SEM images of templates prepared by Bwana are shown in Figure 3.36.[188]

The templates are immersed in a TiO$_2$ precursor solution which contains titanium isopropoxide as a Ti source, and then removed and

(a)   (b)

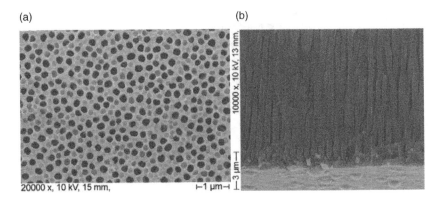

Figure 3.36   Top (a) and side (b) SEM images of anodic alumina templates showing monodisperse nanopores running the full length of the alumina film. Reprinted from Bwana, 2008[189] with kind permission of Springer Science and Business Media

sintered in a furnace at 400 °C. The alumina template is then removed by immersing the samples in 6 M NaOH solution. Nanotubes and rods can be prepared in this way, as shown in Figure 3.37.

The effective electrode area was 1.5 cm × 1.5 cm with roughness factor of ca 27.2 for the nanorods and ca 62.6 for the nanotubules. The current–voltage characteristics of the N719-sensitised DSSCs are shown in Figure 3.38. The nanorod-based electrodes gave an overall conversion efficiency of 5.4%, the nanotubes 4.5% and the nanoparticle films 4.7%. The V_{OC} values were high for the nanorod systems, considered to be a result of the higher density of packed TiO_2 particles making up the structure and the reduced surface area exposed to the electrolyte where recombination with triiodide can take place, compared with the

(a)   (b)

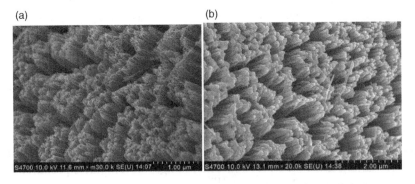

Figure 3.37   SEM images of (a) TiO_2 nanotubules and (b) nanorods synthesised in anodic alumina templates after removal of the templates. Reprinted from Bwana, 2008[189] with kind permission of Springer Science and Business Media

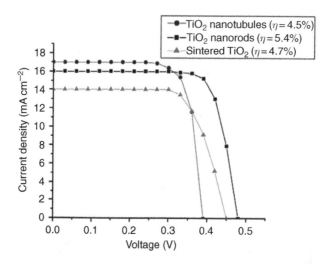

Figure 3.38 Photocurrent–voltage curves of DSSCs based on different TiO$_2$ working electrodes. The electrolyte [Iodolyte TG-50 (Solaronix)] contained 50 mmol l^{-1} of triiodide in tetraglyme, 0.5 M LiI, 0.05 M I$_2$, 0.6 M N-methylbenzimidazole, 0.10 M guanidinium thiocyanate and 0.5 M *tert*-butylpyridine in MPN. Reprinted from Bwana, 2008[189] with kind permission of Springer Science and Business Media

nanoparticle and nanotube electrodes. The higher photocurrents for the organised structures were attributed to higher charge collection efficiencies because of the vectorial electron transport and reduced charge hopping (in random directions) across grain boundaries which is necessary for nanoparticle systems. This allows thicker films to be prepared, increasing light absorption without sacrificing electrical properties.

Alternatively, nanorods, wires or fibres can be grown by plasma-enhanced chemical vapour deposition using a nickel catalyst.[189] An electric field is applied during the deposition process which aligns the growth vertically from the substrate to give a free-standing, brush-like array of anatase nanowires with diameters 50–100 nm and length <40 μm (Figure 3.39). Chemical vapour deposition from titanium isopropoxide can then be used to grow 'nanoneedles', 10–15 nm in diameter, 50–100 nm in length, from the vertically aligned nanowire cores. The surface area for dye adsorption is increased compared with the smooth nanorods alone, however substantial scattering effects (reduction in transmittance) were observed, proportional to the length of the wires. The optical-penetration depth was estimated from transmittance spectra to be 5 μm, *i.e.* the average length of the longest wires used for DSSCs.

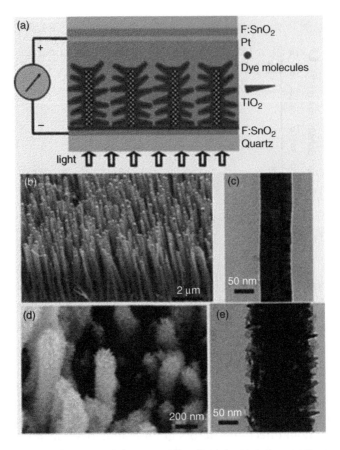

Figure 3.39 (a) Schematic of the vertically aligned nanowire arrays coated with anatase TiO_2 nanoneedles for DSSCs. (b) SEM image at a 45° perspective view. (c) Transmission electron microscopy (TEM) image of an as-grown nanowire. (d) SEM image at a 45° perspective view of the nanowire array after being coated with anatase TiO_2 nanoneedles. (e) TEM image of a nanowire after being coated with anatase TiO_2 nanoneedles. Reprinted with permission from Liu et al., 2009[190]. Copyright (2009) American Chemical Society

The overall conversion efficiency of DSSCs assembled from N719-sensitised 'nanobrushes' was 1.09% ($J_{SC} = 2.91$ mA cm^{-2}, $V_{OC} = 0.640$ V, $FF = 0.582$) and the IPCE at 515 nm was 34%. This was comparable with the efficiency ($\eta = 1.86\%$) of the equivalent nanoparticle-based device.

Inverse Opal   Inverse opals have the opposite morphology to nanoparticle networks, where large voids are created in a three-dimensional, continuous shell of small crystallites.[190] The open structure results

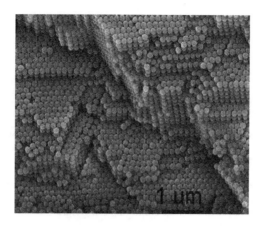

Figure 3.40 The face-centred cubic close packing of polystyrene spheres used as templates. Reprinted from Kuo and Lu, 2008 reference[192] with permission from IOP Publishing Ltd

in interesting optical properties which will be discussed in more detail in the section on Photonic Crystals. Typically, a TiO_2 precursor solution containing a metal alkoxide or organic-layer-coated TiO_2 nanoparticles is added to a colloidal 'opal' template, such as polystyrene or silica (Figure 3.40).[190]

Dense packing within the interstitial volume of the face-centred cubic, colloidal assembly is important to minimise shrinkage-induced cracking of the inverse opal structure and so non-aggregated, monodisperse TiO_2 particles are used or processes such as chemical vapour deposition[190] or electrochemical deposition. Hydrolysis of the precursor and thermal decomposition of the template (or dissolution in HF for silica templates) gives a compact, three-dimensional TiO_2 array of the target material, as illustrated in Figures 3.41 and 3.42.

These structures are advantageous for DSSC applications because the highly ordered, open structure reduces the mass transfer resistances of electrolyte transport and dye impregnation within the voids and results in efficient electrolyte–dye interactions. The thin shells can speed up electron transport because the geometric confinement and structural regularity effects provide direct transport paths for electrons. N719-sensitised DSSCs based on electrodes with 1 μm diameter voids, gave a photoconversion efficiency of 3.47% ($V_{OC} = 0.726$, $J_{SC} = 6.42$, $FF = 0.744$) under AM 1.5 illumination. From impedance measurements, it was indicated that slower recombination occurs with films with large rather than small pores.

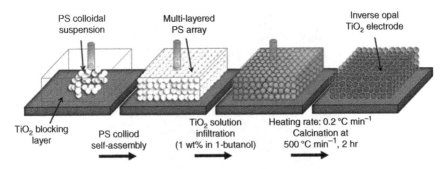

Figure 3.41 Schematic diagram showing the preparation steps of the three-dimensional colloidal-array-templated TiO$_2$ electrode. Adapted from Kwak *et al.*, 2009[142]. Copyright (2009) Wiley-VCH Verlag GmbH & Co. KGaA

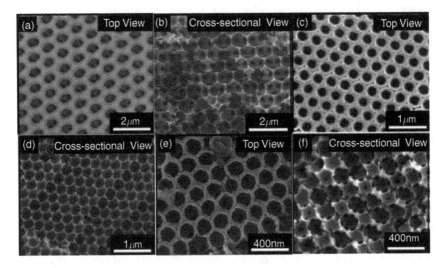

Figure 3.42 SEM images of the templated inverse-opal TiO$_2$ films: (a, b) from 1000 nm polystyrene; (c, d) from 400 nm polystyrene; and (e,f) from 240 nm polystyrene microspheres. The three black spots within the hexagons indicated in (b) show the holes connecting the spherical cavities to the inner layer. Adapted from Kwak *et al.*, 2009[142]. Copyright (2009) Wiley-VCH Verlag GmbH & Co. KGaA

Kuo and Lu coated the surfaces of the spherical voids with 10–15 nm TiO$_2$ nanoparticles by TiCl$_4$ treatment to increase the surface area for dye adsorption.[191] Figure 3.43 shows the SEM and high resolution transmission electron microscopy (HRTEM) images of the structures. The voids were 100 nm in diameter with transport channels of 30–50 nm in between. The film was 25 μm in thickness. The V_{OC} decay when switching from AM 1.5 to dark was recorded for a 13 μm film and the τ_r calculated. The inverse opal structure was found to result in longer

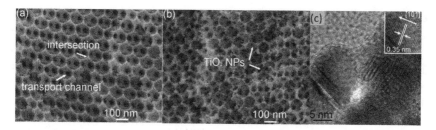

Figure 3.43 SEM images of the TiO_2 inverse opal (a) before and (b) after the $TiCl_4$ post-treatment. (c) HRTEM image taken at a junction of three spherical voids. Reprinted from Kuo and Lu, 2008[192] with permission from IOP Publishing Ltd

electron lifetimes than the equivalent nanoparticle-based electrodes. The conversion efficiency for the DSSC was 4% ($J_{SC} = 8.7$ mA cm^{-2}, $V_{OC} = 0.76$ V) under AM 1.5 illumination.

Aerogels Another way of fabricating high-surface area electrodes is to coat inert, low-density, high-surface-area silica aerogel films with TiO_2 using atomic layer deposition (ALD). ALD enables controlled layer-by-layer deposition of materials such as metal oxides onto templates such as silica aerogels, but also onto nanoparticle assemblies to form core-shell structures. High-surface-area mesoporous aerogel films were prepared on conductive glass substrates. ALD was employed to coat the aerogel template with various thicknesses of TiO_2 with subnanometre precision. The porosity of aerogels is known to be controlled by the concentration of the sol and can exceed 99.5%. In the example by Hamann et al., a porosity of more than 90% was chosen for the template, allowing volume for growth of the TiO_2 layer without clogging the pores.[192] The thickness of the aerogel film (25 μm) was determined by the volume of sol drop cast onto the FTO substrate.

The TiO_2 electrodes were sensitised and assembled in DSSCs. At 100 mW cm^{-2}, the efficiency of the devices reached 4.3%. The charge diffusion length was found to increase with increasing thickness of TiO_2 leading to increased current for thicker films. The V_{OC} was lower than for nanoparticle-based electrodes because of the increase in dark current resulting from the larger surface area and so the overall conversion efficiencies were similar.

Nanostructured ZnO Anodes

ZnO is an attractive material for nanoscale optoelectronic devices, as it is a wide band gap semiconductor with good carrier mobility and can be doped both n- and p-type.[193] The electron mobility is much higher in

ZnO than in TiO_2, while the conduction band edge of both materials is located at approximately the same level. ZnO shows more flexibility in synthesis and morphology than TiO_2. A plethora of ZnO materials with different particle sizes, morphologies and compositions are at present being reported and there are a growing number of studies on the use of these materials for DSSC applications.

Early work on dye-sensitised ZnO photoelectrochemical cells was reported by Gerischer and Tributsch[194–196] in the late 1960s when monocrystalline ZnO was sensitised with ruthenium pyridyl complexes. The use of a porous ZnO electrode with an enhanced surface area was first reported in 1976 by Tsubomura et al.[199] At present, solar cell efficiencies of dye-sensitised ZnO have reached 6.6% at 1000 W cm^{-2}, showing promise for the future. Table 3.2 is a summary of overall efficiencies reached with the dye-sensitised, nanostructured ZnO solar cell (Figure 3.44). DSSCs based on ZnO nanowires have not yet been able to give better performance than those based on spherical particles. Although the electron transport is expected to be improved in the

Table 3.2 Compilation of the overall efficiencies reached in sensitised nanostructured ZnO solar cells

Size/structure	Sensitiser	Intensity (Wm^{-2})	Efficiency (%)	Reference
150 nm	N719	100	5	[198]
20 nm	N719	1000	6.6	[199]
11 nm	N719	1000	4.1	[200]
10–20 nm	Mercurochrome	1000	2.5	[201]
5 nm	MDMO-PPV[a]	710	1.6	[202]
14 nm	Eosin Y	1000	1.11	[203]
Porous crystal	Eosin Y	1000	2.3	[204]
Porous crystal	Squaraine	1000	1.5	[205]
Wires 100–150 nm	CdSe	360	2.3	[206]
Wires 130–200 nm	N719	1000	1.5	[207]
ZnO sheets	N719	1000	3.9	[208]
Aggregates <500 nm (15 nm)	N3	1000	5.4	[209]
Nanorods 30 nm × several µm	N3	1000	4.7	[210]
Nanotubes 210 nm × 64 µm	N719	1000	1.6	[211]
Nanosheets 50 nm thick	N719	1000	3.3	[212]
Nanotetrapods 40 nm, 500–800 nm	N719	1000	3.27	[213]
Nanoflowers 4 µm	N719	1000	1.9	[214]

[a] MDMO-PPV, poly[2-methoxy-5-(3″,7″-dimethyloctyloxy)-1,4-phenylenevinylene].

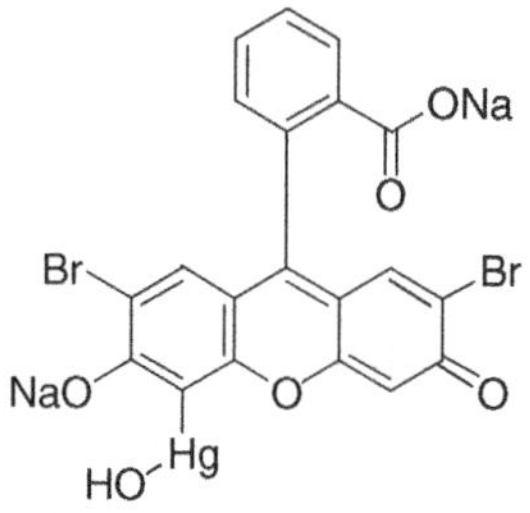

Mercurochrome 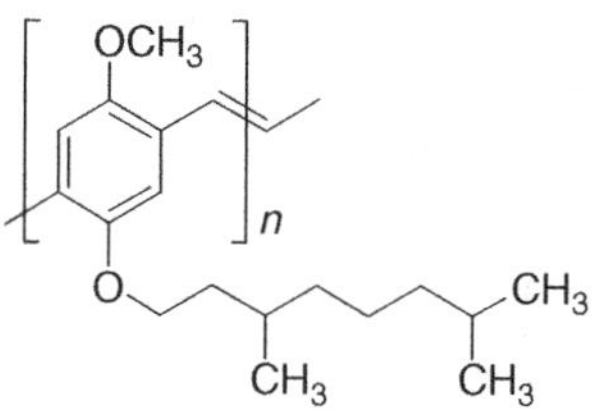 **MDMO-PVP**

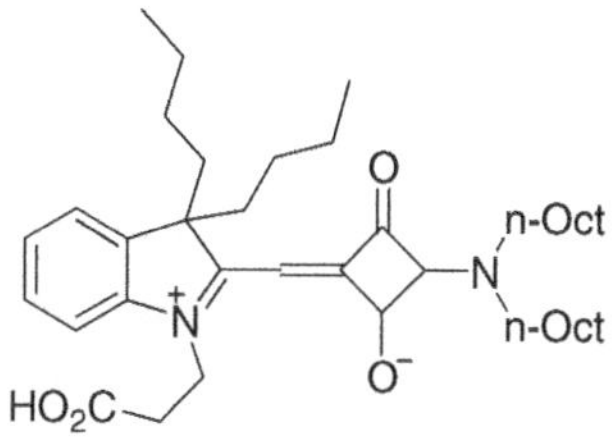

Squaraine 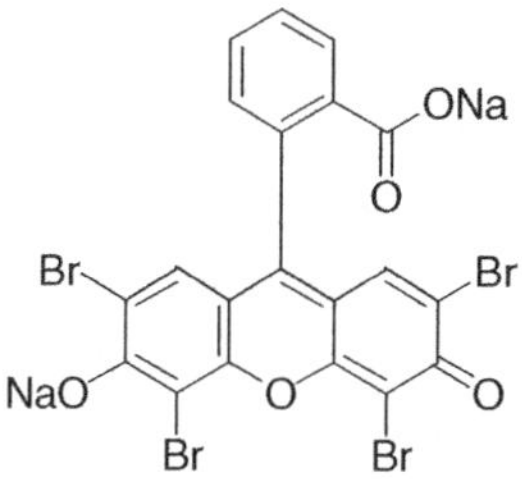 **Eosin Y**

Figure 3.44 Structures of dyes successfully used as photosensitisers in ZnO DSSCs listed in Table 3.2

nanowires, the light absorption of the film is insufficient so far. The performance of dye-sensitised ZnO solar cells is, thus, so far lower than that of TiO_2. This can, at least partly, be attributed to the much lower research activity on the use of ZnO. The main problem with ZnO at present is identified and is related to a complex dye-sensitisation process discussed below.

Material Properties Historically, ZnO has been used for a long time as a pigment and protective coatings on metal surfaces. Its wide band gap of 3.2 eV at room temperature has rendered it useful as a protective UV-absorbing additive in skin cream to advanced plastic and rubber composites. In the era of nanotechnology, the potential applications of ZnO are spanning the vast fields of nanoelectronics and acousto-optics. For example, ZnO has been used in photocopy machines, transparent conducting layers, varistors, as optical wave guides, surface acoustic wave transducers and thin-film transistors.[215] The wide band gap has also made it suitable for short-wavelength optoelectronic devices, including UV detectors, photocatalysts, laser diodes and light emitting diodes (LEDs).[193] For a comprehensive survey of ZnO materials and devices, we refer to the reviews in references.[199-217]

A summary of material properties for ZnO can be found in the ZnO DSSC review by Boschloo *et al.*[218] Zinc oxide for DSSC applications has normally a hexagonal (wurtzite) structure with the lattice parameters a, $b = 3.25$ and $c = 5.12$ Å. ZnO is a direct band gap semiconductor whose electrical conductivity is determined by defects in the material that are present intrinsically or incorporated on purpose. ZnO is intrinsically doped *via* oxygen vacancies and/or zinc interstitials, which act as n-type donors.[219] Recently it has been shown by experiments[220-222] and first-principle calculations[223] that the incorporation of hydrogen produces shallow states 30–40 meV below the conduction band edge. Furthermore, ZnO is frequently doped with aluminum or fluorine to obtain highly n-type conducting films. The electron mobility in single-crystalline ZnO (Hall mobility of 200 cm^2 V^{-1} s^{-1}) is much higher than that of TiO_2 anatase (30 cm^2 V^{-1} s^{-1}). Mobilities tend to decrease upon doping due to scattering of the electrons at the impurities, with a more substantial decrease when going from single crystals to polycrystalline and nanocrystalline materials, due to scattering at grain boundaries and energy barriers at these boundaries. The conduction band edge positions in ZnO and TiO_2 are very similar, about -0.5 V *vs* SCE at pH 5,[115,224-226] and show a Nernstian shift of 59 mV pH^{-1}. The point of zero charge (pzc) for ZnO is determined at pH 8–9 depending on the preparation method and the experimental conditions.[227] The static dielectric constant of ZnO is 7.9[228] and the optical dielectric constant is 3.7.[229] Quantum confinement in ZnO is expected to occur for particle sizes less than 4–5 nm.[218] For these particles, an increase in band gap energies is seen arising from the confinement of the electronic states (see Section 3.2.3.4).[230-232]

Preparation of ZnO for DSSC Applications   Nanostructured ZnO electrodes can either be synthesised from ZnO nanoparticles that are prepared in a separate procedure, or produced in a single synthetic step. Crystalline ZnO nanoparticles can be prepared at room temperature in non-aqueous solutions.[233-237] In a typical preparation, a solution of zinc acetate in alcohol is mixed with an equimolar amount of hydroxide. The dehydrating properties of the solvent prevent the formation of zinc hydroxide and promote formation of crystalline ZnO. Transparent ZnO colloidal solutions are easily prepared. The particles tend to be approximately spherical and their size depends strongly on the preparation temperature. Quantum-sized particles are obtained at low temperatures. Prolonged reflux of the colloidal solution results in formation of ZnO nanorods.[238]

In aqueous preparations, zinc hydroxide rather than zinc oxide tends to be formed, which decomposes at 125 °C to form zinc oxide. Several preparation methods have been developed in which crystalline ZnO is directly formed in aqueous solution by controlled precipitation at temperatures below 125 °C.[239,240] In these methods, elevated temperatures are used and chelating agents are used to prevent formation of zinc hydroxide. Alternatively, zinc hydroxide or carbonate nanoparticles can be precipitated from aqueous solution and converted to ZnO by heat treatment.[241,242] Hydrothermal treatment at 180 °C has been used to form ZnO nanorods with a diameter of about 50 nm.[243] Other methods to prepare ZnO nanoscale powders are spray pyrolysis and other gas-phase methods. To prepare the nanostructured ZnO electrodes, a nanoparticle dispersion is deposited onto a conducting substrate, followed by heat treatment. The conducting substrate is usually a conducting glass, consisting of glass coated with a thin layer of a TCO, such as F-doped SnO_2 or Sn-doped In_2O_3 (ITO). Typical application methods for the dispersion are doctor blading, screen printing, spin coating, dip coating and spraying.

For direct syntheses of nanostructured ZnO films onto the conducting substrates various options are available and reviewed in Boschloo et al.[218] The different methods are summarised below.

Chemical bath deposition (CBD): This is a low-temperature method (<150 °C) in which a substrate is immersed in a semi-stable precursor solution.[244–247] The deposition rate can be controlled by adjusting the temperature, the pH and the relative concentration of the reactants in the solution. Vayssieres et al.[245,246] described the formation of ordered crystalline ZnO microstructures (rods and tubes) and nanorods on several types of substrate. Peterson et al.[248] prepared uniform arrays of nanorods grown epitaxially onto a sputtered ZnO film. Figure 3.45 shows a SEM image of ZnO nanorods grown with similar methods in our laboratory.

Electrodeposition: Electrodeposition is a low-temperature method, where the ZnO is formed directly on the substrate. Film morphology and thickness can be controlled by applied potential, current density, reactant concentrations, temperature and deposition time. Nano-columns of ZnO were prepared by electrodeposition on several substrates.[249,250] Using non-aqueous conditions, deposition of

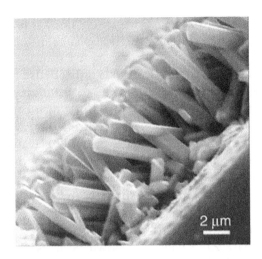

Figure 3.45 SEM image of ZnO rods grown on a FTO layer (compact film in the lower right corner). Reprinted with permission from Vayssieres *et al.*, 2001[247]. Copyright (2001) American Chemical Society

nanostructured ZnO films was achieved.[251,252] In addition, successful electrodeposition of ZnO in the presence of dye has been reported.[253] Cathodic electrodeposition of ZnO from aqueous zinc salt solutions in the presence of oxidants and water-soluble dye molecules provides a nanostructured photoanode to be used in DSSCs.[253–257] Dye molecules with attachment groups such as $-COO^-$ and $-SO_3^-$ adsorb to the growing oxide film and, besides being used as sensitisers the dyes, are also influential in formulating the structure and morphology of the nanostructured film. Highly porous ZnO has been obtained especially in the presence of O_2 and the red xanthene dye eosin Y. The solar cell efficiencies are hampered by dye aggregation in the pores of the as-deposited films originating from the use of acidic anchoring groups. Thus, quenching of the photogenerated electrons resulting in low-electron injection efficiency has been observed. This problem has to a large extent been overcome by desorption and re-adsorption of the dye, leading to the formation of a dye monolayer on the ZnO surface and a significant improvement of the electron injection efficiency being as high as 90%.[206] Although the eosin Y dye has a relatively narrow absorption band, an overall solar-to-electrical energy conversion efficiency of 2.3% was achieved.

Controlled oxidation of Zn in solution: In this method, similar to CBD but also electrochemical in nature, a Zn substrate is immersed in an aqueous formamide solution at slightly elevated temperatures.[258] Zn is oxidised by oxygen to form soluble Zn^{2+}/formamide complexes, which in turn can deposit and react to ZnO on the Zn substrate. Aligned ZnO nanorods are formed. Deposition can also take place on other substrates in close vicinity to the Zn plate.

Gas-phase methods: Nanorods and wires have been produced by vapour–liquid–solid deposition.[259] The substrate is first coated with a thin layer of a catalyst such as gold. A ZnO/graphite mix is placed close to the substrates in a furnace tube and heated at about 900 °C under an argon flow. Zn and CO vapour are formed, and the Zn forms an alloy with the Au particles. ZnO starts to grow when the alloy becomes saturated with Zn. Epitaxial growth of oriented nanowires on silicon substrates has been reported.[260]

Dye-sensitisation of ZnO Electrodes   Compared with TiO_2, the dye-sensitisation process of ZnO is more complicated due to reduced stability towards acidic media. Carboxyl groups are commonly used as anchoring groups for chemisorption of dye molecules onto metal oxide surfaces. Protons from this anchoring group cause dissolution of Zn surface atoms and the formation of Zn^{2+}/dye complexes in the pores of the mesoporous film, which gives rise to a filter effect (inactive dye molecules). Consequently, the net yield for charge carrier injection is decreased, whereas the light harvesting efficiency is increased during the sensitisation process due to the large number of dye molecules in the film. A schematic drawing of the Zn^{2+} dissolution and Zn^{2+}/dye complexation processes can be seen in Figure 3.46.

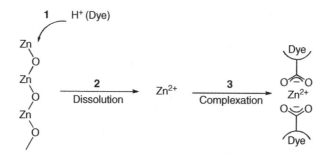

Figure 3.46   A schematic drawing of the problems associated with the dye-sensitisation process of protonated anchoring groups on ZnO surfaces. Reprinted with permission from Boschloo *et al.*, 2006[218]. Copyright (2006) with permission from Elsevier

Thus, careful control of the dye composition, concentration, pH and sensitisation time is necessary in order to avoid the Zn^{2+}/dye complex formation and to achieve high-efficiency solar cells based on ZnO. Preferably, an anchoring group without protons, or other ions which cause dissolution of the ZnO surface, should be developed. So far, very few approaches have been reported to circumvent this problem. Thus, a large potential exists to improve on the performance of dye-sensitised ZnO solar cells by learning how to use new types of anchoring groups and controlling the chemistry at the oxide/dye/electrolyte interface.

SnO_2 Photoanodes

SnO_2 has a larger band gap ($E_g = 3.8$ eV) than TiO_2 and, therefore, offers the advantage of better long-term stability of the solar cell in the presence of UV light.[61] SnO_2 has a conduction band approximately 0.4 V more positive than TiO_2 and therefore the maximum photovoltage is lower with the use of triiodide/iodide as the redox system.[261–264] However, since there is more driving force for injection of an electron from the excited state dye, more photocurrent is possible and dyes with a LUMO level lower than the conduction band of TiO_2 may inject into SnO_2. Nonetheless, efficiencies of SnO_2-based DSSCs lag substantially behind those of TiO_2-based devices. Recently, efficiencies have been improved significantly by coating the SnO_2 with another insulating or semiconducting oxide in a core-shell structure.[265–267] It is thought that the insulating shell blocks recombination between the electrolyte and the electrons in the SnO_2, increasing the photovoltage. Other positive effects could be doping or shifting of the flat band potential of the semiconductor and an increase in the basicity of the surface increasing the affinity of the surface for carboxylic acid functionalised dye molecules.

Nb_2O_5 Photoanodes

The conduction band potential of Nb_2O_5 is ca 100 mV more negative than for TiO_2.[140] Therefore, solar cells built incorporating a sensitised Nb_2O_5 photoanode are expected to have an increased V_{OC} over those using TiO_2. In addition, it has a lower photocatalytic activity than TiO_2 under UV irradiation because of the slightly larger band gap. Sayama et al. prepared an 8 μm nanoporous Nb_2O_5 electrode from $Nb(OH)_5$, calcined in air, post-treated with a niobium alkoxide and sensitised with N3 dye.[140] Whilst the photovoltage was higher than the cells using TiO_2, the photocurrent was lower because of the lower driving force for injection. The efficiency was 2% ($J_{SC} = 4.9$ mA cm^{-2}, $V_{OC} = 0.63$ V, $FF = 0.66$ at 100 mW cm^{-2}) and the IPCE was above 30% at 515 nm (Figure 3.47).

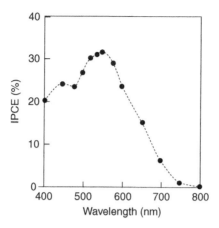

Figure 3.47 Photocurrent action spectrum for the dye-sensitised Nb_2O_5 cell with $Nb(OC_2H_5)_5$ treatment. The spectrum was obtained using a thin-layer cell (1 cm^2) containing 0.5 M tetra-*n*-propylammonium iodide and 0.05 M iodine in ethylene carbonate and dry acetonitrile mixed solvent (60:40 v/v) under monochromatic light passed through a bandpass filter (3–5 mW cm^{-2}). Reprinted with permission from Sayama *et al.*, 1998[141]. Copyright (1998) American Chemical Society

3.2.3.2.6   Cathode Materials

Counter Electrodes

In the optimal Grätzel cell, the material chosen for the counter electrode is Pt on account of its stability, low sheet resistance and high catalytic activity for the 'regeneration' step: reduction of I_3^- to I^-.[268] However, for large-scale manufacture, cheaper, more abundant materials are preferred and for applications such as building integration and electronics, alternative, flexible substrates such as steel and plastic are also advantageous. A number of materials have been used to prepare counter electrodes for DSSCs including carbon counter electrodes using graphite, activated carbon, carbon black, or single-wall carbon nanotubes, and polymer-based counter electrodes using poly(3,4-ethylenedioxythiophene) (PEDOT), polypyrrole, or polyaniline (Figure 3.48).[269]

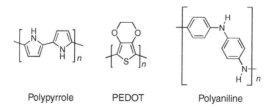

Polypyrrole   PEDOT   Polyaniline

Figure 3.48   Conducting polymers typically used in solid-state DSSCs

In some applications it is necessary to illuminate through the counter electrode and therefore transparent materials, such as a thin Pt layer or transparent polymer on conducting glass or plastic are necessary.

The rate of reaction at the counter electrode has a significant affect on the *FF* and therefore the efficiency of the cell. If the reaction is slow, I^- will not be regenerated quickly enough to suppress the back reaction of I_3^- with the electrons in the TiO_2. This increases the losses in the cell and adds to the internal resistance. The ability of the counter electrode to catalyse a reaction can be observed from cyclic voltammetry.

Figure 3.49 shows a typical voltammogram for the I_3^-/I^- electrolyte. Two pairs of waves are observed: one pair is assigned to the redox reaction 1 and the positive pair is assigned to redox reaction 2:

$$(1) \ I_3^- + 2e^- = 3I^- \tag{3.32}$$

$$(2) \ 3I_2 + 2e^- = 2I_3^- \tag{3.33}$$

The peak separation between the oxidation wave and the reduction wave is a measure of the overpotential for the reaction.

Counter electrodes are also compared in terms of their charge-transfer resistance (R_{ct}), a measure of the catalytic activity, or exchange current density (J_0).[268]

$$R_{ct} = RT/nFJ_0 \tag{3.34}$$

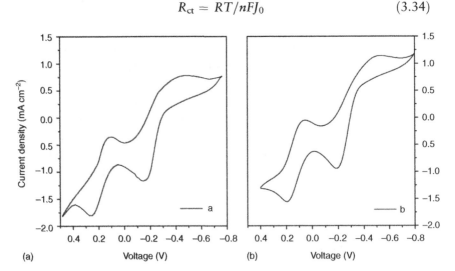

Figure 3.49 Cyclic voltammograms for (a) a platinum nanoparticle electrode and (b) a platinum plate electrode in 10 mM LiI, 1.0 mM I_2 acetonitrile solution containing 0.1 M $LiClO_4$ as the supporting electrolyte: $[I^-]/[I_2] = 10/1$. Reprinted with permission from Li *et al.*, 2008[270]. Copyright (2008) Elsevier

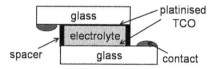

Figure 3.50 The electrochemical cell used for the impedance measurements. Reprinted with permission from Hauch and Georg, 2001[271] Copyright (2001) Elsevier

where R is the gas constant, T is temperature, n is number of electrons transferred in the elementary electrode reaction ($n = 2$ for I_3^- reduction) and F is the Faraday constant. R_{ct} can be determined from impedance spectra.[271]

Hauch and Georg used sandwich cells as illustrated in Figure 3.50 to compare R_{CT} for Pt counter electrodes prepared using different deposition methods (sputtering, evaporation and thermal) on conducting glass with a sheet resistance of 10 Ω sq^{-1}. For champion cells, the counter electrode is prepared by thermal deposition of a $H_2[PtCl_6]$ film on a conducting glass substrate to give a catalytic, microcrystalline Pt layer less than 10 nm thick. A sandwich cell using this deposition method and filled with 0.5 M tetrapropyl ammonium iodide and 0.05 M I_2 in acetonitrile electrolyte produced a low R_{CT} of 1.3 Ω cm^{-2}.[271]

Figure 3.51 shows the impedance spectra for a Pt/FTO sandwich cell at different bias potentials reported by Hauch and Georg.[271] The impedance element on the left of the spectrum (high frequencies) is the Pt/ electrolyte interface (charge-transfer resistance and double layer capacitance); on the right (low frequencies) there is the Nernst diffusion impedance. The diameter of the high frequency semicircle in the impedance

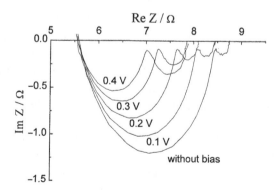

Figure 3.51 Impedance spectra for different bias potentials. Reprinted with permission from Hauch and Georg, 2001[271]. Copyright (2001) Elsevier

spectra is the sum of the charge-transfer resistances of the anode and cathode for symmetrical sandwich cells such as these. The equation for the charge-transfer resistance above is defined at equilibrium with 0 V bias. When a bias potential is applied, the cell is no longer at equilibrium because there are different overpotentials and different ion concentrations at the anode and cathode and R_{CT} is interpreted as a differential resistance (there are different values of R_{CT} at the different electrodes):

$$R_{CT} = \delta\eta/\delta I \qquad (3.35)$$

The characteristics of the impedance spectra at different potentials correspond to the kinetics at the electrodes and can be explained by the Butler–Volmer equation.[272]

Hauch and George have suggested the mechanism below for oxidation of I^- at the Pt electrode from analysis of the impedance measurements and Langmuir isotherms for the Pt/electrolyte interface.[271] The rate-determining step of the suggested mechanism is the oxidation of iodide at the Pt surface.

$$\text{(a)} \quad I^- + Pt \leftrightarrow I^-(Pt) \quad \text{(adsorption of } I^- \text{on Pt)} \quad \text{fast} \qquad (3.36)$$

$$\text{(b)} \quad I^-(Pt) \leftrightarrow I(Pt) + e^- \quad \text{(oxidation of } I^-) \quad \text{slow} \qquad (3.37)$$

$$\text{(c)} \quad I(Pt) + I(Pt) \leftrightarrow I_2 + 2Pt \quad \text{(production of } I_2) \quad \text{fast} \qquad (3.38)$$

$$\text{(d)} \quad I_2 + I^- \leftrightarrow I_3^- \quad \text{(production of } I_3^-) \quad \text{fast} \qquad (3.39)$$

Pt-Based Counter Electrodes   Rather than incurring the cost of the catalytic metals, Pt or for example Ru, FTO-glass is one of the major contributors to the materials and energy cost of DSSCs. For building integration, flexible, metal substrates are preferred to glass. Because the iodine-based electrolyte is corrosive to metal substrates such as steel and nickel, anti-corrosion materials such as carbon or fluorine-doped SnO_2 are used to cover the metal substrates before the Pt is deposited. Efficiencies above 5% for DSSCs fabricated on Ni and steel substrates have been achieved. Another advantage of metal substrates is the low sheet resistance of the substrate which means high FF can be achieved.

DSSCs based on plastic substrates have also received attention in recent years because they are lightweight and suitable for roll-to-roll manufacturing processes. Ma et al. have prepared DSSCs using Pt deposited on plastic substrates, achieving 5.4% efficiency using a polyethylene naphthalate film coated with tin-doped indium oxide (ITO-PEN), compared with 5.39% efficiency obtained for Pt/FTO in the same

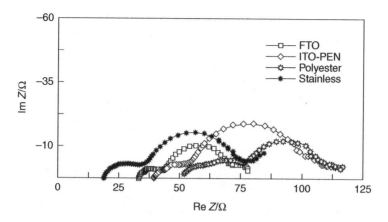

Figure 3.52 Electrical impedance spectra of the DSSC with different counter electrode substrates. Reprinted with permission from Ma *et al.*, 2004[273]. Copyright (2004) Elsevier

study.[273] The impedance spectra of the cells are shown in Figure 3.52. For asymmetric sandwich cells such as these, a third semicircle is present at intermediate frequencies due to the resistance at the anode.[274] Ito *et al.* also used ITO-PEN as the CE for flexible DSSCs.[275] They deposited Pt by the electrochemical deposition method with an electrolyte of 5 mM $H_2[PtCl_6]$ aqueous solution and achieved 7.2% efficiency ($J_{SC} = 13.6$ mA cm^{-2}, $V_{OC} = 780$ mV, $FF = 0.68$) for the cell irradiated through the CE.

In an effort to increase the surface area and reduce the quantity of Pt on the counter electrode and hence reduce cost, a number of groups have distributed Pt in other materials such as acetylene black, NiO or $SnO_2:Sb$.[276–278] A light-to-electricity energy conversion efficiency of 8.6% was achieved for the DSSCs with the Pt/acetylene black electrode. Wei *et al.* employed polyvinylpyrrolidone (PVP)-capped Pt nanoparticles as a counter electrode in DSSCs to reduce the Pt-loading (4.89 μg cm^{-2}) while maintaining a low charge-transfer resistance (5.66 Ω cm^{-2}). A 2.84% conversion efficiency was obtained for the device ($V_{OC} = 0.66$ V $J_{SC} = 10.5$ mA cm^{-2}, $FF = 0.41$).[279]

Carbon-Based Counter Electrodes Whilst carbon is a cheaper, abundant alternative to Pt, the catalytic activity is far lower, resulting in a lower efficiency when used in the DSSC.[280] However, from a manufacturing perspective, carbon electrodes are favourable since they do not require the heating step that is necessary for adhesion of Pt to the substrate and

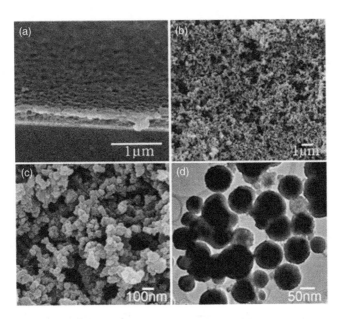

Figure 3.53   (a–c) SEM micrographs of AB thin films deposited on FTO glass and (d) TEM image of the porous AB spheres. Reproduced from Cai *et al.*, 2006[282] with permission from The Chemical Society of Japan

therefore can be deposited on plastic and other flexible substrates for roll-to-roll processing. A number of methods have been used to improve the efficiency of the reaction at the counter electrode by reducing the resistance and increasing the surface area or roughness.[281] Figure 3.53 shows the photocurrent–photovoltage characteristics for DSSCs assembled with different carbon electrodes with different roughness factors reported by Imoto *et al.*[281] Their report demonstrated that activated carbon with a sufficiently high roughness factor (above 10 400) can perform as well or better than Pt. Likewise, Cai *et al.* used porous acetylene-black (AB) spheres (Figure 3.53), prepared by the pyrolysis of acetylene, as the counter electrode material.[282] The BET specific surface area of the porous spheres was very high ($280 \, m^2 \, g^{-1}$) and the R_{CT} at the counter electrode/electrolyte interface was low ($7.9 \, \Omega \, cm^{-2}$). A thin AB film with a thickness of 20 μm gave a conversion efficiency of 5.76%.

Another way to improve the catalytic surface area while not increasing the resistance is to use carbon nanotubes (Figure 3.54). Carbon nanotubes have a high longitudinal conductivity and are good triiodide reduction catalysts.[283] Suzuki *et al.* used single-wall carbon nanotube counter electrodes, deposited on both FTO-glass and a Teflon membrane filter,

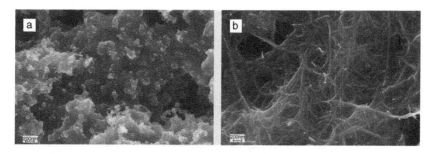

Figure 3.54 SEM images of nanocarbon ink. (a) SuperP and (b) multi-wall carbon nanotubes ink, both prepared by doctor blade deposition. Reproduced with permission from Gagliardi et al., 2009[283]. Copyright (2009) Elsevier

achieving conversion efficiencies of 3.5% and 4.5%, respectively.[284] The sheet resistance of the nanotubes on the Teflon membrane (1.8 Ω sq^{-1}) was four times lower than that of FTO-glass (ca 8–15 Ω sq^{-1}) rendering them an attractive alternative.

Carbon black is commonly used as printing ink but is electrically conducting and has also been applied as catalysts for I_3^- reduction as counter electrodes for DSSCs. Murakami et al. achieved an efficiency of 9.1% ($J_{SC} = 16.8$ mA cm^{-2}, $V_{OC} = 0.790$ V, $FF = 0.685$) for a 14.5 μm thick carbon layer.[269] The contribution of the high surface area (163.9 m^2 g^{-1} from BET measurements) in the thick carbon layer considerably improved the cell performance. By using carbon black on steel, Murakami et al. have fabricated a low cost substrate/low cost catalyst counter electrode.[285] The photocurrent–photovoltage curves are shown in Figure 3.55. Efficiencies similar to the FTO glass/Pt system were achieved (the SUS-304 steel substrate gave $J_{SC} = 15.4$ mA cm^{-2}, $V_{OC} = 780$ mV, $FF = 0.737\%$, $\eta = 8.86\%$, the SUS-316 support gave $J_{SC}, = 16.3$ mA cm^{-2}, $V_{OC} = 785$ mV, $FF = 0.714\%$, $\eta = 9.15\%$).

Conducting Polymers A number of conducting polymers have been used as counter electrode materials, such as poly(3,4-ethylenedioxythiophene) (PEDOT) doped with perchlorate (PEDOT-ClO$_4$), p-toluenesulfonate (PEDOT-TsO) or polystyrenesulfonate (PEDOT-PSS).[286-295] Cyclic voltammetry experiments have confirmed that the polymer electrodes show an electro-catalytic effect for I_3^- and values of R_{CT} close to that for Pt/FTO have been measured. As for the carbon-based counter electrodes, the efficiency of polymer-based counter electrodes can be improved by increasing the surface area e.g. by blending the polymer with TiO$_2$.[288] With a TiO$_2$ loading amount of ca 15 g m^{-2} in a TiO$_2$:ITO/PEDOT-PSS

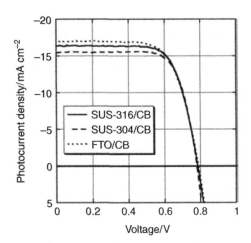

Figure 3.55 Typical photocurrent–voltage curves of cells using a carbon layer on a stainless steel substrate as counter electrodes. Reprinted with permission from Murakami and Grätzel, 2008[269]. Copyright (2008) Elsevier

film (4 wt% TiO_2), the composite film had an interfacial resistance close to that of a sputtered Pt coated on a glass electrode. The energy conversion efficiency of a full-plastic cell using the ITO-PEN counter electrode coated with TiO_2:ITO/PEDOT-PSS reached 4.38% (5.41% Pt/FTO).

Lee *et al.* incorporated a poly(3,3-diethyl-3,4-dihydro-2H-thieno-[3,4-b][1,4]dioxepine)(PProDOT-Et$_2$) counter electrode, prepared by electrodeposition, in a plastic DSSC and obtained a conversion efficiency of 5.20% ($J_{SC} = 11.22\,mA\,cm^{-2}$, $V_{OC} = 0.740\,V$, $FF = 0.627$), higher than an equivalent cell fabricated with sputtered-Pt (5.11%).[290] The root-mean-square (RMS) roughness value of the electrode was *ca* 83.5 nm, compared with that of *ca* 23.6 nm for the Pt film. The characteristic voltammograms of the films in the presence of I_3^-/I^- are shown in Figure 3.56. The anodic and cathodic peaks and associated charge capacities are similar for the Pt and the PProDOT-Et$_2$ electrodes. The PProDOT-Et$_2$-based plastic DSSC showed better cell performance due to the higher surface area from its wire-like structure and good catalytic behaviour for reducing I_3^-. The *FF* showed an optimal value of 0.627.

These examples demonstrate that replacement of the platinised glass counter electrode in favour of cheaper catalysts and flexible substrates should not be a substantial hurdle in the upscale of DSSC manufacture in terms of maintaining reasonable efficiencies. However, issues with the long-term stability arising from reactions with the corrosive electrolyte will need to be overcome.

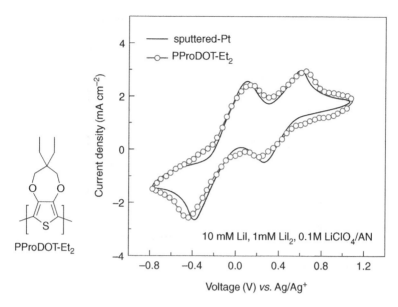

Figure 3.56 Structure of PProDOT-Et$_2$ and cyclic voltammograms for PProDOT-Et$_2$ film and Pt electrode at a scan rate of 100 mV s^{-1}. Reproduced from Lee *et al.*, 2009[290] by permission of the PCCP Owner Societies

p-Type and Tandem Dye-Sensitised Solar Cells

To make DSSCs a third generation technology, efficiencies well in excess of 20% are necessary.[29] Despite the attention paid to improving the performance, the record efficiency is still 11.1% for cells and 8.4% for modules.[43] One way of improving the overall efficiency is to assemble a tandem solar cell where both electrodes are photoactive, one harvesting the high-energy photons, the other harvesting the low-energy photons, as shown in Figure 3.57.[296] The theoretical upper limit for a cell with one photoactive electrode is around 30% whereas for a tandem device, with both electrodes photoactive, the theoretical limit is 43%.

In the tandem cell, the V_{OC} is the difference between the quasi-Fermi levels above the valence band of the photocathode and below the conduction band of the photoanode and is therefore the sum of the individual n- and p-type devices. The total current is limited by the lower performing electrode and the currents from both sides must be matched. In order to exploit the possible improvements from tandem DSSCs, the efficiency of the photocathode must be brought in line with that of the photoanode. Unlike the n-type devices, there have been few reports of p-type DSSCs.

The differences between n-type and p-type photocathodes can be illustrated using the examples of pyrite (FeS$_2$) and pyrrhotite (FeS) reported

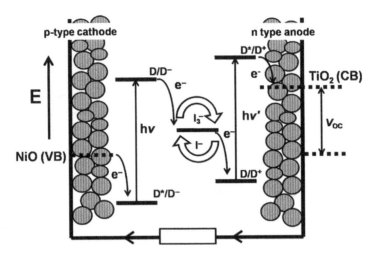

Figure 3.57 Schematic illustration of the relative energy levels and reactions in a tandem DSSC

by Hu *et al.*[297] FeS$_2$ has a band gap suitable for absorption of solar light ($E_g = 0.95$ eV) and a high absorption coefficient (5×10^5 cm^{-1} at 700 nm), two orders of magnitude bigger than crystalline silicon. FeS has interesting electrical and magnetic properties. Figure 3.58 shows the FeS and FeS$_2$ nanosheet films prepared from hydrothermal treatment of iron and sulfur with and without hydrazine.

The films were assembled with N3-sensitised TiO$_2$ nanorod photoanodes in DSSCs [Figure 3.59(a)]. The *J–V* characteristics of the DSSCs are shown in Figure 3.59(b), and a number of interesting observations can be

Figure 3.58 SEM images of the as-prepared FeS$_x$ ($x = 1, 2$): (a) FeS; (b) FeS$_2$. Reprinted with permission from Hu *et al.*, 2008[297]. Copyright (2008) American Chemical Society

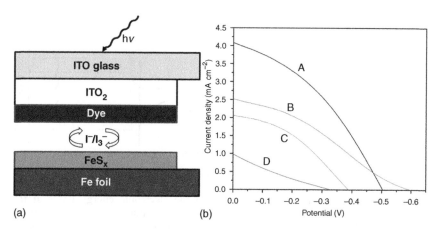

Figure 3.59 (a) Schematic illustration of the sandwich-type double-photoelectrode DSSCs. (b) Current–voltage characteristics of dye-sensitised TiO_2 solar cell with different photocathodes: (A) Pt; (B) FeS; (C) FeS_2; (D) Fe. Reprinted with permission from Hu et al., 2008[297]. Copyright (2008) American Chemical Society

made. The V_{OC} of the FeS/TiO_2 DSSC was higher than that of the Pt/TiO_2 DSSC. The FeS-based photocathode was better than that of the FeS_2-based nanosheet film photocathode, with a conversion efficiency of $\eta = 1.32\%$ ($J_{SC} = 2.53$ mA cm^{-2}, $V_{OC} = 0.60$ V, $FF = 0.31$ under 100 mW cm^{-2} illumination).

The reason for the difference is that FeS is a p-type semiconductor and FeS_2 is an n-type semiconductor. In both devices the usual processes occur at the TiO_2 anode (the electrons are injected into the conduction band of the TiO_2 from the photoexcited dye and the corresponding dye cation is reduced by redox mediator), generating a potential difference (U_1) between the quasi-Fermi level close to the conduction band in the TiO_2 [$E_{CB}(TiO_2)$] and the potential of the redox mediator ($U_{red/ox}$). If a passive, Pt counter electrode is used the V_{OC} is the difference between these, i.e. $E_{CB}(TiO_2) - U_{red/ox}$.

If the cathode is photoactive then, simultaneously with the reactions at the anode, electron–hole pairs form in the conduction and valence band in the semiconductor at the cathode, generating a potential difference (U_2) between the Fermi level in the cathode semiconductor and $U_{red/ox}$. Figure 3.60 shows the changes in energy level in the bulk for dense (e.g. single crystal) n- and p-type semiconductors in equilibrium with an electrolyte, in the dark and upon illumination.

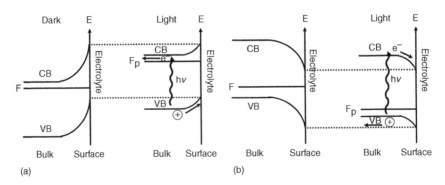

Figure 3.60 Scheme showing the change of the electronic energy levels at the interface between an (a) n-type and (b) p-type semiconductor before and under illumination. Reprinted with permission from Hu *et al.*, 2008[297]. Copyright (2008) American Chemical Society

In the case of n-type FeS_2, the Fermi level is close to the conduction band ($[E_{CB}(FeS_2)]$ and the maximum open circuit voltage (U_A) will be:

$$U_A = U_1 - U_2 = \left[E_{CB}(TiO_2) - U_{red/ox}\right]$$
$$-\left[E_{CB}(FeS_2) - U_{red/ox}\right] = E_{CB}(TiO_2) - E_{CB}(FeS_2) \quad (3.40)$$

Since $E_{CB}(FeS_2)$ is more negative than $U_{red/ox}$ in the case of I^-/I_3^-, U_A is lower than the V_{OC} of the Pt/TiO_2 DSSC.

In the case of p-type FeS, U_2 is the difference between the Fermi level close to the valence band of FeS and $U_{red/ox}$

$$U_B = U_1 - U_2 = \left[E_{CB}(TiO_2) - U_{red/ox}\right]$$
$$-\left[E_{VB}(FeS) - U_{red/ox}\right] = E_{CB}(TiO_2) - E_{VB}(FeS) \quad (3.41)$$

Since $E_{VB}(FeS)$ is more positive than $U_{red/ox}$ in the case of I^-/I_3^-, U_B is higher than U_A and the V_{OC} of the Pt/TiO_2 DSSC. An illustration of the processes in the case of FeS and FeS_2 are shown in Figure 3.61.

Early studies on dye sensitisation of Cu-based p-type semiconductors were carried out by Tennakone *et al.*[298,299] Typically, SCN^- salts of cationic dyes such as methyl violet (Figure 3.62) were chosen to sensitise 1.3 μm CuSCN films and the standard, triiodide/iodide redox couple was used. Sensitisers used in the p-type systems must have HOMO levels which lie below the top of the semiconductor valence band and LUMO

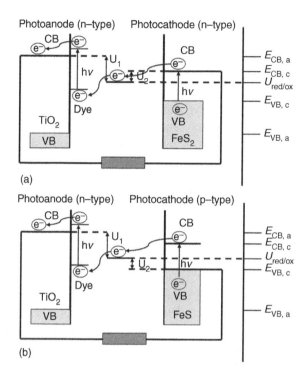

Figure 3.61 Schematic illustration of the band structure alignment and charge-transfer in the PV cell with different photocathodes, (a) n-type and (b) p-type, under illumination. Reprinted with permission from Hu *et al.*, 2008[297]. Copyright (2008) American Chemical Society

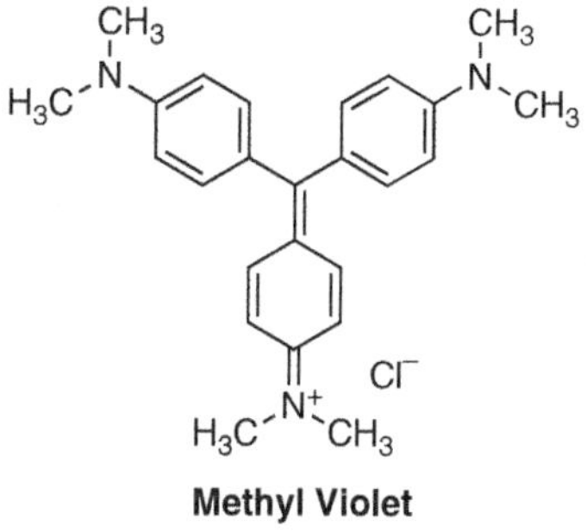

Methyl Violet

Figure 3.62 Structure of a cationic dye, methyl violet, used to sensitise CuSCN

levels which lie above the energy level of the triiodide/iodide redox system and below the semiconductor conduction band. In contrast to TiO_2-based photoanodes, electron transfer occurs from the semiconductor to the dye on illumination, *i.e.* holes are injected into the valence band of the semiconductor and the dye is reduced. The dye is regenerated to its initial neutral ground state by the redox couple which travels to the counter electrode where it is re-oxidised, completing the circuit. Low photocurrents in the order of 100 $\mu A\ cm^{-2}$ were obtained when the films were illuminated by a 60 W tungsten filament lamp, because, due to the low porosity of the CuSCN, the dye was located on the film surface instead of penetrating through the semiconductor film. CuSCN has since been more successfully used as an inorganic hole conductor in solid-state DSSCs[300] and *eta* cells (see later).

NiO-Based Photocathodes   The most successful material used for p-type photocathodes to date is NiO.[296,301–317] In 1999 He *et al.* prepared a dye-sensitised, nanostructured NiO cathode.[301] NiO is a transparent ($E_g \sim 3.6$ eV) p-type semiconductor with a cubic (bunsenite) structure. The p-type character is believed to arise from Ni^{2+} vacancies.[302] The hole mobility $\mu = 0.43\ cm^2\ V^{-1}$, and $m_h{}^* = 0.8 - 1.0 m_0$ at 1373–1673 K. The valence band lies around 0.54 V *vs* NHE at pH 7.[303,304]

He *et al.* prepared the electrodes by depositing a $Ni(OH)_2$ slurry on an FTO substrate and sintering in air at 500 °C to give 1 μm thick porous, nanostructured NiO films.[301] Erythrosin B and tetrakis(4-carboxyphenyl)porphyrin (TPPC) (Figure 3.63) were used as photosensitisers and under illumination a cathodic current was generated (in the opposite direction to the Grätzel-type cells) at wavelengths corresponding to the

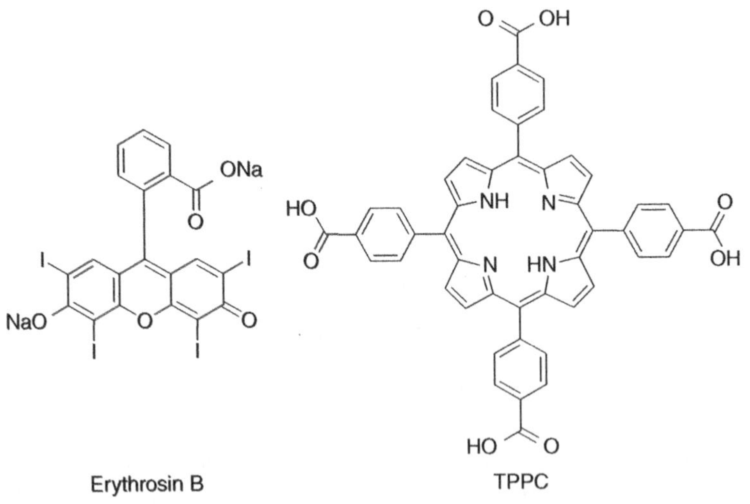

Erythrosin B   TPPC

Figure 3.63 Structures of erythrosin B and tetrakis(4-carboxyphenyl) porphyrins (TPPC) used as dyes in p-type DSSCs

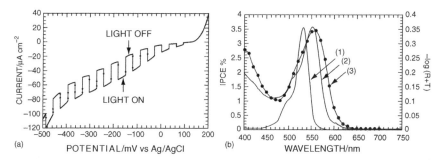

Figure 3.64 (a) Current *vs* potential curves for a nanostructured NiO film coated with erythrosin B with a 0.1 M LiI/0.01 M I$_2$ in propylene carbonate electrolyte, using intermittent light from a solar simulator (light intensity: 85 mW cm^{-2}). (b) Absorption spectra [$-\log(R+T)$] of erythrosin B in ethanol solutions (curve 1, 3.4 × 10^{-6} M) and on the nanostructured films measured in an integrating sphere (curve 2). Action spectra (curve 3) of nanostructured NiO films coated with erythrosin B in sandwich-type measurements. Reprinted with permission from He *et al.*, 1999[301]. Copyright (1999) American Chemical Society

absorption spectrum of the dyes when irradiated with monochromatic light (Figure 3.64).

Sandwich cells were prepared by assembling the sensitised electrodes with a platinum foil counter electrode and using 0.5 M LiI/0.05 M I$_2$ in 1:1 ethylene carbonate, propylene carbonate as the redox couple (*ca* 0.44 V *vs* NHE). The devices were illuminated with sun-simulated light at 68 mW cm^{-2} and for the TPPC cell, the J_{SC} = 0.079 mA m^{-2}, V_{OC} = 98.5 mV, *FF* 0.285, and overall conversion efficiency η = 0.0033%. For the erythrosin B cell, the J_{SC} = 0.232 mA cm^{-2}, V_{OC} = 82.8 mV, *FF* = 0.270, and overall conversion efficiency η = 0.0076%. The V_{OC} was limited by the small energy difference between the NiO valence band and the redox system, and the current by the low light absorption by the thin film.

This paper was quickly followed by the report by He *et al.* of a tandem DSSC assembled from an erythrosine B-sensitised NiO cathode and a N3-sensitised TiO$_2$ anode.[296] The electrolyte was 0.5 M LiI/0.05 M I$_2$ in 1:1 ethylene carbonate/propylene carbonate. The V_{OC} of the tandem cell was 732 mV, approximately the sum of the V_{OC} values from the n-type device (650 mV) and the p-type device (83 mV). The overall efficiency, η = 0.39%, was low because of the low current on the p-side and the *J–V* curve was S-shaped because of the mismatch in current between the anode and cathode (Figure 3.65).

A large improvement in the performance was reported by Nakasa *et al.* who used a sol-gel method to prepare NiO.[305] A precursor

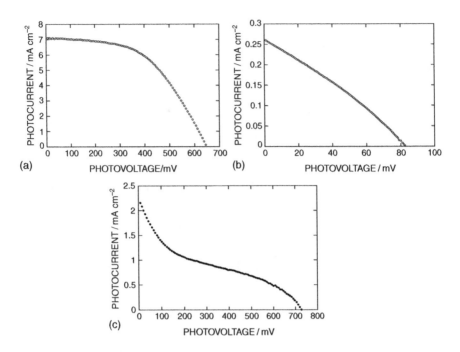

Figure 3.65 *J–V* characteristics of different types of sandwich cells, approximately 0.25 cm², illuminated with 85 mW cm⁻² light from a solar simulator. (a) N3-coated nanostructured, 4.4 µm thick, TiO₂ film with a platinised transparent conducting oxide as counter electrode. The electrode was treated with 4-*tert*-butylpyridine before measurements. (b) Erythrosin B-coated nanostructured, 1 µm thick, NiO electrode. Counter electrode as in (a). (c) A tandem DSSC of the dye-coated TiO₂ film in (a) with erythrosin B-coated NiO film in (b). The cell was illuminated from the erythrosin B-coated NiO film side. Reprinted with permission from He *et al.*, 2000[296]. Copyright (2000) Elsevier

solution containing a triblock copolymer template (HO(OCH₂-CH₂)₂₀(OCH(CH₃)CH₂)₇₀(OCH₂CH₂)₂₀OHCH₂, P123) and NiCl₂ in ethanol and water was spin-coated onto FTO followed by sintering at 450 °C to give crack-free, 1 µm thick, nanostructured (30–40 nm), NiO films. N3 or 3-carboxymethyl-5-[2-(3-octadecyl-2-benzothiazolinyldene) ethylidene]-2-thioxo-4-thiazolidine (Figure 3.66) was used as the sensitiser. The *J–V* characteristics are shown in Figure 3.67.

A tandem DSSC was prepared by assembling the cathode with a N3-sensitised TiO₂ anode (prepared using the P123-template in the same way) and the efficiency was 0.78% when illuminated through the cathode. The shape of the *J–V* curve was still S-shaped, however, since the currents from each side were still not matched. As for the device prepared by He *et al.*, the J_{SC} was substantially higher than that for the p-type

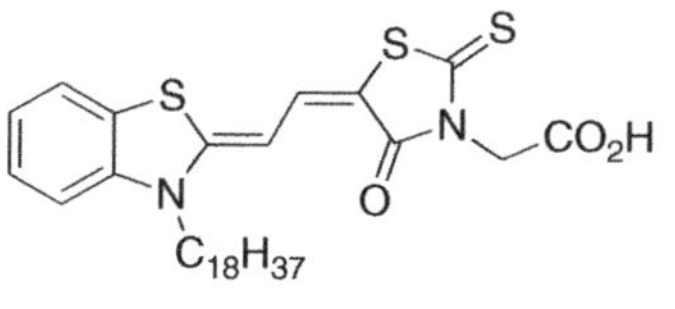

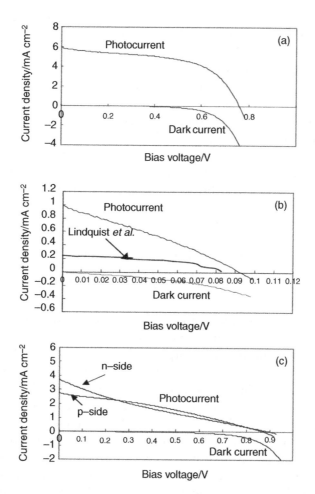

Figure 3.66 Structure of 3-carboxymethyl-5-[2-(3-octadecyl-2-benzothiazolinyldene) ethylidene]-2-thioxo-4-thiazolidine

Figure 3.67 The *J–V* characteristics of (a) n-DSSC, (b) p-DSSC and (c) tandem DSSC, irradiated at 100 mW cm⁻² with a light source simulating AM 1.5 global solar radiation; n-side and p-side indicates the illumination direction. Reprinted from Nakasa *et al.*, 2005[305] with permission from the Chemical Society of Japan

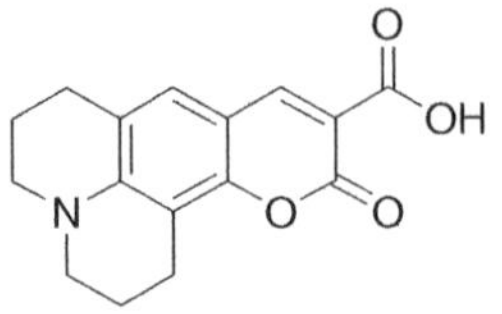

Figure 3.68   Structure of coumarin dye C343 used in p-type DSSCs

device alone, suggesting not all the current was photogenerated at the cathode.

Coumarin C343 (Figure 3.68) is a commercially available dye which has been studied as a sensitiser for TiO_2 and now is commonly used as the dye in NiO-based DSSCs. Despite its unfavourable optical properties (absorbing towards the UV) C343 has been shown to inject holes efficiently into NiO and reasonable currents have been reported. Morandeira *et al.* reported the charge-transfer dynamics of C343-sensitised NiO and showed that whilst charge injection (Equation 3.42) occurred in around 200 fs (similar to electron injection into the TiO_2 conduction band), charge recombination (Equation 3.43) occurred in around 20 ps.[317] For regeneration of the dye to compete with this fast process, it is postulated that the redox mediator (probably I_3^-) must be pre-associated with the dye.

$$C343/NiO + h\upsilon \rightarrow C343^{-\cdot}/NiO^+ \qquad (3.42)$$

$$C343^{-\cdot}/NiO^+ \rightarrow C343/NiO \qquad (3.43)$$

$$C343^{-\cdot}/NiO^+ + I_3^- \rightarrow I_2^{-\cdot} + I^- + C343/NiO^+ \qquad (3.44)$$

This fast charge recombination process is characteristic for organic sensitisers on NiO. The same group have also reported the charge-transfer kinetics of a phosphorous porphyrin bound to NiO.[310] In this case, charge injection was also fast (2–20 ps) and likewise followed by rapid recombination within 1 ns. Unlike C343, however, the IPCE was very low (2.5%), possibly as a result of an absence of pre-associated I_3^-.

The highest reported IPCE for a p-type DSSC was 64% using the 'push-pull' dye P1 (Figure 3.69).[316] 'Push-pull' dyes have a donor–acceptor character designed to improve the distance between the holes in the NiO and the unpaired electron in the reduced dye. The calculated frontier orbitals are illustrated in Figure 3.69. The HOMO level of the dye is evenly distributed over the whole molecule, including the carboxylic acid anchoring group, whereas the LUMO is located at the acceptor 'arms', further away from the NiO, pointing towards the electrolyte.

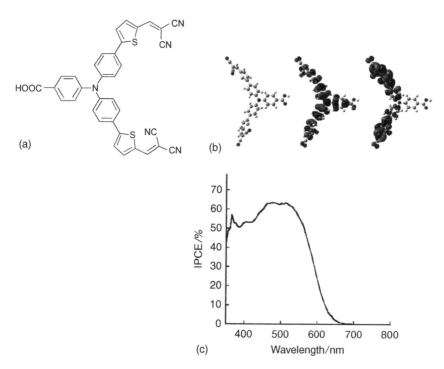

Figure 3.69 (a) Structure of P1. (b) The optimised structure (left) of P1, the frontier molecular orbitals of the HOMO (middle) and LUMO (right) calculated with DFT at the B3LYP/6-31+G(d) level. (c) IPCE of P1-sensitised NiO DSSC using 1.0 M LiI, 0.1 M I_2 in acetonitrile as the redox electrolyte. Reproduced with permission from Li *et al.*, 2009[316]. Copyright (2009) Wiley-VCH verlag Gmbh & co. KGaA

The previously reported IPCE was 35% for P1-sensitised NiO, prepared by the dehydration of $Ni(OH)_2$.[303,315] The improvement to the cell performance was attributed to the improvements made to the NiO when the F108-templated method described by Sumikura *et al.* was used.[310]

Charge Transport in NiO   NiO is one of the few metal oxides that are p-type in nature and is also studied for applications such as smart windows and supercapacitors.[303] As prepared, the nanostructured material has some surface colouration attributed to Ni(III) sites on the surface. At negative potentials, or in the presence of a mild chemical reducing agent, the material can be bleached. On scanning to positive potentials the colour turns brown then black. The spectroelectrochemistry was studied by Boschloo and Hagfeldt, and two surface redox reactions were observed (Figure 3.70) attributed to oxidation of Ni^{II} to Ni^{III} coupled

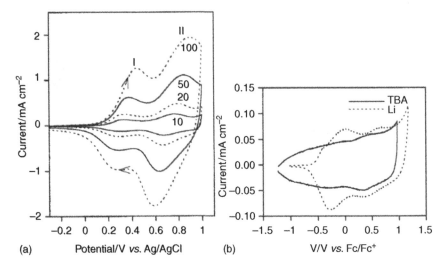

Figure 3.70 Cyclic voltammograms of (a) a nanostructured NiO electrode (0.75 μm) in aqueous electrolyte (0.2 M KCl + 0.01 M KH$_2$PO$_4$ + 0.01 M K$_2$HPO$_4$, pH 6.8). Scan rates (mV s^{-1}) are indicated. (b) A nanostructured NiO (0.75 μm) electrode in 3-methoxypropionitrile containing either 0.2 M lithium triflate or 0.2 M TBA triflate. Scan rate 10 mV s^{-1}. Reprinted with permission from Boschloo and Hagfeldt, 2001[303]. Copyright (2001) American Chemical Society

with desorption of protons at the surface in aqueous solution or adsorption of cations in the supporting electrolyte in nonaqueous solvents (Figure 3.71).[303,304]

Zhu *et al.* have studied the effects of cations in the electrolyte on hole transport times (τ_{tr}) and lifetimes (τ_h) in C343-sensitised NiO, prepared form Ni(OH)$_2$ colloids, using small-modulation transient photocurrent and photovoltage measurements.[309] Figure 3.72 shows the measured time constants plotted against J_{SC}.

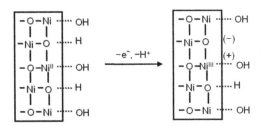

Figure 3.71 Surface oxidation of NiO in aqueous electrolyte. Reprinted with permission from Boschloo and Hagfeldt, 2001[303]. Copyright (2001) American Chemical Society

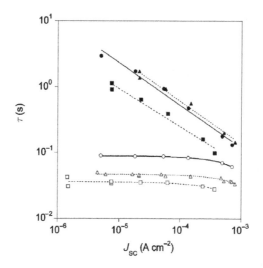

Figure 3.72 Hole transport times (open symbols) and lifetimes (filled symbols) in C343-sensitised mesoporous NiO solar cells as a function of short-circuit current density. The electrolyte was 0.5 M of LiI (circles), NaI (triangles), or TBAI (squares) and 0.1 M I_2 in 3-methoxypropionitrile. Reprinted with permission from Zhu *et al.*, 2007[309]. Copyright (2007) American Chemical Society

The J_{SC} was found to vary linearly with light intensity. τ_h (measured under open circuit conditions) was found to vary with light intensity. The transport times were found not to vary significantly with light intensity. This is contrary to the transport of electrons in TiO_2 which is light intensity dependent (see earlier). τ_{tr} was shown to vary significantly with the type of cation in the electrolyte solution, decreasing in the order $Li^+ > Na^+ >$ TBA^+. The authors attributed this to adsorption of the cations affecting the hopping of charge at the NiO/electrolyte interface, suggesting that the holes are Ni^{3+} at the surface. These observations could partly explain the extremely fast recombination between the dye radical anions and the holes in the NiO observed spectroscopically.[313,317]

Mori *et al.* studied the transport properties of F88 polymer-templated NiO, sensitised with different dyes, including C343, also using photocurrent and photovoltage transient measurements.[314] Diffusion coefficients (D) were calculated from τ_{tr} using Equation 3.45 where w is the film thickness.

$$D = w^2/(2.77\tau_{tr}) \tag{3.45}$$

The trends matched those reported by Zhu *et al.* The values of D were three orders of magnitude lower than the typical values for electrons in dye-sensitised TiO_2 solar cells but the hole lifetimes were comparable. Variations in the I^-/I_3^- ratio did not appear to change the hole lifetime or D. Interestingly, increasing the concentration of I_3^- did not improve the IPCE at wavelengths corresponding to the dye absorption but did increase at wavelengths corresponding to I_3^- absorption. This suggests that despite the short charge-separated state lifetime, increasing $[I_3^-]$ did not speed up the regeneration as would be expected if the process was kinetically limited.

Alternative Redox Mediators   So far, much of the research into p-type DSSCs has focused on improvement in the photocurrent. One of the limiting factors to the efficiency, however, is the very low V_{OC} caused by the small difference in energy between the quasi-Fermi level above the valence band in the NiO and the energy level of the redox couple. Because the Grätzel-type DSSCs are optimised for the I_3^-/I^- electrolyte, it would be a likely choice for use in tandem devices and therefore has been used in p-type devices as well. A number of problems arise, however; the first being the mismatch in energy with the NiO valence band. The second is that the triiodide is strongly coloured, with the absorption maximum at around 360 nm, tailing into the visible to over 500 nm.[309,318] In the tandem cell, the light must pass through the electrolyte to reach the second electrode and so an optically transparent electrolyte is necessary. A curious effect of using triiodide and NiO was observed by Zhu et al. who noticed that a significant contribution to the IPCE came from the triiodide itself (Figure 3.73).[309]

Under irradiation, triiodide dissociates as in Equation 3.46:

$$I_3 + h\nu \rightarrow I_2^- + I \qquad\qquad (3.46)$$

$$I_2^- + e^- \rightarrow 2I^- \qquad E \approx 0.93\,V\ vs\ NHE\ in\ MeCN \qquad (3.47)$$

The photoproducts can either recombine, or alternatively I_2^- can react with NiO (i.e. inject a hole) since the redox potential is more positive than that of the top of the valence band. The one-electron oxidation of iodide is thought to be the process that occurs in the regeneration of the ground state dye in n-type DSSCs and is thought to be the reason behind the apparent 0.6 V overpotential required to drive the system. The equivalent oxidation of triiodide lies symmetrically to negative potentials of the formal redox potential of I_3^-/I^- (0.32 V vs NHE in MeCN) according to Equation 3.48:

$$I_3^- + e^- \rightarrow I_2^- + I^- \qquad E \approx -0.35\,V\ vs\ NHE\ in\ MeCN \qquad (3.48)$$

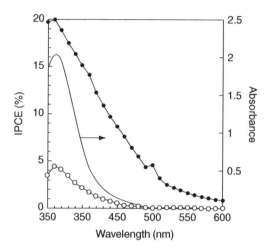

Figure 3.73 IPCE spectra of bare mesoporous NiO in iodide/triiodide electrolyte: 0.5 M LiI, 0.1 M or 0.5 M I$_2$ (open and closed symbols, respectively) in 3-methoxypropionitrile. Also shown is the absorption spectrum of diluted electrolyte (line). Reprinted with permission from Zhu *et al.*, 2007[309]. Copyright (2007) American Chemical Society

Analogous to the n-type system, this is expected to be the reaction involved in regeneration of the ground state dye in the p-type system. A consequence of this is a limit to the LUMO energy or more importantly the reduction potential of the sensitiser and therefore the range of wavelengths available.

In order to avoid these limitations, there is a need for alternative electrolytes which work in both n-type and p-type systems. One such class of species is the cobalt polypyridyl complexes such as Co$^{II/III}$ tris(4,4″-di-*tert*-butyl-2,2″-dipyridyl) perchlorate.[319–323] Gibson *et al.* employed this redox couple in a NiO-based p-type device in conjunction with two perylene-based dyes, PI and PINDI (Figure 3.74).[324]

The PINDI 'dyad' contained a coupled naphthalene diimide acceptor unit which had a more positive reduction potential than the PI absorber (PI$^{0/-}$ = −0.66 V *vs* NHE, NDI$^{0/-}$ = −0.19 V *vs* NHE). On illumination an electron-transfer cascade from the NiO to the PI to the NDI occurred, extending the distance between the NiO(h+) and the unpaired electron in the reduced dye compared with the PI alone, and increasing the lifetime of the charge-separated state by five orders of magnitude. Because the potential of the NDI acceptor was more positive than the reduction potential of the triiodide (Equation 3.48), little difference between the performance of devices using PI and PINDI was observed when

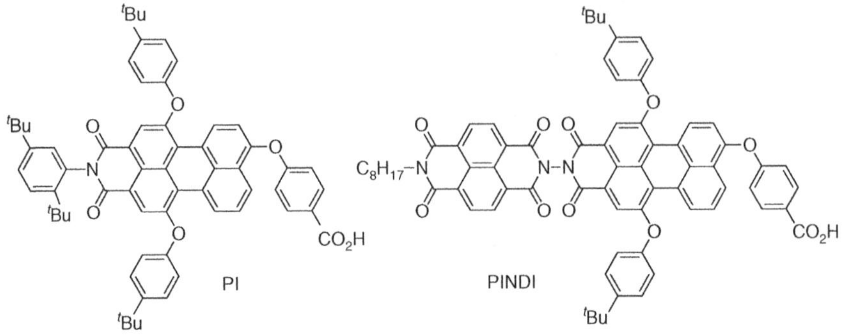

Figure 3.74   Structures of PI and PINDI sensitisers[313,324]

the triiodide redox couple was used.[313] However, when the cobalt redox mediator was used, the PI-sensitised DSSC performed poorly because of the short charge-separated state lifetime whereas a threefold increase in photovoltage was observed for the PINDI-sensitised DSSC [Figure 3.75(a)].

A tandem cell was assembled with an N719-sensitised TiO_2 anode where the current was matched to the cathode by tuning the TiO_2 film thickness [Figure 3.75(b)]. The FF (0.62) was improved compared with the previously reported tandem DSSCs described above and the V_{OC} was approximately the sum of the p-type and n-type devices (0.91 V). The current was still low because of the substantial overlap of the absorption spectra of the dyes used and the relatively low ionic strength of the electrolyte because of the low solubility of the bulky cobalt complex and the efficiency was 0.55%.

Alternative p-Type Semiconductors   Because the valence band potential of NiO lies only 100 mV more positive than the optimum redox couple for n-type devices, a substantial increase in voltage for the p-type/tandem systems should be achieved if the p-type semiconductor has a valence band with a much lower energy than the triiodide/iodide redox couple. The other requirements for the material include optical transparency ($E_g > 3$ eV), mechanical and electrochemical stability, good electronic properties (high charge-carrier mobility) and a convenient means of anchoring the dye (*e.g.* metal oxides and carboxylic acids). Whilst other p-type semiconductors exist, few combine all the properties required and as yet there have been no p-type semiconductors reported that perform better than NiO in a p-type DSSC.[325–327] For example, several

(a)   (b)

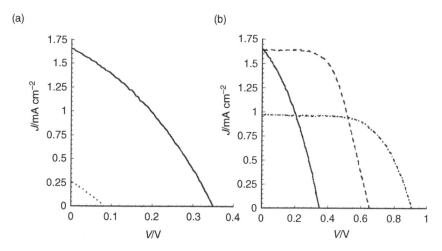

Figure 3.75 *J–V* characteristics for: (a) a NiO-based p-type DSSC sensitised with PI (dotted line) or PINDI (solid line); (b) a tandem-DSSC (dot-dashed line) assembled with the PINDI-sensitised NiO-based photocathode (solid line) and an N719-sensitised TiO_2-based photoanode (dashed line). The electrolyte was 0.1 M Co^{II} tris(4,4′-di-*tert*-butyl-2,2′-dipyridyl) perchlorate, 0.1 M Co^{III} tris(4,4′-di-*tert*-butyl-2,2′-dipyridyl) perchlorate and 0.1 M $LiClO_4$ in propylene carbonate. Reproduced with permission from Gibson *et al.*, 2009[324]. Copyright (2009) Wiley-VCH Verlag GmbH & Co. KGaA

Cu-based semiconductors have been investigated.[326] Fast green and NK-3628 were shown to inject holes into CuO by Sumikura *et al.*, however the band gap is narrow ($E_g = 1.4$ eV) and CuO absorbs light itself.[328] Nonetheless, a fast green-sensitised 0.34–0.52 µm thick CuO electrode when assembled in a device with Pt/Pd counter electrode and a triiodide/iodide electrolyte gave an overall efficiency of 0.011% ($J_{SC} = 0.30$ mA cm^{-2}, $V_{OC} = 0.115$ V, $FF = 0.31$). The CuO alone gives a poor photocurrent because of fast charge recombination between the excited electron–hole pairs in the semiconductor. The V_{OC} of this device, and others using Cu-based semiconductors, are low as a result of the small difference in energy between the valence band potential of the CuO (which is similar to NiO) and the triiodide/iodide redox couple. The development of alternative p-type semiconductors with the required properties for a DSSC photocathode is a crucial target in DSSC research.

3.2.3.2.7 *Alternative Architectures*

In the laboratory, the 'standard' sandwich cell architecture described above is used. This can be expensive and impractical to assemble for modules with a large area and novel geometries are being developed, driven by cost and ease of upscale or to improve efficiency.

Monolithic Solar Cells

In 1996, Kay and Grätzel introduced the concept of monolithic DSSCs, where the device is assembled on one substrate rather than two in the sandwich cell.[329] The assembly process and the device structure is shown in Figure 3.76.

Three porous layers are deposited on a transparent, conducting substrate: a TiO_2 (anatase) photoelectrode, a TiO_2 (rutile) electrically insulating, light-reflecting spacer and a graphite powder and carbon black counter electrode. The annealed device is sensitised by soaking in a dye solution, the redox electrolyte containing iodide is applied and finally the device is sealed to prevent evaporation of the electrolyte and intrusion of humidity or oxygen. The spaces between the adjacent cells can be filled with an insulating polymer to prevent short-circuiting.

The primary advantage is the reduction in materials cost since most of the cost of manufacturing DSSCs comes from the glass substrate. Also, the sheet resistivity of the conducting SnO_2 limits the width to less than 1 cm for the sandwich structure. Other advantages include the shorter distance over which the redox mediator has to travel to the counter

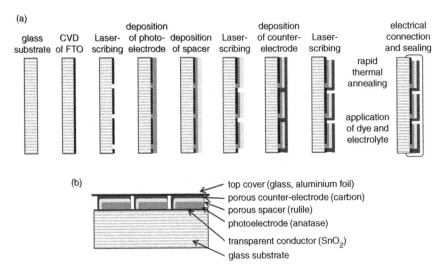

Figure 3.76 (a) Continuous process for the fabrication of monolithic series connected dye-sensitised PV modules. (b) Schematic cross-section of the monolithic series connected dye-sensitised PV modules. Reprinted with permission from Kay and Grätzel, 1996[329]. Copyright (1996) Elsevier

electrode, because the counter electrode is deposited directly on top of the working/photoelectrode. The two electrodes are separated by a 10 μm, insulating rutile spacer which also reflects light, improving the efficiency. Each of these layers is porous or sponge-like and so absorbs the electrolyte by capillary forces. This means that the active area is maximised so the catalytic efficiency is increased and there is no 'dead space' between the electrodes where ohmic losses from diffusion occur, as in the free-flowing layer in sandwich cells. Also, since the sensitisation and electrolyte-filling occurs through these porous materials after assembly and heat treatment (the dye soaks through the porous counter electrode and insulating layers and is adsorbed on the active anatase layer), heat and humidity-sensitive dyes and electrolytes can be applied. The standard printing techniques used to build the active layers on the single, transferrable substrate are suitable for large-scale manufacture in a continuous process, and allow for patterning of the electrodes to increase the surface area. It is easy to connect the individual cells in series by overlapping the counter electrodes with the back-contact of adjacent cells in the printing process. Contacts, therefore, only need to be made at the first counter electrode and last photoelectrode.

The size of the rutile particles (0.3–0.5 μm) in the insulating/scattering layer is chosen to be optimum for scattering red light. A binder must be added because the temperature for adhesion of the large particles is much higher than for the small anatase particles. The particles used are 20 nm ZrO_2 because ZrO_2 is insulating ($E_g = 5$ eV), it prevents electrical contact between the rutile particles, and it sinters at 450 °C. There is a trade-off between the thickness for optimum reflectance and insulation vs voltage losses from diffusion of the electrolyte. The 10 μm layer has a diffuse reflectance of 70% at 700 nm after exposure to the dye solution and filling of the pores with electrolyte.

The counter electrode must be a good conductor and exhibit a low overvoltage for the redox couple and be porous to allow the dye and then the redox electrolyte to permeate through to the underlying layers. Carbon is chosen because it has good conductivity, heat and corrosion resistance and sufficient catalytic activity for triiodide reduction. Note, 20% carbon black is added to graphite flakes to increase the catalytic activity by increasing the surface area and filling the gaps between the flakes. Pt is unsuitable not only because of cost but also because dyes can bind to the surface and poison the catalyst. TiO_2 nanoparticles less than 20 nm in size are used as the binder which provide good adhesion and scratch resistant films on sintering at 450 °C. The resulting 50 μm thick electrode has a roughness factor in excess of 1000 and a sheet resistance of 5 Ω, which doubles on soaking with electrolyte due to swelling.

Kay and Grätzel reported an energy conversion efficiency of 6.67% for a cell under simulated sunlight of 1000 W m^{-2}. A small module comprising six series connected elements of 4.7 × 0.7 cm^2 each gave an efficiency of 5.29% (V_{max} = 3.90 V, J_{max} = 28.55 mA, FF = 0.614) at 1000 W m^{-2} with respect to the total surface area of 21.06 cm^2 (5.65% with respect to the active surface). The long-term stability of the monolithic dye-sensitised cells was tested by illuminating the array continuously with simulated sunlight, whilst operating close to the maximum power point. The results are shown in Figure 3.77. The conversion efficiency was fairly constant over 100 days, corresponding to about 2 years of outdoor illumination in middle Europe.

Petersson *et al.* have modified the structure to improve the conductivity, catalytic activity and adhesion of the counter electrode by developing a quadruple-layer monolithic device.[330] The structure is shown in Figure 3.78 and consists of a layer of nanocrystalline/nanoporous TiO$_2$, an insulating ZrO$_2$ layer, a carbon layer promoting good adhesion to the substrate (Carbon 1), and a platinised carbon counter electrode with good catalytic and conducting properties (Carbon 2).

The cells are assembled on a glass substrate in a multicell, array which is shown in Figure 3.79. The manufacturing technique used is a modified version of that described by Kay *et al.*, *i.e.* sequential screen printing onto

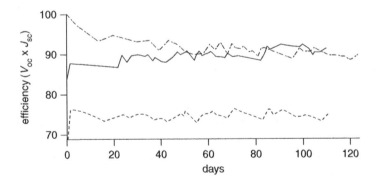

Figure 3.77 Evolution of open-circuit photovoltage, short-circuit photocurrent and the normalised product of these two representing conversion efficiency, for three monolithic DSCCs under xenon lamp illumination of 800 W m^{-2} with cut-off below 435 nm. Each cell was connected to a 100 Ω resistor for operation near the maximum power point; dye, N3; electrolyte, 1 M TBAI and 0.1 M I$_2$ in acetonitrile. Reprinted with permission from Kay and Grätzel, 1996[329]. Copyright (1996) Elsevier

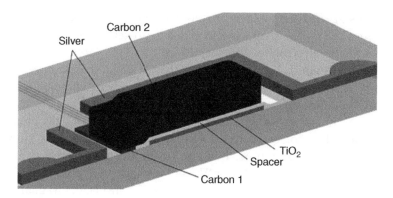

Figure 3.78 Schematic cross-section of the four-layer monolithic dye PV structure, consisting of dye-sensitised TiO_2, porous ZrO_2 (Spacer), an adhesion layer (Carbon 1), and platinised carbon (Carbon 2) on structured TCO glass. Reprinted from Petersson *et al.*, 2007[330] with permission from John Wiley & Sons, Ltd

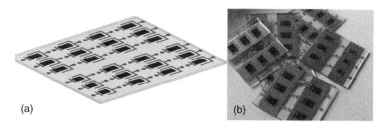

(a)   (b)

Figure 3.79 (a) Monolithic multicell with 24 individual cells; (b) photograph of divided multicells. Reprinted from Pettersson *et al.*, 2007[330] with permission from John Wiley & Sons, Ltd

laser-structured transparent conducting oxide (TCO) glass plates of 12 μm TiO_2 (followed by sintering), then 10 μm thick silver current collectors, followed by the 10 μm spacer, then the 10 μm Carbon 1 layer and finally 50 μm of Carbon 2. The spacer layer is 1.0 mm broader than the TiO_2 layer in all directions. The Carbon 1 layer is placed next to the spacer layer whereas the Carbon 2 layer is placed on top of the spacer layer and the Carbon 1 layer. The silver lines are placed 1.0 mm from the external contours of the cell, on three sides for the working electrode and on one side for the counter electrode. The Carbon 2 layer had a sheet resistance less than 5 Ω sq^{-1} before applying the electrolyte (*i.e.* conducts better than the TCO substrate). The assembled layers are sintered at 390 °C before soaking in N3 solution and application of the electrolyte. The multicells were encapsulated for long-term testing in a vacuum/heat process using the thermoplastic material Surlyn®. The space between the silver lines and the electrodes is also sealed to prevent contact between the silver and the electrolyte. Efficiencies up to 6.8% (area $= 0.48$ cm^2, $J_{SC} = 14.1$ mA cm^{-2}, $V_{OC} = 780$ mV, $FF = 0.62$) at a light intensity of 100 mW cm^{-2} (up to 7.5% at 25 mW cm^{-2}) have been obtained with an electrolyte solution based on γ-butyrolactone.

Figure 3.80 shows the results from stability testing of cells incorporating the γ-butyrolactone-based electrolyte, where the cells showed some degradation over time. Long-term stability was obtained at cell efficiencies close to 5% at 100 mW cm^{-2} with an electrolyte containing 0.8 M 1-propyl-3-methylimidazolium iodide, 50 mM I$_2$, 0.1 M guanidine thiocyanate and 0.5 M N-methyl-benzimidazole in glutaronitrile, with good reproducibility.

Han *et al.* have developed a solid-state, triple-layer monolithic structure, comprising of a nanoporous TiO$_2$ working electrode layer, a ZrO$_2$/TiO$_2$ insulating/scattering layer, and a carbon counter electrode layer.[331] A vacuum filling method was used to improve the filling of the nanoporous structure by iodine-containing, polymer-based electrolytes. The counter electrode (Figure 3.81) contained graphite particles coated

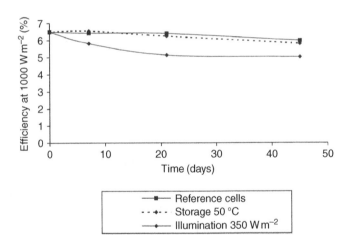

Figure 3.80 Efficiency mean value at 100 mW cm^{-2} of six cells using electrolyte containing 0.6 M 1-butyl-3-methyl-imidazolium iodide, 0.25 M N-methyl-benzimidazole and 0.05 M I$_2$ in γ-butyrolactone *vs* time for cells exposed to accelerated testing: continuous illumination at 35 mW cm^{-2} using a sodium lamp (open-circuit condition) protected by a UV cut-off filter (cut-off at 400 nm): storage in darkness at 50 °C; or storage in darkness at room temperature (reference cells). Reprinted from Pettersson *et al.*, 2007[330] with permission from John Wiley & Sons, Ltd

Figure 3.81 SEM image of the carbon film. Reprinted with permission from Han et al., 2009[331]. Copyright (2009) American Institute of Physics

with 0.5 wt% Pt and used ZrO_2 nanoparticles as a binder. The sheet resistance of a 50 μm thick carbon film was 13 Ω sq^{-1}, which is similar to that of the platinised fluorine-doped tin oxide conducting glass used as counter electrode in conventional DSSCs.

The cell efficiency was 3.65%. After illuminating for 3192 h, the short-circuit photocurrent density of the monolithic solid-state DSSC decreased from 7.44 to 6.83 mA cm^{-2}. The major loss in short-circuit photocurrent occurred within the first 1272 h of the stability test (5.9%). Photocurrent losses over the following 1920 h were less than 2.4%, indicating that the initial photocurrent losses had plateaued. However, the open-circuit voltage increased steadily from 524 to 665 mV for the same period. As a result, the overall energy conversion efficiency of the monolithic device increased from 2.73 to 3.09% after 3192 h.

Dye Cocktails and Co-sensitisation

A number of groups have attempted to increase the absorption over the full solar spectrum in order to increase the photocurrent from DSSCs and reduce the thermalisation losses. The record DSSC efficiency was measured for a cell sensitised with the 'red dye', N719. For this device, the light-harvesting efficiency is approximately unity over the range of wavelengths absorbed by the dye and the current obtained (ca 20 mA cm^{-2}, 90% IPCE) is close to the theoretical limit for this dye (estimated by integrating the total number of photons available across the absorption weighted solar spectrum). To increase the current, and therefore conversion efficiency, the range of absorption of the DSSC must be extended to longer wavelengths. The most successful panchromatic dye is the 'black dye'.[332] The drawback of increasing the absorption towards the IR is that the gap between the HOMO and LUMO levels in the dye decrease and, therefore, so do the driving forces. A lowering of the LUMO energy means a lowering of the driving force for injection, which leads to a reduction in the photocurrent. The injection yield can be increased by lowering the conduction band edge of the TiO_2 by adding lithium cations but this reduces the V_{OC} of the device.

Another way to increase the range of wavelengths absorbed is to use a dye cocktail.[333–335] However, problems arise due to competition for binding sites and unfavourable interactions between the dye molecules, such as energy transfer competing with charge injection. Inakazu et al. avoided the problem of dyes competing for binding sites by exploiting the phenomenon that large ruthenium-based dyes such as the black dye adsorb from the surface downward towards the substrate.[336] The authors firstly sensitised the porous TiO_2 film with the black dye under pressurised CO_2. The dye adsorption was stopped at a certain time and the rest of the unstained TiO_2 layer was coated with NK3705, as illustrated in Figure 3.82. The solar cell was assembled in the standard manner and the measured efficiency was 9.16% ($J_{SC} = 21.8$ mA cm^{-2}, $V_{OC} = 0.60$ V, $FF = 0.70$). This was higher than either of the cells sensitised with one dye because of the additive effects of the dyes on the light harvesting, as illustrated by the IPCE spectra shown in Figure 3.83.

Lee et al. have developed an elegant way of controlling the desorption and adsorption depth of different coloured dyes to layer yellow (2-cyano-3-(5-(4-ethoxyphenyl)thiophen-2-yl)acrylic acid, P5; Figure 3.84), red (N719) and green [black dye (N749)] dyes within a TiO_2 film.[337] They noticed that the rate of flow of the dye solution in the nanoporous film can be slowed by inserting a polymer into the pores. First the yellow dye was adsorbed, followed by infiltration of the pores with styrene which was polymerised in situ to cover the surface and reduce the pore size. The dye was selectively desorbed, using an aqueous solution of NaOH and polypropylene glycol, from the upper region of the film. The polypropylene glycol retards the flow of the solution and solvated ions in the film so only the upper region is penetrated and the dye is desorbed selectively. The red dye is then adsorbed at the upper surface only, because the dye molecules are too big to penetrate the lower part of the film, and partly desorbed at the uppermost region, and then the green dye is adsorbed near the surface.

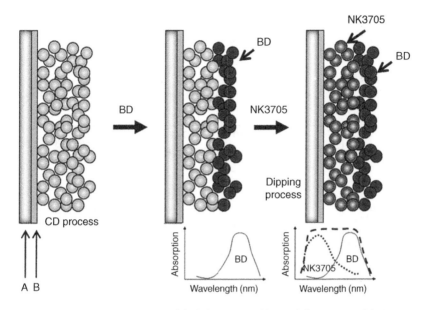

Figure 3.82 Fabrication process of the bilayer TiO_2 film and illustration of the increase in wavelength range absorbed by the two-dye system. A, glass; B, F-doped SnO_2; BD, black dye, NK3705; CD process, pressurised CO_2 conditions. Reprinted with permission from Inakazu et al., 2008[336]. Copyright (2008) American Institute of Physics

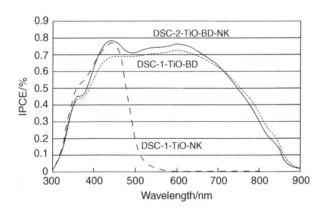

Figure 3.83 IPCE curves for bilayer and single-layer cells. Reprinted with permission from Inakazu et al., 2008[336]. Copyright (2008) American Institute of Physics

The photocurrent density (J_{SC}) of the three-dye cell was 10.6 mA cm^{-2}, which was higher than that for the single-dye cells: P5 (2.5 mA cm^{-2}), N719 (5.4 mA cm^{-2}) and N749 (7.5 mA cm^{-2}). The V_{OC} (619 mV) for the three-dye cell was close to the average (623 mV) of the V_{OC} values for the

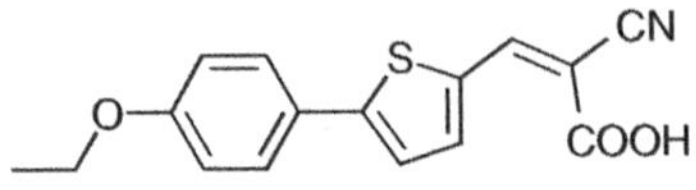

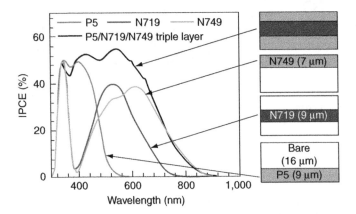

Figure 3.84 Structure of 2-cyano-3-(5-(4-ethoxyphenyl)thiophen-2-yl)acrylic acid, P5

Figure 3.85 IPCE spectra of the selectively positioned three-dye cell and those of the single-dye cell with P5 on the bottom, N719 in the middle and N749 on the top of the TiO_2 film. Reprinted with permission from Lee *et al.*, 2009[337]. Copyright (2009) Macmillan Publishers Ltd

three single-dye cells. The IPCE spectra of the individual and multilayered devices are shown in Figure 3.85.

Another way to prepare multilayers of dye is to use an Al_2O_3 blocking layer to separate the dyes adsorbed to the surface of TiO_2.[338,339] Two organic dye sensitisers (Figure 3.86) were employed in the example of the layered assembly shown in Figure 3.87: 3-[5'-{N,N-bis(9,9-dimethyl-fluorene-2-yl)phenyl}-2,2'-bisthiophene-5-yl]-2-cyanoacrylic acid (JK-2), which absorbs in the blue part of the visible spectrum, and 5-carboxy-2-[{3-[(1,3-dihydro-3,3-dimethyl-1-ethyl-2H-indol-2-ylidene)methyl]-2-hydroxy-4-oxo-2-cyclobuten-1-ylidene}methyl]-3,3-trimethyl-1-octyl-3H-indolium (SQ1), which absorbs in the red region. The secondary Al_2O_3 layer was coated by the hydrolysis of an aluminium isopropoxide on a JK-2-sensitised TiO_2 electrode. The Al_2O_3-coated JK-2-sensitised TiO_2 film was then dipped in the SQ1 solution in ethanol for the second sensitisation step. The energetics of the two dyes were such that a charge-transfer cascade was formed, whereby electron transfer occurred from JK-2 to the TiO_2 conduction band. The resultant hole was channelled from JK-2 to

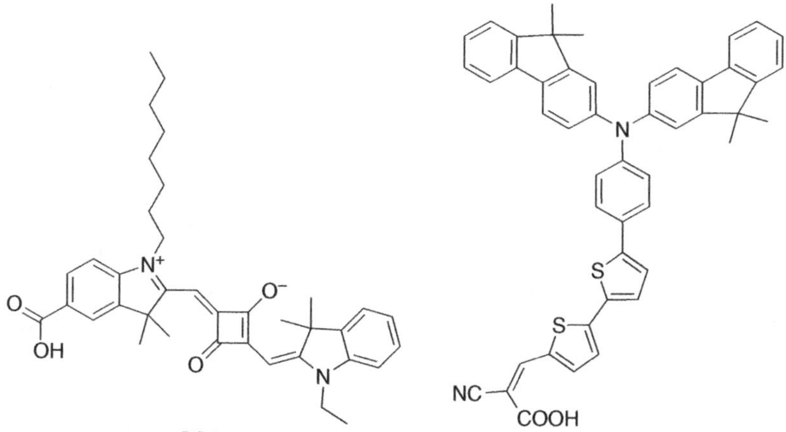

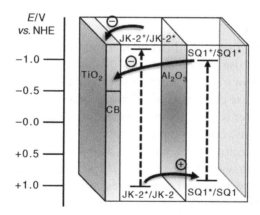

Figure 3.86 Structures of 3-[5'-{N,N-bis(9,9-dimethylfluorene-2-yl)phenyl}-2,2'-bisthiophene-5-yl]-2-cyanoacrylic acid (JK-2) and 5-carboxy-2-[{3-[(1,3-dihydro-3, 3-dimethyl-1-ethyl-2H-indol-2-ylidene)methyl]-2-hydroxy-4-oxo-2-cyclobuten-1-ylidene}methyl]-3,3-trimethyl-1-octyl-3H-indolium (SQ1) used in the multilayer co-sensitised device described by Clifford *et al.*[338] and Choi *et al.*[339]

Figure 3.87 Charge-transfer processes in multilayer co-sensitised nanocrystalline TiO_2 films. CB, conduction band; NHE, normal hydrogen electrode. Reproduced with permission from Choi *et al.*, 2008[339]. Copyright (2008) Wiley-VCH Verlag GmbH & Co. KGaA

SQ1, which resided further away from the TiO_2 surface, thus increasing the distance between injected electron in the TiO_2 film and the oxidised dye.

The IPCE for the JK-2/Al_2O_3/SQ1 cell reached *ca* 85% at 453 nm, and extended to 700 nm, with IPCEs of 79% at 660 nm. The cell resulted in an overall conversion efficiency of $\eta = 8.65\%$ ($J_{SC} = 17.6$ mA cm^{-2}, $V_{OC} = 0.696$ V, $FF = 0.70$). In the assembled devices, a more than 30 mV increase in V_{OC}, compared with the single-dye devices and devices in which simultaneous co-sensitisation using a dye cocktail had been employed, was attributed to the suppression of the charge recombination reaction by the insulating Al_2O_3 layer, which increases the separation between the injected electrons and oxidised dye. This insulating layer led to suppression of the dark current, an increase in electron lifetime (τ_e), and the lower series resistance[j] in the device [JK-2/SQ1 (52.6 Ω) > JK-2/Al$_2$O$_3$/SQ1 (45.0 Ω)], indicating improved charge generation and transport. In 1:1 mixed solutions of JK-2 and SQ1, the excitation of JK-2 resulted in energy transfer to SQ1 and is expected to occur on the surface of TiO$_2$ films co-sensitised with both dyes in the normal manner. In the presence of the Al_2O_3 layer, energy transfer was not observed and, instead, the hole-transfer mechanism shown above dominated. This multistep electron transfer mechanism is important for maintaining the photocurrent in the presence of an insulating layer, since electron-transfer reactions such as charge injection or regeneration of the oxidised dye through this layer would otherwise be impaired.

Photonic Crystals

Instead of tuning the dye, a different method to enhance the red response of the DSSC is to use photonic crystals. Nishimura *et al.* reported the first example of a device incorporating a photonic crystal, fabricated using a two-layer structure consisting of submicrometre spheres and a nanoparticulate TiO$_2$ layer to enhance light collection by scattering the incoming light.[340] Photonic crystals have interesting optical properties resulting from different regions with different dielectric constants located in a repeating fashion within the crystal. They have been exploited in low threshold lasers, waveguides, waveplates, reflective polarisers and dielectric mirrors. These properties include the reflection of certain wavelengths of light (those that fall in the photonic 'band gap' or stop band), birefringent and anisotropic dispersion and nonlinear dispersion (or the superprism phenomenon). Light is localised in different regions of the photonic crystal depending on the energy and according to the refractive index of the crystal at different regions. The photocurrent of solar cells can be increased by coupling a photonic crystal to a conventional photoelectrode.[341–343]

---

[j]Measured by impedance spectroscopy.[334]

The TiO_2 inverse opal structure described earlier in the chapter has a wide band gap, so it has no appreciable absorbance in the visible part of the spectrum. With a relatively high refractive index, it provides good dielectric contrast with air or common liquid electrolytes so strongly diffracts light at wavelengths near its stop band maxima. The large cavities result in diffraction of light at wavelengths relative to the size and refractive index of the materials. Bragg's law can be used to determine the optical characteristics of the material.

$$\lambda_p = 2d_{hkl}\sqrt{[n_s^2 f + n_f^2(1-f)]} \qquad (3.49)$$

where λ_p is the maximum wavelength of the reflected peak (the position of the photonic band gap), d_{hkl} is the interplanar spacing between hkl planes, n_s is the refractive index of the spheres and n_f is the refractive index of the framework.

The ordered, periodic structure results in periodic modulation of the dielectric constant perpendicular to the substrate surface, along the (111) direction of the film, which produces sharp, characteristic reflection peaks. Figure 3.88 shows the reflectance spectrum for 200 nm colloidal polystyrene particles at each of the stages from the preceding scheme. The reflection peak changes from 822 nm for the template assembly

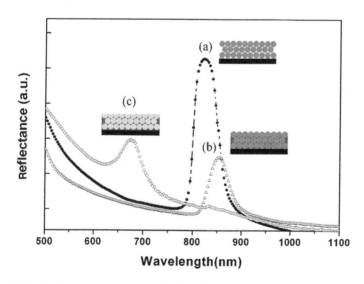

Figure 3.88 Reflectance spectra of the (a) 400 nm polystyrene colloidal array, (b) TiO_2 nanoparticle-infiltrated polystyrene colloidal array and (c) liquid-electrolyte-infiltrated TiO_2 inverse opal structure. Adapted from Kwak *et al.*, 2009[141]. Copyright (2009) Wiley-VCH Verlag GmbH & Co. KGaA

(corresponding to 345 nm polystyrene particles) to 854 nm when infiltrated with TiO_2 nanoparticles, to 675 nm when the calcined film is surrounded by liquid electrolyte, corresponding to the change in refractive index of the materials. Unlike typical scattering layers formed by large TiO_2 crystals, which scatter light in random directions, the transparency of the electrode outside of the stop-band is maintained.

In the example by Nishimura et al., inverse opal structured TiO_2 photonic crystals were made by the templating method introduced above. A conventional nanocrystalline TiO_2 layer was applied to a 3 μm thick photonic crystal layer and the resulting ordered, mesoporous TiO_2 framework was then sensitised with N719. The wavelength dependence of the photocurrent in the bilayer cell was compared with that of a conventional nanoparticle-based electrode of the same thickness (10 μm). The IPCE for cells assembled with each electrode are shown in Figure 3.89. The overall gain, integrated over the visible spectrum (400–750 nm), was about 30%.

Colodrero et al. have studied the effects of photonic crystals on the LHE and photocurrent of DSSCs closely.[344] They have attributed the improvement in current in the region of the stop band of the photonic crystal as due to the dielectric mirror phenomenon, a result of the periodic repeating structure. For a structure such as that shown above, where a photonic crystal is assembled above a dye-sensitised, nanoparticle-based electrode, the current is increased or decreased depending on the direction of illumination. When the light reaches the electrode through

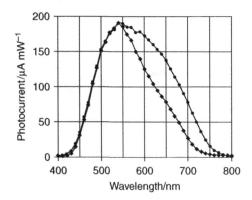

Figure 3.89 Wavelength dependence of the short-circuit photocurrent in the bilayer electrode (upper curve) and the conventional nanocrystalline TiO_2 photoelectrode (lower curve). The position of the stop band maximum in the bilayer electrode was 610 nm. Reprinted with permission from Nishimura et al., 2003[340]. Copyright (2003) American Chemical Society

the counter electrode side, first passing through the photonic crystal, the light in the region of the photonic band gap is reflected, causing a reduction in the IPCE in that range. When the light reaches the electrode first and then the photonic crystal, the IPCE is increased in the 'forbidden' range of the photonic band gap. An important finding is that the photonic crystals do not need to absorb light and do not themselves contribute to the photocurrent. A 15–30% increase in efficiency was achieved by a photonic crystal formed from repeating layers of SiO_2 and TiO_2.[344]

Multiple Junction DSSCs

Another way simultaneously to harvest the full solar spectrum is to assemble multiple subcells in a tandem device. This section will compare tandem solar cells consisting of a series of complete solar cells (unlike the tandem cells described above) stacked in the direction of the incident light. Each cell harvests a portion of the incident spectrum. To minimise thermalisation losses, the optimum configuration is to assemble the highest band gap solar cell at the top and the lowest band gap cell at the bottom, as indicated in Figure 3.90. A number of series-connected devices have been reported, typically using the red dye, N3 or N719, to collect the higher energy photons and the black dye for the longer wavelengths.

Bremner et al. have calculated the potential efficiencies for stacked arrangements of different numbers of solar cells. Assuming all the photons are absorbed, each photon produces one electron–hole pair, and there are no overlaps between the absorption spectra of the cells.[345] For the unconstrained array, each solar cell is connected to a separate circuit and allowed to operate at its maximum power point. For a single cell

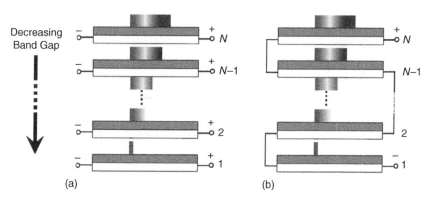

Figure 3.90 Tandem solar cell stack in the (a) unconstrained and (b) constrained cases. Adapted from Bremner et al., 2008[345] with permission from John Wiley & Sons, Ltd

under AM1.5 irradiation, operating at the optimum band gap (1.34 eV), the maximum efficiency is 33.7%. This increases to 46.1% for two cells operating with band gaps $E_1 = 0.94$ and $E_2 = 1.73$ eV and to 62.3% for an array of eight cells with band gaps ranging from $E_1 = 0.51$ to $E_8 = 2.57$ eV. For the constrained case, the cells are connected in a circuit in parallel or in series. When the cells are connected in series the currents need to be matched. The maximum efficiency becomes 45.7% for two cells with band gaps $E_1 = 0.94$ eV and $E_2 = 1.60$ eV, and 61.4% for an array of eight cells with band gaps ranging from $E_1 = 0.51$ to $E_8 = 2.35$ eV. With concentrated light, the maximum efficiency can reach the thermodynamic limit and for an array of eight cells in an unconstrained solar cell stack, Bremner et al. calculated a value of 77.6%.[345]

Parallel Connection   In the parallel connection (Figure 3.91), the current density is the sum of the current densities of the top and bottom cells, but the photovoltage is supplied by the lower-voltage cell. Typically, the top cell is sensitised with N719, to absorb the shorter wavelength light, and the bottom cell is sensitised with the black dye, to absorb the longer wavelength light transmitted by the top layer. Large, light-scattering rutile particles may be incorporated in the lower cell to increase the response at longer wavelengths. An example of the IPCE spectra and $J–V$ characteristics for a parallel tandem device along side the individual DSSCs are shown in Figure 3.92.[346] The tandem cell shows an extended spectral response compared with the individual cells, leading to about 20% higher photocurrent. In this example the V_{OC} of the tandem cell was similar to that

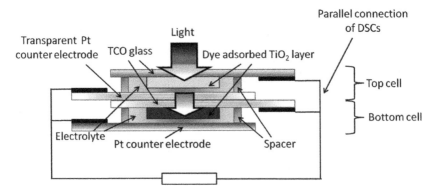

Figure 3.91 Structure of a parallel-connected tandem DSSC. Reprinted with permission from Kubo et al., 2004[346]. Copyright (2004) Elsevier

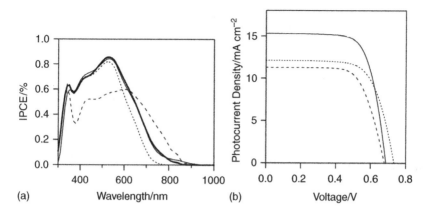

Figure 3.92 (a) Measured (bold solid curve) and calculated (solid curve) IPCE spectrum for the tandem DSSC using N719 and black dye and measured IPCE spectra for single N719 (dotted curve) and black dye (dashed curve) cells. (b) J–V characteristics for the tandem DSSC (solid curve) and the single DSSCs, N719 (dotted curve) and black dye (dashed curve) measured under AM 1.5, 100 mW cm^{-2} irradiation. Reprinted with permission from Kubo *et al.*, 2004[346]. Copyright (2004) Elsevier

of the black-dye-sensitised cell. The overall conversion efficiency for this tandem device was 7.6% (J_{SC} = 15.9 mA cm^{-2}, V_{OC} = 0.69 V, FF = 0.70).

Dürr *et al.* used different sized TiO$_2$ particles to optimise the performance of a tandem cell, with smaller particles (14 nm) in the transparent layer facing the incoming light and larger particles (20 nm and 400 nm) in the lower layer to scatter the light and lengthen the optical path for the incoming light.[347] The transparency of the first Pt counter electrodes was 70%. The second counter electrode was made of a Pt mirror. For this tandem cell, the V_{OC} = 690 mV was lower than that for the higher performing cell (740 mV), but the J_{SC} = 21.1 mA cm^{-2}, approximately the sum of the J_{SC}s of two separate cells (top: J_{SC} = 16.3 mA cm^{-2}; *bottom*: J_{SC} = 5.4 mA cm^{-2}). The conversion efficiency of the tandem cell was 10.5% at V_{max} = 545 mV.

Series Connection Whilst a parallel-connected tandem solar cell can successfully be used to increase the photocurrent, they cannot utilise the higher energy of the shorter-wavelength photons at the same time as harvesting the longer-wavelength photons. In series-connected tandem DSSCs, the photovoltage becomes the sum of photovoltages of the top and bottom cells, but the current density of both the top and bottom cells should be matched. The top cell should absorb the higher-energy photons, and therefore have a high voltage, and the bottom cell should absorb the lower-energy photons and add to the photovoltage.

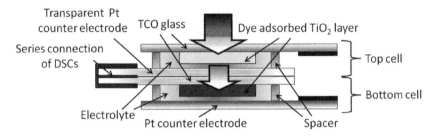

Figure 3.93   Structure of series-connected tandem DSSC. Reprinted with permission from Yamaguchi *et al.*, 2009[348]. Copyright (2009) Elsevier

Figure 3.93 shows an example of a series-connected tandem DSSC, reported by Yamaguchi *et al.*[348] As for the parallel-connected tandem cells described above, the top cell was sensitised with N719, and the bottom cell was sensitised with the black dye. The photoconversion efficiency was 10.4% (J_{SC} = 10.8 mA cm^{-2}, V_{OC} = 1.45 V, FF = 0.67).

The J-V characteristics and the IPCE spectra of the component cells are shown in Figure 3.94. The IPCE spectrum for the tandem cell is approximately the superimposed spectra of the individual N719-sensitised and black-dye-sensitised devices. The photocurrents of the individual cells (N719 J_{SC} = 11.3 mA cm^{-2}, black dye J_{SC} = 10.1 mA cm^{-2}) were approximately matched and were maintained in the tandem cell (J_{SC} = 10.8 mA cm^{-2}). The V_{OC} of the tandem cell (1.45 V) was almost the sum of the V_{OC}s of the individual cells (N719 V_{OC} = 0.77 V, black dye V_{OC} = 0.66 V).

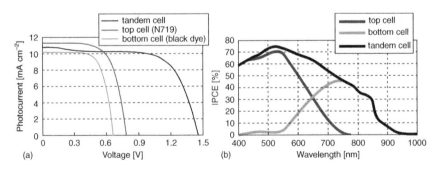

Figure 3.94   (a) J-V curves and (b) IPCE spectra for the tandem DSSC and individual cells. Reprinted with permission from Yamaguchi *et al.*, 2009[348]. Copyright (2009) Elsevier

A Dye-Sensitised Solar Tube An entirely different tandem cell configuration was reported by Usagawa *et al*.[349] The cells, consisting of a dye-stained porous TiO_2 layer, a gel electrolyte and two titanium charge collectors (anode and cathode), were assembled around a glass rod as shown in Figure 3.95. Light was introduced through the glass rod and absorbed by the dye-sensitised electrodes. The operating principle is the same as for flat electrodes. TiO_2 paste was coated on a glass rod (diameter 9 mm, length 30 mm) and the glass rod was baked at 450 °C for 30 min. The process was repeated to prepare a porous TiO_2 electrode with 6 μm thickness. A mixture of ZnO and TiO_2 particles in ethanol was sprayed on the porous TiO_2 layer, followed by another sintering step at 450 °C. Ti was sputtered on the layer and then the ZnO crystals were removed by dipping the rod into a solution consisting of 40% acetylacetone and 60% methanol to make a porous Ti electrode. The rod was stained and then covered with a 35 μm thick porous poly(tetrafluoroethylene) (PTFE) film and a 50 μm thick Ti sheet. The electrolyte (500 mM LiI, 50 mM I_2, 580 mM t-butylpyridine, 6:4 w/w methylpropylimidazolium iodide and ethylmethylimidazolium dicyanoimide in acetonitrile) solution was injected into the porous PTFE film.

An N719-senstised solar rod gave a conversion efficiency of 1.33% (J_{SC} = 5.42 mA cm^{-2}, V_{OC} = 0.65 V, FF = 0.37), much lower than cells fabricated on a flat substrate because almost all the light passed through

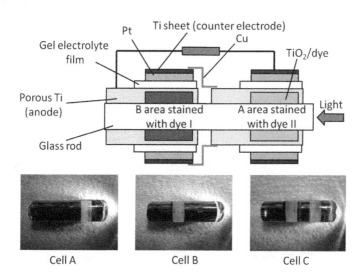

Figure 3.95 Tandem cell structure fabricated on a glass rod. Reproduced from Usagawa *et al*., 2009[349] with permission from the Japan Society of Applied Physics

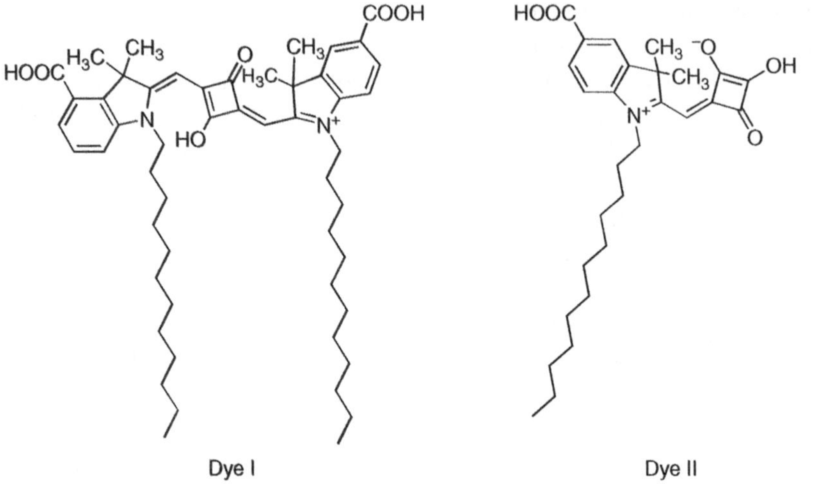

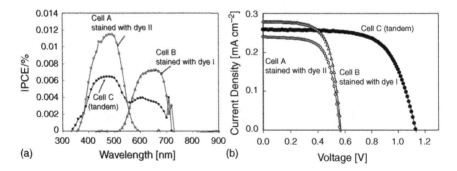

Figure 3.96 Structures of the dyes used by Usagawa *et al.* to sensitise a 'solar tube'[349]

the glass rod. A tandem cell was fabricated by staining two cells, A and B, with 'dye II' (λ_{max}: 429 nm) and 'dye I' (λ_{max}: 646 nm), respectively (Figure 3.96). Cell A was connected to cell B by using a Cu sheet. The IPCE spectra of the individual and tandem cells are shown in Figure 3.97. The IPCE spectrum for the tandem DSSC contained peaks corresponding to the absorption spectra of both the dyes. Also shown in Figure 3.98 are the PV properties of single and tandem cells. The V_{OC} of the tandem cell (1.13 V) was the sum of the V_{OC} values of cell A (0.57 V) and cell B (0.57 V).

Figure 3.97 (a) IPCE curves for single and tandem cells. (b) PV performances for single and tandem cells. Reproduced from Usagawa *et al.*, 2009[349] with permission from the Japan Society of Applied Physics

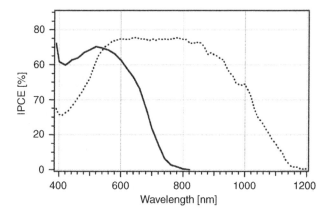

Figure 3.98 Photocurrent action spectra measured separately for a DSSC (solid line) and a CIGS cell (dotted line). Reprinted with permission from Liska *et al.*, 2006[350]. Copyright (2006) American Institute of Physics

Whilst the efficiency was lower than for tandem cells using the stacked array of flat DSSCs, the photovoltage is impressive, and the architecture is interesting and demonstrates the versatility of the DSSC assembly which can be adapted for different substrate geometries. Filling the tube with scattering material may improve the light harvesting. Alternatively, it may be more useful to have the absorption on the outside of a tube. This would further remove the dependence of the efficiency on the angle of incidence of the radiation and increase the active area.

$TiO_2/Cu(In, Ga)Se_2$ Thin-Film Tandem Cells

A thin-film tandem cell combining a DSSC top cell with a thin-film CIGS bottom cell in a series-connected, double-junction device has been reported giving conversion efficiencies greater than 15%.[350] High-energy photons are absorbed by the top DSSC and transmitted low-energy photons can be absorbed in an underlying CIGS cell. As mentioned above, the band gap of the CIGS semiconductor can be tuned but the high-efficiency devices have a band gap of *ca* 1.25 eV, which complements the HOMO-LUMO gap of the N719 sensitiser, 1.65 eV. The IPCE spectra of the individual cells are shown in Figure 3.98. The DSSC shows a strong response in the UV, blue, and green wavelength domains. By contrast the CIGS cell exhibits high external quantum efficiencies in the red and near-infrared parts of the spectrum extending from 700 to 1150 nm where the DSSC is insensitive to light.

The DSSC was prepared by screen printing a transparent single layer 12 μm thick consisting of 20 nm-sized TiO_2 particles on conducting (FTO) glass. The CIGS cell was fabricated by depositing polycrystalline layers of ZnO:Al/ZnO/CdS/CIGS/Mo on a soda-lime glass substrate. When illuminated with simulated AM 1.5 sunlight, the CIGS cell alone gave a solar-to-electric power conversion efficiency $\eta = 13.9\%$ ($J_{SC} = 26.8$ mA cm^{-2}, $V_{OC} = 699$ mV, $FF = 0.744$). When the cells were stacked above each other, the CIGS cell gave a solar-to-electric power conversion efficiency $\eta = 7.28\%$ ($J_{SC} = 14.3$ mA cm^{-2}, $V_{OC} = 0.65$ mV, $FF = 0.77$) and the DSSC gave a solar to electric power conversion efficiency $\eta = 8.18\%$ ($J_{SC} = 13.66$ mA cm^{-2}, $V_{OC} = 0.798$ mV, $FF = 0.75$). When the two cells were connected in series, the photovoltaic efficiency was 15.09% ($J_{SC} = 14.05$ mA cm^{-2}, $V_{OC} = 1.45$ V, $FF=0.74$) (Figure 3.99).

In this and each of the tandem cells assembled on flat substrates described above, optical losses arise at the interfaces such as the glass (FTO or Pt/FTO) substrates. In this example, the FTO glass absorbs in the near infrared wavelength region above 1000 nm, thereby reducing significantly the photocurrent drawn by the bottom cell. Also, if the assembly was to be scaled up the cost of the multiple layers of glass would be

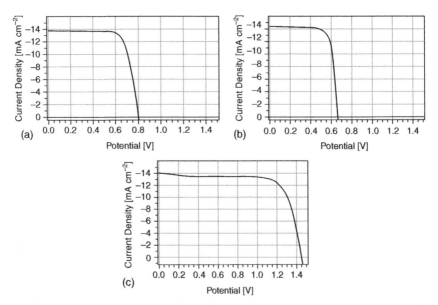

Figure 3.99 Photocurrent density–voltage characteristics under AM 1.5 full sunlight (100 mW cm^{-2}) for a two-wire tandem DSSC/CIGS cell. (a) J–V curve for the DSSC top cell, (b) J–V curve for the CIGS bottom cell and (c) J–V curve for the two-terminal DSSC/CIGS tandem cell. Reprinted with permission from Liska et al., 2006[350]. Copyright (2006) American Institute of Physics.

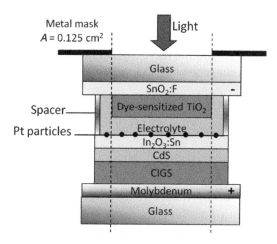

Figure 3.100 Schematic of the monolithic device structure using a dye-sensitised mesoporous TiO$_2$ film as top absorber of visible light and a CIGS bottom absorber of transmitted near-infrared light. Reprinted with permission from Wenger *et al.*, 2009[351]. Copyright (2009) American Institute of Physics

significant. Therefore the design was modified and a monolithic device was fabricated.[351] Figure 3.100 shows the structure of the monolithic DSSC/CIGS tandem device.

A mesoporous C101 dye-sensitised TiO$_2$ film was directly sandwiched with a platinised CIGS solar cell using a spacer, avoiding the back glass electrode commonly used in the DSSC.[352] The void was filled through a hole in the top electrode with an acetonitrile-based electrolyte containing the I$^-$/I$_3^-$ redox couple. The front contact of the CIGS cell, a 600 nm thick layer of In$_2$O$_3$:Sn (ITO), was covered with a transparent layer (1 nm) of sputtered Pt particles. The charges generated in the subcells recombine at the catalytic Pt particles on the electrolyte/ITO interface, *i.e.* the oxidised I$_3^-$ ions react with electrons from the bottom cell. Therefore, the current densities of the subcells must be matched to minimise electronic losses. The current density of the DSSC can be tuned with choice of the sensitiser, by variation in the optical band gap, film thickness and by variation in the optical path length. The current density of the CIGS cell can be tuned with variation of the band gap by changing the In/Ga ratio in the absorber.

The conversion efficiency of the monolithic device, $\eta = 12.2\%$, slightly exceeded the performance of the CIGS cell (11.6%), but was a little lower than the stacked cell described above. The V_{OC} of the tandem device was close to the sum of the V_{OC}s of the individual DSSC and CIGS cells, confirming the series connection of the subcells. The device suffered from electric shunts and the performance of the device (V_{OC} and *FF*) degraded quickly due to a corrosion of the CIGS cell by the electrolyte percolating through pinholes. Nonetheless, these examples demonstrate again the versatility of the DSSC and the increase in efficiency that can be achieved by combining different types of PV devices.

3.2.3.3 Extremely Thin Absorber (eta) Solar Cells

The *eta* (extremely thin absorber) solar cell uses the concepts behind DSSCs but is made only from solid-state inorganic compounds. The concept was described by Kaiser *et al.* in 2001.[353] A semiconductor absorber, one to tens of nanometres thick, is sandwiched between inter-penetrating, nanostructured electron and hole conductors to form a p-i-n junction (Figure 3.101). The absorber is located in the electric field between the two semiconductors. By using very thin layers, the transport path for excited charge carriers in the absorber before they are separated can be reduced substantially and there are lower restrictions on the purity or electron transport properties of the material. Moreover, the increase of the photon path length as a result of scattering at the internal surfaces improves the light absorption.[355] Because less material is required, materials that would otherwise be expensive or not abundant may be used at reasonably low cost.

n-Type windows are typically made from nanostructured TiO_2 or ZnO.[356] A number of absorbers including Se (IPCE = 16.1%, $\eta = 0.13\%$, V_{OC}= 600 mV, J_{SC} = 3.0 mA cm^{-2} at 800 W m^{-2}),[357] CdS (0.4%),[357] In_2S_3 (3.4%),[358,359] $CuInS_2$ (IPCE = 80%, $\eta = 4\%$, J_{SC} = 18 mA cm^{-2}, V_{OC} = 0.49 V, FF = 0.44),[353,360] PbS,[361] a-Si:H (1%),[361] CdTe,[362,363] CdSe (IPCE = 60%, 400–500 nm, APCE $\sim$ 100%, $\eta = 1.3\%$, J_{SC} = 2.3 mA cm^{-2}, V_{OC} = 0.86 V, FF = 0.651 sun),[364]

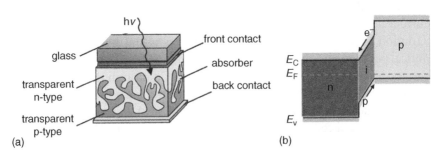

Figure 3.101 (a) Scheme of the *eta* solar cell and (b) band diagram. Reproduced from Tena-Zaera *et al.*, 2006[354] with permission from Elsevier

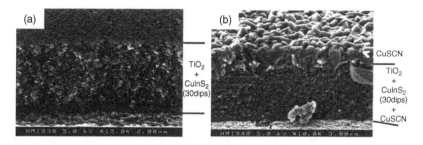

Figure 3.102 Cross-section SEM micrographs of devices on the base of nanoporous TiO_2 taken at a tilted angle. (a) TiO_2 substrate covered by CIS prepared by 30 dipping cycles; (b) the whole structure including CuSCN. Reprinted with permission from Kaiser *et al.*, 2001[353]. Copyright (2001) Elsevier

$HgCdTe$,[365] $Cu_{2-x}S$(0.06%),[366] and Sb_2S_3[367] have been used. p-Type materials such as β-CuSCN and ZnTe fill the voids and complete the device. The absorbers are typically deposited by electrodeposition or chemical deposition. SEM images of some examples are shown in Figures 3.102 and 3.103.

The working mechanism of the *eta* cell, can be summarised by four main processes: light harvesting, by light excitation of the absorber; charge separation, induced by quenching of the excited state through electron injection in the conduction band of a metal oxide; charge collection, by electron percolation through the n-type metal oxide; and absorber regeneration, the photo oxidised absorber is regenerated by capture of an electron from the valence band of a p-type hole conductor. The two charges, separated by the internal electric field and the band alignment diffuse through the respective semiconductors to reach the front and back contacts. In order for an *eta* cell to function efficiently, the band alignment must be optimised so that charge separation and transport occur

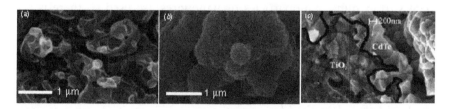

Figure 3.103 Electron micrographs of (a) the TiO_2 substrate and (b) the TiO_2 substrate covered with a 150 nm CdTe layer deposited in electrodeposition and (c) a cross-sectional micrograph at the TiO_2/CdTe interface. Reprinted from Ernst *et al.*, 2003[368] with permission from IOP Publishing Ltd

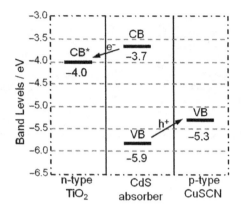

Figure 3.104 Energy band levels of components of the TiO_2/CdS/CuSCN cell. Reprinted with permission from Larramona *et al.*, 2006[364]. Copyright (2006) American Chemical Society

quickly to limit recombination of the charge carriers (Figure 3.104). This can be achieved by tuning the particle size (see Section 3.2.3.4) or by doping or alloying the absorber semiconductor.[361,363,368]

The *eta* thin film approach reduces internal recombination by optimising the ratio of the electron diffusion length (L_n) to the absorber thickness ($1/\alpha$). L_n is given by:

$$L_n = (\tau_n D_n)^{0.5} \tag{3.50}$$

where τ_n is the electron lifetime and D_n is the electron diffusion coefficient.

If $L_n/(1/\alpha) < 1$, the charge carriers will recombine before they reach the collector electrode. If $L_n/(1/\alpha) > 1$, most of the charge carriers produced by light will reach the collector electrode, producing electrical current. Since this condition can easily be fulfilled by varying the light absorption length, α, an absorber with low electronic quality can be used.[369] As for DSSCs, thin films, nanoparticles or nanowires are used so there is no band bending in the semiconductors. An alternative transport mechanism occurs similar to that in the DSSC.

A relatively high efficiency device was prepared by Itzhaik *et al.* using Sb_2S_3 (stibnite) as the absorber.[367] The band gap of Sb_2S_3, 1.7–1.8 eV, is similar to that of CdSe. The device was assembled with a compact TiO_2 blocking layer, deposited by spin coating; a 1–2 µm TiO_2 porous layer (25 nm particles); a 1 nm buffer layer of $In_x(OH)_yS_z$ to prevent oxidation of the absorber by TiO_2, deposited by chemical bath deposition (CBD) from $In_2(SO_4)_3$ and thioacetamide; a 5–10 nm layer of Sb_2S_3 deposited by

CBD from $SbCl_3$, $Na_2S_2O_3$ and annealed at 300 °C in a nitrogen atmosphere and cooled in air to allow a passivation layer of S_2O_3 to form; CuSCN, deposited from a solution in 1:1 di-n-propyl sulfide and PrS infiltrated into the pores at 65 °C; a 80 nm gold contact was deposited by evaporation. KCN treatment prior to CuSCN deposition was found to increase the short-circuit current as a result of doping with excess SCN^-. The device was aged for a few days. The IPCE matched the optical spectra of Sb_2S_3 and the external quantum efficiency was as high as 80% between 450 nm and 520 nm. The reported efficiency was 3.37% at 1 sun (J_{SC} = 14.1 mA cm^{-2}, V_{OC} = 490 mV, FF = 0.488).

    eta cells have also been prepared using ZnO nanowires (density 10^9 cm^{-1}) as the n-type window and CdTe (E_g = 1.54 eV as deposited) and CdSe as the absorber.[206,354,370] ZnO is transparent (E_g = 3.35 eV) exhibits good transport properties (carrier concentration = 10^{20} cm^{-3}, mobility = 23 cm^2 V^{-1} s^{-1}) and can be grown as free-standing, single crystal, ZnO nanowires, which are not in contact with each other and so there is no lateral conduction, enlarging the surface area 10 to 100 times compared with flat substrates. This means they require 10 to 100 times lower absorber thickness to harvest the same amount of light, making them an ideal choice as an anode material for *eta* solar cells. Moreover they have energy levels in good alignment with CuSCN, typically used as the p-type cathode (Figure 3.105).

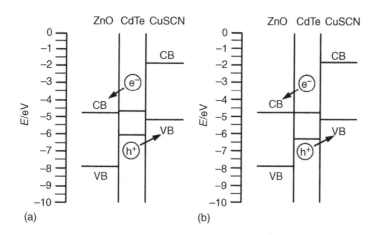

Figure 3.105 Band diagram of the (a) ZnO/CdTe/CuSCN and (b) ZnO/CdTe/CuSCN heterostructures. CB and VB are the conduction and valence bands, respectively. Reproduced with permission from Tena-Zaera *et al.*, 2006[354]. Copyright (2006) Elsevier and Tena-Zaera *et al.*, 2005[370]. Copyright (2005) Elsevier

The band alignment at the interfaces promotes the electron injection from absorber (CdSe or CdTe) to ZnO and hole transport from the absorber to CuSCN, separating the charge carriers. The glass/SnO_2:F substrate was covered by a 150 nm thin continuous layer of ZnO, deposited by spray pyrolysis (ZnOsp) of zinc acetate and acetic acid at 450 °C, which prevents short-circuiting between SnO_2:F and CuSCN. ZnO nanowire films (150 nm in diameter, 2 μm in height) were deposited by electrodeposition from $ZnCl_2$ (ZnOed). About 65 nm thick CdTe was deposited on the ZnO by metal organic chemical vapour deposition (MOCVD)[354] or 30–40 nm CdSe by electrodeposition from an aqueous selenosulfate solution.[206] If the surface reaction rate is slower than the arrival of the precursor molecules, the precursor molecules reach the bottom of the film before the surface reactions take place and the CdTe deposition is uniform in all parts of the nanowires. A CuSCN layer by a chemical solution deposition technique fills the voids, and a graphite contact to this completes the cell.[370] SEM images of the coated substrate, nanowires and the complete device are shown in Figure 3.106.

The J–V characterisation for the CdSe and CdTe devices are shown in Figure 3.107. For the CdTe device, a promising open-circuit voltage (V_{OC} = ca 200 mV) was observed but the short circuit current density (J_{SC} = ca 0.03 mA cm^{-2}) and the fill factor (FF = ca 0.28) were very low. The CdSe device performed much better, at 1/3 sun: V_{OC} = 460 mV, FF = 0.42, J_{SC} = 2.6 mA cm^{-2}, η = 1.5%. After 1 week the current increased and an efficiency of η = 2.3% was measured. The quantum efficiency (Φ) matched the absorption spectrum of the absorbers. For CdSe the internal Φ = 28% and external Φ = 25% at ca 500 nm. The electron lifetime measured from transient photovoltage decay under a bias light of 1 sun was 14 μs. The decay time of the transient photocurrent at short circuit was also 14 μs, implying that recombination was limiting charge collection in the devices.[206]

The highest reported efficiency for an *eta* device was obtained by Belaidi *et al.* using an In_2S_3 absorber layer on ZnO nanorods with CuSCN as the p-type hole conductor and a compact ZnO blocking layer as for the devices described above.[359] Gold contacts were deposited on the CuSCN. However, the 100 nm compact layer was deposited by sputtering and the 0.1 μm diameter nanorods by CBD from an alkaline solution of $Zn(NO_3)_2$ at 80 °C. The In_2S_3 layer was deposited by spray pyrolysis of $InCl_3$ followed by reaction with H_2S. The performance of the device was optimised by altering the nanorod length and morphology and the thickness of the absorber layer, and a maximum efficiency of 3.4% was reported.

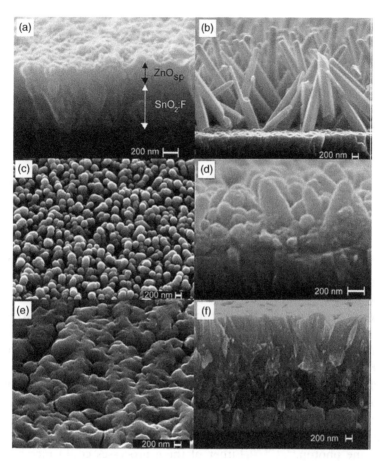

Figure 3.106 SEM images of: (a) cross-section of glass/SnO$_2$:F/ZnOsp sample; (b) cross-section of glass/SnO$_2$:F/ZnOsp/ZnOed sample; (c, d) glass/SnO$_2$:F/ZnOsp/ZnOed/CdTe sample (c, planar view; d, cross-section); (e, f) glass/SnO$_2$:F/ZnOsp/ZnOed/CdTe/CuSCN sample (e, planar view; f, cross-section). Reprinted with permission from Tena-Zaera *et al.*, 2005[370]. Copyright (2005) Elsevier

3.2.3.4 Quantum-Dot-Sensitised Solar Cells

Quantum confinement is manifested when carriers in a semiconductor are confined by potential barriers to regions of space less than or equal to the de Broglie wavelength or the Bohr radius of excitons in the bulk material.[231] Quantum dots (QDs) are confined in all three dimensions. For example, bulk CdSe has a band gap of 1.7 eV, the effective mass of an electron is 0.44 m_0, the effective mass of a hole is 0.13 m_0 and the Bohr radius of the CdSe exciton is 5.6 nm. Particles

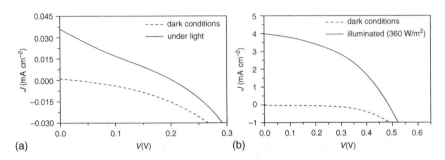

Figure 3.107 (a) *J–V* curves of ZnO/CdTe/CuSCN.; Reprinted with permission from Lévy-Clément *et al.*, 2005[206]. Copyright (2005) John Wiley & Sons, Ltd. (b) *J–V* curves of a ZnO/CdSe/CuSCN (after ageing) *eta* solar cell. Reprinted with permission from Tena-Zaera *et al.*, 2006[354]. Copyright (2006) Elsevier

of a similar diameter will exhibit quantum confinement and the absorption spectrum will be significantly blue-shifted relative to that of the bulk material.

Several properties of QDs that make them suitable for applications in solar cells include high extinction coefficients from quantum confinement effects, large intrinsic dipole moments which encourage rapid charge separation and size tuneable band energies which can be adjusted to match solar spectrum or optimise band alignment for charge transfer between semiconductor nanostructures. Much of the recent interest comes from the anticipation that QDs can be used to capture 'hot electrons' and the observation that multiple charge carriers can be generated from one photon.[371–380] Either of these processes could increase the maximum theoretical efficiency to 66%. The following sections will contain examples of each of these processes which demonstrate the progress made towards QD-sensitised solar cells.

3.2.3.4.1 Charge Transfer and QD Sensitisation of Transparent Semiconductors

Nanostructured electrodes of wide band gap semiconductors can be coated with QDs by a number of methods; direct adsorption of QDs on nanostructured semiconductor networks can be achieved by CBD, electrodeposition or spin coating. Alternatively, they can be anchored using linkers, such as mercaptopropionic acid (MPA), which have an acid function which binds to metal oxides, such as TiO_2 and ZnO, and a thiol (or amine) group, which binds to CdSe or CdS. When linkers such as MPA are used, a uniform coverage throughout the internal and external

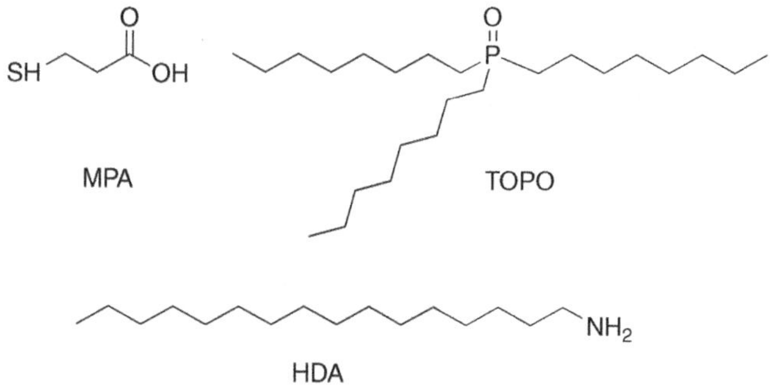

Figure 3.108 Examples of surface-active molecules used in QD synthesis and sensitisation

surface of the film is obtained. QDs are usually capped with TOPO (tri-n-octylphosphine oxide) or hexadecylamine (HDA) to control particle growth, passivate the surface or increase solubility in organic solvents (Figure 3.108). These can be exchanged with linkers with thiol anchors such as MPA.

The loading of the QDs on the nanostructured film can easily be quantified by UV-visible absorption spectroscopy. The spectrum of the sensitised film should match that of the QDs in solution, super-imposed on the spectrum of the substrate, if the native quantisation property of the CdSe nanocrystals is maintained after assembly on the TiO_2 surface. The electrodes can be sandwiched with a counter electrode and filled with a redox mediator or assembled with a solid-state hole conductor to make devices in the same way as dye-sensitised analogues (Figure 3.109).

As in the examples of the inorganic absorbers described in Section 3.2.3.3 on *eta* solar cells, QDs can inject charge into the conduction band of wide band gap metal oxides, such as TiO_2, SnO_2 and ZnO, providing the photoexcited electron in the conduction band of the QD is higher in energy than that of the semiconductor substrate.[381–388] CdSe sensitisation of TiO_2 has been well documented.[374,375] QDs are strongly emissive compared with bulk semiconductors where the excitons are not confined and tend to relax non-radiatively. On deposition on a semiconductor such as TiO_2, quenching of this emission occurs, providing evidence for charge transfer (Figure 3.110). If the QD is deposited on an insulator such as ZrO_2, charge transfer cannot occur, since the conduction band of the bulk material is higher than that of the QD, and no quenching is observed.[389]

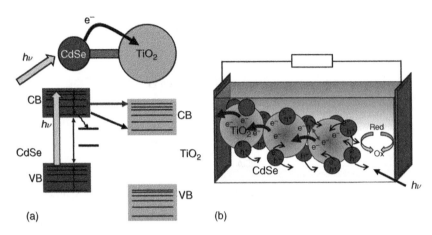

Figure 3.109 Principle of operation of the quantum-dot-sensitised solar cell (QDSSC). Charge injection from excited CdSe QDs into TiO_2 nanoparticles (a) is followed by collection of charges at the electrode surface (b). The redox electrolyte (*e.g.* sulfide/polysulfide) scavenges the holes and thus ensures regeneration of the CdSe. Reprinted with permission from Kamat, 2008[372]. Copyright (2008) American Chemical Society

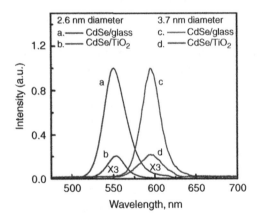

Figure 3.110 Emission spectra of 2.6 and 3.7 nm diameter CdSe QD film deposited on glass and FTO/TiO_2 (nanoparticle) films. Reprinted with permission from Kongkanand *et al.*, 2008[381]. Copyright (2008) American Chemical Society

The rate of charge transfer between QDs and nanostructured semiconductors can be studied by transient absorption spectroscopy in the same way as for dyes in DSSCs (Figure 3.111).[372] Band-gap excitation results in a bleach for the excitonic band which represents charge separation in

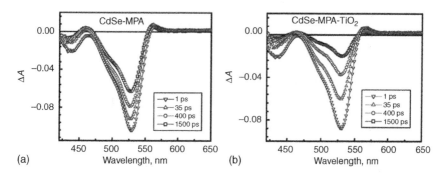

Figure 3.111 Time-resolved spectra recorded following laser pulse excitation of 3 nm CdSe QDs in the (a) absence and (b) presence of TiO$_2$. Adapted with permission from Robel *et al.*, 2006[383]. Copyright (2006) American Chemical Society

semiconductor nanocrystal as electrons and holes accumulate in the conduction band and valence band.

$$CdSe + h\upsilon \rightarrow CdSe(e_p + h_p) \rightarrow CdSe(e_s + h_s) \quad \text{Charge separation}$$

$$(3.51)$$

$$CdSe(e_s + h_s) \rightarrow CdSe + h\upsilon' \quad \text{Charge recombination (radiative)} \quad (3.52)$$

$$CdSe\ (e_s + h_s) + TiO_2 \rightarrow CdSe\ (h) + TiO_2(e) \quad\quad (3.53)$$

where s is s state and p is p state of holes and electrons.

Recombination results in luminescence or the recovery of the bleach. When in contact with another semiconductor, *e.g.* TiO$_2$, the emission is quenched as electrons are transferred from the conduction band of the CdSe to the conduction band of the semiconductor providing the conduction band is energetically lower than that of the CdSe QD. The recovery of the bleach occurs at a faster rate than when the QD is in solution in the absence of TiO$_2$. The electron-transfer rate constant (k_{et}) can be estimated from the differences between these two bleaching recovery lifetimes.

$$k_{et} = 1/\tau_{(CdSe+TiO_2)} - 1/\tau_{CdSe} \quad\quad (3.54)$$

The performance of the QD-based devices can be compared in terms of the IPCE (Figure 3.112). For the devices illustrated, the IPCE

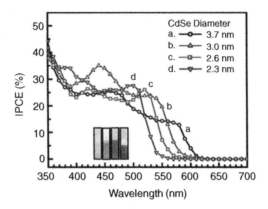

Figure 3.112 Photocurrent action spectra recorded in terms of incident photon-to-charge carrier generation efficiency (IPCE) of FTO/TiO$_2$(NP)/CdSe (electrodes are shown in the inset). The individual IPCE responses correspond to (a) 3.7, (b) 3.0, (c) 2.6 and (d) 2.3 nm diameter CdSe QDs anchored on nanostructured TiO$_2$ films. 0.1 M Na$_2$S solution was used as redox electrolyte. Reprinted with permission from Kongkanand et al., 2008[381]. Copyright (2008) American Chemical Society

spectrum matches that of the absorption spectrum of the QD. Since smaller particles have a higher energy conduction band than smaller particles, the electrons in their excited state are more energetic and therefore are injected into TiO$_2$ at a faster rate. It is hoped that QDs will enable 'hot electron' (electrons at higher energy than the lowest energy state) to be collected in this way. However, as yet no evidence for hot electron injection has been published. While the increase in driving force increases the rate of electron transfer to TiO$_2$, increasing the photocurrent, the spectral blue shift results in a reduction in the total light harvesting efficiency because fewer low-energy photons can be utilised.

Charge transfer between semiconductor nanostructures has been compared for a number of variables such as the nature and morphology of the materials, the method of attachment (e.g. deposition time for chemical deposition), the linker molecule, the environment (e.g. electrolyte composition) and the surface properties. The efficiency of QD-sensitised films prepared by CBD of the material directly onto the semiconductor substrate has been shown to be strongly dependent on the deposition time.[378,384] The tendency is for the IPCE to increase until an optimum is reached, after which the efficiency falls for longer deposition times. This is accompanied by an increase in intensity and a red shift of the absorption or IPCE spectra as the material is deposited.

In the case of PbS on TiO_2, (optimum IPCE = 81% at 1 mW cm^{-2}) the decrease in efficiency is attributed to an increasing crystal size and corresponding lowering of the conduction band energy with increasing material deposition, until the driving force becomes too low for charge transfer to occur ($E_{CB} < 1.3$ eV). For CdSe on SnO_2 the driving forces are larger since the conduction band energy is lower for SnO_2 than for TiO_2. However, as more material is deposited, the photons are absorbed closer to the surface of the film, further away from the charge collector. This means the electrons have to travel further in the nanostructured film, and so have a higher probability of recombination with the redox mediator. Increasing the thickness of the QD film also leads to increased losses in the electrode since the increased distances for the carriers to travel means there is a higher probability of exciton recombination in the QDs before the electron and hole are separated; more CdSe-CdSe interfaces means there is more chance of trapping at the interfaces; electrons have to be transported through the QD network before injection into the SnO_2 increasing the probability of electrons recombining with the holes in the CdSe or the redox mediator.

To avoid these problems, organic linker molecules have been used to assemble pre-formed QDs on bulk semiconductor substrates. Guijarro *et al.* compared the effect of adsorption time for CdSe QDs directly assembled on TiO_2 and those attached *via* a linker (MPA) on the IPCE of QD-sensitised devices (Figure 3.113).[387] For the QDs deposited directly on TiO_2-the trend as described above was observed. However, for QDs attached *via* an organic linker, the IPCE did not decrease after an optimum deposition time since aggregation was inhibited. However, the IPCEs were not as high for the linked QDs possibly due to a slower injection rate.

For CdSe QDs assembled on TiO_2 using organic linkers, the electron-transfer rate ($k_{et} = 0.073$–1.95×10^{11} s^{-1}) is proportional to the linker length and therefore tunnelling distance between the semiconductors.[383,386] The nature of the linker also affects the charge-transfer rate. Mora-Seró *et al.* obtained the highest IPCE in their study with a cystein-linked system (23%) followed by thioglycolic acid (10%) then the longer MPA (4%).[386] Cystein (Figure 3.114) has an amine function which can either bind to the CdSe, shortening the distance further, or the electron-donor property can mediate charge transfer. Evidence for the cystein bridge trapping the electron was observed by way of a signal for the cystein radical in the EPR spectrum. In a three-electrode set-up, measured at 1 sun, the efficiency $\eta = 0.92\%$ ($V_{OC} = 0.620$ V, $J_{SC} = 2.67$ mA cm^{-2}, FF = 0.54). In a closed cell, the efficiency was lower

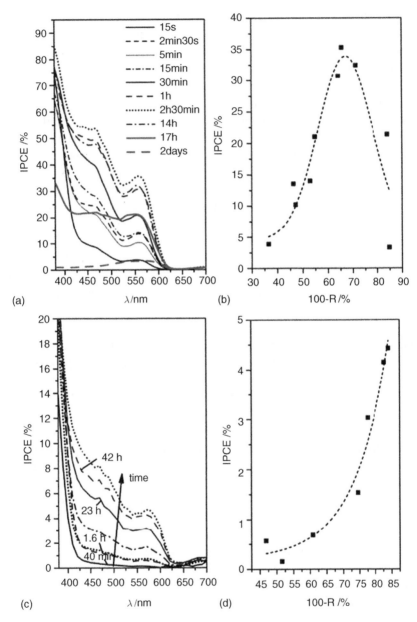

Figure 3.113 (a) IPCE spectra of TiO$_2$ nanoporous electrodes sensitised through direct adsorption of CdSe QDs with different coverage values. (b) IPCE at the excitonic peak vs 100-R/% at the excitonic peak. (c) IPCE spectra of TiO$_2$ nanoporous electrodes sensitised with MPA-attached CdSe QDs with different coverage values. (d) IPCE at the excitonic peak vs 100-R% at the excitonic peak. R, reflectance. Adapted with permission from Guijarro *et al.*, 2009[387]. Copyright (2009) American Chemical Society

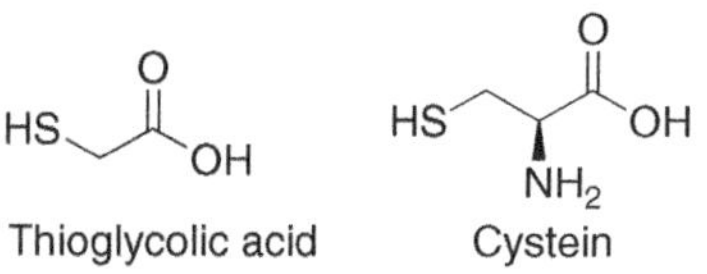

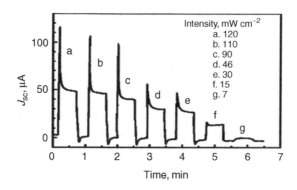

Thioglycolic acid   Cystein

Figure 3.114 Structure of thioglycolic acid and cystein used as linkers for QDs to TiO_2

(0.40–0.55%) because of increased charge-transfer resistance at the counter electrode, possibly as a result of adsorption of sulfur species.

3.2.3.4.2 Bulk Semiconductor Morphology

Sensitisation by CdSe QDs has been studied on a range of TiO_2 morphologies including nanoparticles, nanotubes and inverse opal.[383] Robel *et al.* observed an intensity-dependent, sharp rise in photocurrent on illumination followed by stabilisation at lower current, as shown in Figure 3.115 for nanoparticle TiO_2/CdSe electrodes. This was indicative of slow transport (ms–μs) of the electrons through the nanoparticle network compared with the ultrafast charge injection and accumulation in the TiO_2 nanoparticles.[383]

To improve the charge transport in the bulk semiconductor network, alternative materials and morphologies have been employed. Kongkanand *et al.* have shown that the maximum IPCE for CdSe of the same size (3 nm) assembled on TiO_2 nanoparticles and nanotubes differed by

Figure 3.115 Effect of excitation intensity on the stability of photocurrent generation following the visible excitation (>400 nm) of a FTO/TiO_2/CdSe electrode. The Na_2S concentration was 0.1 M. Excitation intensity was maintained at (a) 120, (b) 110, (c) 90, (d) 46, (e) 30, (f) 15 and (g) 7 mW cm^{-2}. Adapted with permission from Robel *et al.*, 2006[383]. Copyright (2006) American Chemical Society

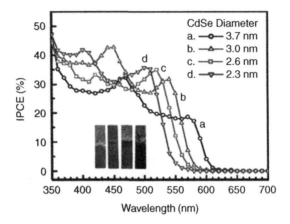

Figure 3.116 Photocurrent action spectra recorded in terms of incident photon-to-charge carrier generation efficiency (IPCE) of Ti/TiO$_2$(NT)/CdSe electrodes (electrodes are shown in the inset). The individual IPCE responses correspond to (a) 3.7, (b) 3.0, (c) 2.6 and (d) 2.3 nm diameter CdSe QDs anchored on nanostructured TiO$_2$ films. 0.1 M Na$_2$S solution was used as redox electrolyte. Reprinted with permission from Kongkanand et al., 2008[381]. Copyright (2008) American Chemical Society

around 10% (Figures 3.113 and 3.116). This was matched by a conversion efficiency for the electrodes using Na$_2$S electrolyte and a platinum counter electrode of 0.6% for the electrodes using TiO$_2$ nanoparticles and 0.7% for nanotubes at 80 mW cm^{-2}. The V_{OC} also differed, 600 mV for nanoparticles and 580 mV for nanotubes ($FF \sim 0.4$). They attribute the improvement in efficiency to better electron transport and longer electron lifetime in the TiO$_2$ nanotubes due to fewer grain boundaries and less surface defects than the nanoparticle system.[381]

Leschkies et al. substituted TiO$_2$ for ZnO nanowires which are known to posses fast charge-transfer properties by avoiding the particle–particle hopping of nanoparticle assemblies.[388] In this system, all the photogenerated electrons have a direct connection with the charge collector. The MPA linker was used to assemble a monolayer of 3 nm CdSe QDs over the nanowires. The triiodide/iodide redox mediator was used despite the instability of the QDs in this electrolyte. The ZnO nanowires were grown on FTO from Zn(NO$_3$)$_2$ and methenamine at 80–95 °C and were 2 μm in diameter and 12 μm in length (6–40 wires per μm). As for the TiO$_2$-based electrodes, the absorption properties of the films matched those of the QDs in solution and these were preserved on adsorption to ZnO. The IPCE matched the absorption spectrum, with maxima at 470 nm and 540 nm. The APCE was 50–60% at 500 nm, much higher than that reported for the colloidal systems described above and similar to dye-sensitised ZnO analogues. At 1 sun $\eta = 0.4\%$ ($J_{SC} = 2.1$ mA cm^{-2}, $V_{OC} = 0.6$ V, $FF = 0.3$).

3.2.3.4.3 Size Effects on the Band Energies of Semiconductors

Because the energy levels of the QDs can be so easily tuned to match the energy levels of wide band gap semiconductors such as TiO_2, SnO_2 and ZnO, a number of research groups are exploiting the properties of QDs as sensitisers in photoelectrochemical solar cells analogous to DSSCs. The conduction band energy varies with the size of the particle, shifting to higher energy with decreasing size. For example, 7.5 nm CdSe colloids have $E_{CB} = ca$ –0.8 V vs NHE, 3.0 nm particles $E_{CB} = ca$ –1.57 V vs NHE. This gives rise to a blue shift in the absorption spectrum. The conduction band varies more than the valence band because of the small electron effective mass ($m_e = 0.13m_0$) compared with the hole effective mass ($m_h = 1.14m_0$); the higher the energy, the larger the driving force [$-G = E_{CB}(CdSe) - E_{CB}(TiO_2)$] for electron injection into TiO_2 ($E_{CB} = ca$ –0.5 V vs NHE). The crystallite sizes can be measured using X-ray diffraction and calculated using the Scherrer equation, and from the onset wavelength in the absorption spectrum using the effective mass approximation.[390] Figure 3.117 illustrates the principle of different band energies and hence driving force for electron transfer for different size CdSe particles. The electron transfer rate of the smaller CdSe particle (1.2×10^{10} s^{-1}) corresponds to an average lifetime of 83 ps. A logarithmic plot of electron transfer rate against driving force is shown, demonstrating that the electron transfer kinetics follow Marcus normal type behaviour (i.e. k_{et} increases with increasing driving force) similar to dye-sensitised systems.[382]

Vogel et al. demonstrated the possibility of tuning band energies of semiconductor–semiconductor systems for a range of low band gap semiconductors [PbS (0.3 eV), CdS (2.4 eV), Ag_2S (1.0 eV), Sb_2S_3 (1.7 eV), Bi_2S_3 (1.3 eV)] with a number of wide band gap semiconductors (TiO_2, ZnO, SnO_2, Nb_2O_5, Ta_2O_5) to speed up or slow down charge transfer between the two in the presence of a sulfide electrolyte.[379] Sb_2S_3, Bi_2S_3 and Ag_2S decomposed under illumination making them impractical for sensitisation. Ta_2O_5 has the highest energy conduction band and therefore electrodes with Ta_2O_5 gave the highest V_{OC} but required very small CdS for charge injection. The optimum photocurrent quantum yields were obtained with TiO_2; 70% for CdS and PbS on TiO_2 and 15% for CdS on ZnO matching the IPCE reported by Hotchandani and Kamat for colloidal CdS to ZnO at 400 nm.[380]

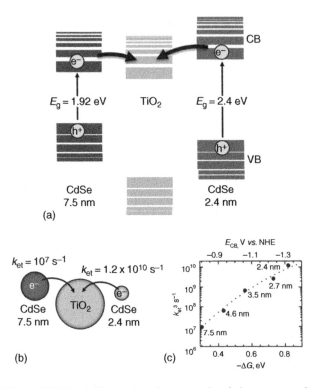

Figure 3.117 (a, b) Scheme illustrating the principle of electron transfer from two different size CdSe QDs into TiO_2 nanoparticles. (c) The dependence of electron transfer rate constant on the energy difference between the conduction bands. Adapted with permission from Robel *et al.*, 2007[382]. Copyright (2007) American Chemical Society

3.2.3.4.4 Redox Mediators for QD-Sensitised Solar Cells

Organic hole conductors have been used in place of the solution-based electrolyte to overcome the photostability problems.[389] Spiro-OMe-TAD is a well-documented hole conductor used in solid-state DSSCs (Figure 3.118).

Plass *et al.*[389] reported a PbS QD-sensitised TiO_2-based solid-state device for which the photoresponse extended above 800 nm, performing with an IPCE of 45% at 420 nm and an overall conversion efficiency of 0.49% at 1/10 sun. Ultrafast time-resolved transient absorption spectroscopy was used to study the charge transfer processes for PbS-sensitised TiO_2 and ZrO_2 using a pump wavelength of 600 nm and probe wavelengths of 778 nm to measure the kinetics of the charge in the PbS and 1400 nm to monitor the electrons in the TiO_2. A nanosecond system with a pump wavelength

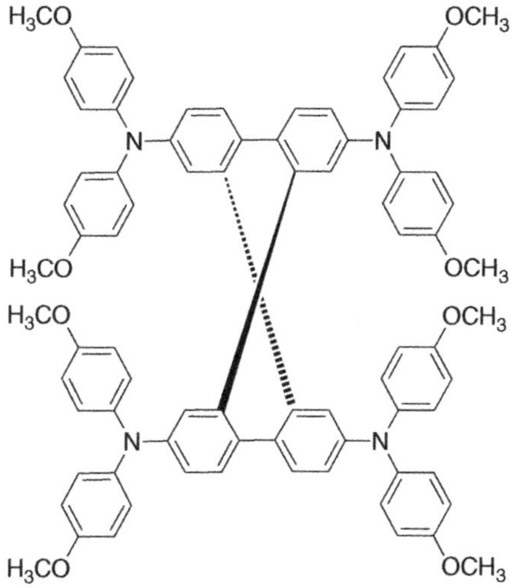

Figure 3.118 Structure of hole conductor Spiro-OMeTAD used in solid-state solar cells

of 450 nm, recording the spectral changes between 480 nm and 680 nm, was used to measure the recombination process. The postulated mechanism (with time constants in brackets) is:

$$PbS + h\nu \rightarrow h^+(PbS) + e^-(PbS) \quad \text{excitation (fs)} \quad (3.55)$$

$$e^-_{(free)}(PbS) \rightarrow e^-_{(trap)}(PbS) \quad \text{charge-separation (1 ps)} \quad (3.56)$$

$$e^-_{(trap)}(PbS) \rightarrow e^-_{CB}(TiO_2) \quad \text{electron injection (20 ps)} \quad (3.57)$$

$$h^+(PbS) + e^-(PbS) \rightarrow \Delta \quad \text{recombination (30 ps)} \quad (3.58)$$

$$h^+(PbS) + Spiro\text{-}OMeTAD \rightarrow h^+(Spiro\text{-}OMeTAD) \quad \text{hole injection (4 ps)} \quad (3.59)$$

$$e^-_{CB}(TiO_2) + h^+(Spiro\text{-}OMeTAD) \rightarrow \Delta \quad \text{recombination (2 μs)} \quad (3.60)$$

A record efficiency, $\eta = 1.2\%$ at 1 sun ($J_{SC} = 3$ mA cm^{-2}, $V_{OC} = 0.61$V, $FF = 0.61$), Cd-Se QD-sensitised solar cell using a cobalt-based redox mediator, [Co(phen)$_3$]$^{2+/3+}$ (Figure 3.119) has been reported.[391] The big difference between this system and those reported before was the stability. QD-based photoelectrochemical systems are typically characterised using

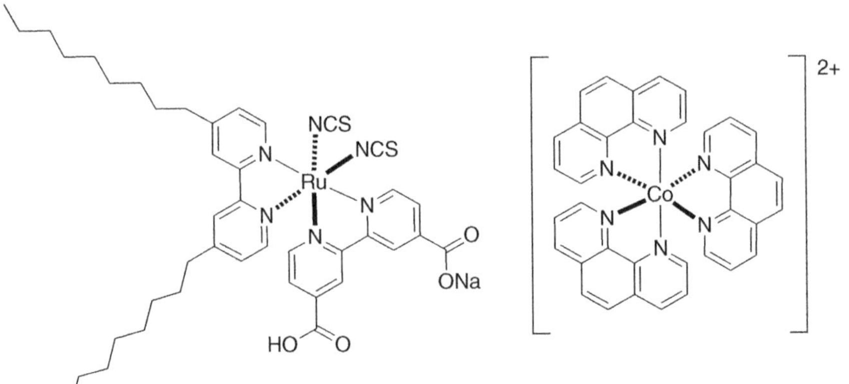

Figure 3.119 Structures of 'hydrophobic' dye Z907Na and redox mediator $[Co(phen)_3]^{2+}$

a hole scavenger rather than a regenerative redox couple or are stable for a couple of hours or days with triiodide/iodide or Na_2S. With the $[Co(phen)_3]^{2+/3+}$ redox couple, the cells maintained 82% of their original efficiency after 42 days. The efficiency of the QD-sensitised TiO_2 was compared with a typical dye sensitiser, Z907Na (Figure 3.119) and was found to be half as efficient. Lifetime, measured using photovoltage decay and charge-extraction techniques, was found to be four times longer with the QD than with the dye, suggesting better surface coverage, steric blocking or fewer pinholes than with the dye-sensitised film, preventing the redox mediator recombining with the electrons in the TiO_2. The authors attribute the lower efficiency to be as a result of smaller optical density of the CdSe compared with the ruthenium-based dye (the extinction coefficient is an order of magnitude higher). Moreover, the absorption of the dye extends to 800 nm whereas the absorption of the CdSe QD only extends to 630 nm.

A cobalt-based redox mediator was also employed by Yu *et al.* with an InAs QD directly adsorbed on TiO_2.[390] As with the system described above, stable (weeks), reproducible cells were prepared. An overall efficiency of 1.7% was obtained at 5 mW cm^{-2}; however at higher light intensities the efficiency was lower (0.3% at 1 sun) as a result of poor mass transport through the TiO_2 nanostructure.

3.2.3.4.5 'Hot' Charge Carriers

Hot carriers are formed from absorption of photons with energies larger than the band gap of a semiconductor which absorbs them. The excess energy $(hv - E_g)$ is divided up as kinetic energy between the corresponding excited electron and hole, depending on the effective masses of each (the lower the effective mass, the more kinetic energy). This excess energy is usually lost as heat through electron–phonon scattering, leading to phonon emission when the carriers relax back to the band edges (bottom of the conduction band for electrons, top of the valence band for holes). The excess energy can also be supplied by applying an electric field. Utilisation of hot carriers could increase the photovoltage.[29,372,373]

Nozik demonstrated a way to slow the hot electron cooling in CdSe and InP QDs which could enable these processes to be exploited.[373] An Auger process speeds up cooling when the hot electrons give the excess energy to a thermalised hole. The hole relaxes more quickly because of the higher effective mass of a hole compared with the effective mass of an electron, and the more closely spaced quantised states. Removal of the holes by a fast hole trap at the surface (in this case sodium biphenyl) blocks the Auger process, slowing the cooling (7 ps with trap, 0.3 ps without).

3.2.3.4.6 Multiple Charge Carriers

The generation of multiple charge carriers from an inverse Auger process arising from a single photon could enable quantum yields in excess of 100%. Collection of the charge carriers would give higher photocurrents. Hot carriers produce a second or more electron–hole pair through impact ionisation, which is when a photon is absorbed by a semiconductor with an energy at least twice that of the band gap. The high energy exciton relaxes to the band edges *via* energy transfer (above E_g) to a valence band electron which is then excited above the band gap itself. The threshold photon energy for impact ionisation in bulk semiconductors is too high to be useful. The process is efficient in quantum confined conditions because of enhanced electron–hole columbic interactions and limitations, which apply to bulk materials, are relaxed *e.g.* of conservation of momentum.

Schaller and Klimov demonstrated multiexciton generation in 4–6 nm PbSe, passivated with oleic acid, using transient absorption spectroscopy.[377] At least two excitons were generated from one photon with energy over three times the band gap energy. The time constants for the bi- (and tri-) excitons were in the 10–100 ps range, suggesting that if charge collection occurred in the *ca* 200 fs timescale, multiple excitons could be utilised. Ellingson *et al.* measured even more efficient multiple exciton generation in PbSe and PbS (Figure 3.120).[378]

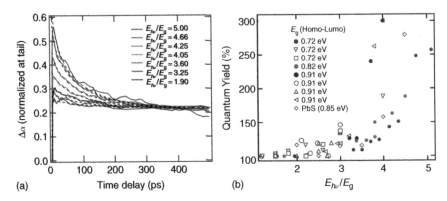

Figure 3.120 (a) Exciton population decay dynamics obtained by probing intraband (intraexciton) transitions in the mid-infrared for a sample of 5.7 nm diameter PbSe QDs. (b) Quantum yield for exciton formation from a single photon *vs* photon energy expressed as the ratio of the photon energy to the QD band gap (HOMO-LUMO energy) for three PbSe QD sizes and one PbS (diameter 3.9, 4.7, 5.4, and 5.5 nm, respectively, and E_g 0.91, 0.82, 0.73 eV, and 0.85 eV, respectively). Adapted with permission from Ellingson *et al.*, 2005[378]. Copyright (2005) American Chemical Society

For 3.9 nm PbSe QDs, excitation with photons with four times E_g gave rise to quantum yields of 300% (3 excitons per absorbed photon). They attribute the high efficiency of multiple exciton generation as due to the small, similar magnitude of the electron and hole effective mass in the PbS(e) which reduced the Auger-like cooling ('phonon bottleneck') which dominates in materials which have a much larger hole effective mass than electron effective mass such as CdSe and InP described above.

3.3 SUMMARY

Photovoltaics is perhaps the fastest growing industry today, with an increase in solar cell production of around 50% in recent years. Silicon type solar cells dominate, holding a market share of about 90%. The so-called second generation, thin-film solar cells are, however, catching up and there are various production lines being set up all over the world. In this chapter, we have reviewed the state-of-the-art performances of all the different solar cell technologies of interest for industrialisation at present, as well as commenting on technical, environmental and economic issues. Almost all of these technologies are inorganic and we have outlined the basic properties of these materials. It is too early to point out a winner for the large scale use of photovoltaics for electricity production. Given the speed of industrialisation and research, it will be highly exciting to follow the different developments over the coming decades.

In this chapter we have paid a special attention to dye-sensitised solar cells (DSSCs). With the discovery of how to achieve high efficiencies with dye-sensitised photoelectrochemical cells in the early 1990s, conventional solid-state photovoltaic technologies are now challenged by devices functioning at a molecular and nano level. The prospect of low-cost investments and fabrication are key features. In addition, DSSCs perform relatively better compared with other solar cell technologies under diffuse light conditions and at higher temperatures. The possibilities to design solar cells with respect to shape, colour and transparency and to integrate them into different products open up new commercial opportunities. From a fundamental viewpoint the DSSC technology also shows some unique features. It resembles nature in that it, like photosynthesis, separates the function of light absorption with charge transport. From a material point of view, it is a hybrid of inorganic/organic and solid-state/liquid materials. Perhaps a key concept for the future success of DSSC in this regard is 'diversity'. The DSSC is a technology to be explored, and the opportunities for materials combination and product design are only limited by imagination.

DSSCs are devices whose essential characteristic behaviour at the molecular level is collective. Conventional approaches are inadequate to describe such complex molecular systems where new and unexpected behaviour emerges in response to the strong and competing interactions among the elementary constituents. For DSSC devices, based on nanostructured TiO_2 electrodes, top efficiencies of 12% on small laboratory samples have been reported. There is potential to boost this performance substantially by imaginative approaches that exploit nanoscience and an improved understanding of complex molecular devices. This will be made by developing new functional material components which are tailor-made for the systems. Since the basic knowledge of how the DSSC molecular system operates still remains unclear, the specifications on how to tailor-make the components are incomplete and there is still a lot of fundamental and exploratory work to be done.

For most of the development of third generation photovoltaic systems, nanotechnology will be essential, utilising, for example, properties in the quantum-size domains. Hence, the DSSC technology provides an interesting launching pad. DSSC devices can be designed by an appropriate choice of the sensitiser to absorb, and quantitatively convert, incident photons to electric current in selective spectral regions of the solar emission, while maintaining high transparency in the remaining wavelength range. A further advantage of sensitised nanostructured electrodes is that their short-circuit photocurrent output can readily be varied by changing the film thickness and effective pore size. This, along with the ease of formation of multilayer structures by simple techniques, such as screen printing, constitutes a great advantage with regard to conventional solar cells, facilitating the fabrication and optimisation of tandem cells. Moreover, the concept of a dye-sensitised nanostructured oxide film can be generalised into mesoscopic injection solar cells (MISCs). With a mesoscopic electrode, we can utilise a very large internal surface area for heterogeneous electron transfer reactions to produce electricity or even chemical fuels. Moreover, a mesoscopic electrode provides a simple way of contacting the electrode interface by an electrolyte. The sensitisers can be dye molecules, supramolecular complexes or semiconductor QDs. A fundamental difference in MISCs compared with conventional semiconductor solar cell technologies is also that we can control and alter the semiconductor/sensitiser/electrolyte interface, rather than the bulk, by designing and preparing well-matched, energetically as well as kinetically, components that optimise the overall device performance.

Thus, research and development of DSSCs is not only important for the technology itself, but will also provide research and development platforms for future generation photovoltaics as well as other optoelectronic applications.

ACKNOWLEDGEMENTS

We would like to thank the Knut and Alice Wallenberg Foundation and the Swedish Energy Agency for financial support and Ulrika Jansson for assistance with the manuscript.

REFERENCES

[1] N. S. Lewis and D. G. Nocera, *Proc. Natl. Acad. Sci. USA*, **103**, 15729 (2006).
[2] D. Hopwood, *Renewable Energy Focus*, **8**, 46–52 (2007).

[3] J. Barber, *Philos. Trans. R. Soc. London, Ser. A*, **365**, 1007 (2007).

[4] G. M. Whitesides and G. W. Crabtree, *Science*, **315**, 796 (2007).

[5] *Department of Energy Report on the Basic Energy Sciences Workshop on Solar Energy Utilization 2005*, http://www.sc.doe.gov/bes/reports/files/SEU_rpt.pdf. Accessed 22-11-2010.

[6] BP Statistical Review of World Energy, June 2010. Available from: www.bp.com. Accessed 22-11-2010.

[7] http://rredc.nrel.gov/solar/spectra/am1.5/. Accessed 22-11-2010.

[8] R. Memming, *Semiconductor Electrochemistry*, Wiley-VCH, Weinheim, 2001.

[9] C. E. Jennings, R. M. Margolis and J. E. Bartlett, *Technical Report*, NREL/TP-6A2-43602, December 2008.

[10] International Energy Agency, Report IEA-PVPS T1-19: 2010. Available from www.eia.pvps.org. Accessed 22-11-2010.

[11] Dyesol Limited, Pegasus Corporate Advisory, DYE Research Note, July 2008.

[12] D. L. Pulfrey, *Solid-State Electron.*, **21**, 519 (1978).

[13] M. Wolf and H. Rauschenbach, *Adv. Energy Conversion*, **3**, 455 (1963).

[14] M. K. Nazeeruddin, A. Kay, I. Rodicio, R. Humphry-Baker, E. Müller, P. Lisker, N. Vlachopoulos and M. Grätzel, *J. Am. Chem. Soc.*, **115**, 6382 (1993).

[15] https://www.nrel.gov/analysis/sam/. Accessed 22-11-2010.

[16] M. A. Green, *Third Generation Photovoltaics: Advanced Solar Energy Conversion*, Springer-Verlag, Berlin, 2004.

[17] M. A. Green, *Energy Policy*, **28**, 989 (2000).

[18] J. Zhao, A. Wang, M. A. Green and F. Ferrazza, *Appl. Phys. Lett.*, **73**, 1991 (1998).

[19] L. M. Gonçalves, V. de Z. Bermudez, H. A. Ribero and A. M. Mendes, *Energy Environ. Sci*, **1**, 655 (2008).

[20] S. W. Glunz, *Adv. OptoElectron.*, Article ID 97370 (2007).

[21] http://www.IEA-pvps-task12.org. Accessed 22-11-2010.

[22] *Nature*, **460**, 677 (2009).

[23] D. Chapin, C. Fuller and G. Pearson, *J. Appl. Phys.*, **25**, 676 (1954).

[24] R. B. Bergmann, *Appl. Phys. A*, **69**, 187 (1999).

[25] M. B. Prince, *Phys. Rev.*, **93**, 1204 (1954).

[26] G. W. Ludwig and R. L. Watters, *Phys. Rev.*, **101**, 1699 (1956).

[27] P. Würfel, *Physica E*, **14**, 18 (2002).

[28] M. A. Green, *Prog. Photovolt: Res. Appl.*, **9**, 123 (2001).

[29] W. Shockley and H. Queisser, *J. Appl. Phys.*, **31**, 510 (1961).

[30] R. M. Swanson, *Prog. Photovolt: Res. Appl.*, **14**, 443 (2006).

[31] International Energy Agency, Experience Curves for Energy Technology Policy, 2000.

[32] R. Bhandariand and I. Stadler, *Solar Energy*, **83**, 1634 (2009).

[33] W. G. J. H. M. van Sark, E. A. Alsema, H. M. Junginger, H. H. C. de Moor and G. J. Schaeffer, *Prog. Photovolt: Res. Appl.*, **16**, 441 (2008).

[34] W. Hoffmann, *Sol. Energy Mater. Sol. Cells*, **90**, 3285 (2006).

[35] P. Mints and D. Tomlinson, *Analysis of Worldwide PV Markets and Five-Year Application Forecast Report*, Navigant Consulting Limited, August 2009.

[36] S. Hegedus, *Prog. Photovolt: Res. Appl.*, **14**, 393 (2006).

[37] C. R. Wronski, *Proc. 28th IEEE Photovoltaic Specialists Conf.*, 1 (2000).

[38] D. Staebler and C. Wronski, *Appl. Phys. Lett.*, **31**, 292 (1977).

[39] A. Romeo, M. Terheggen, D. Abou-Ras, D. L. Bätzner, F.-J. Haug, M. Kälin, D. Rudmann and A. N. Tiwari, *Prog. Photovolt: Res. Appl.*, **12**, 93 (2004).

[40] D. Cusano, *Solid-State Electron.*, **6**, 217 (1963).

[41] R. W. Birkmire and E. Eser, *Annu. Rev. Mater. Sci.*, **27**, 625 (1997).

[42] X. Wu, J. C. Kane, R. G. Dhere, C. DeHart, D. S. Albin, A. Duda, T. A. Gessert, S. Asher, D. H. Levi and P. Sheldon, *Proc. 17th Eur. Photovoltaic Solar Energy Conf. Exhibit.*, 995 (2002).

[43] M. A. Green, K. Emery, Y. Hishikawa and W. Warta, *Prog. Photovolt: Res. Appl.*, **17**, 320 (2009).

[44] A. Morales-Acevedo, *Sol. Energy Mater. Sol. Cells*, **90**, 2213 (2006).

[45] M. I. Alonso, K. Wakita, J. Pascual, M. Garriga and N. Yamamoto, *Phys. Rev. B*, **63**, 075203 (2001).

[46] I. Repins, M. Contreras, M. Romero, Y. Yan, W. Metzger, J. Li, S. Johnston, B. Egass, C. DeHart, J. Scharf, B. E. McCandless and R. Noufi, *Proc. 33rd IEEE Photovoltaic Specialists Conf.*, NREL/CP-520-42539 (2008).

[47] M. Kemell, M. Ritala and M. Leskel, *Crit. Rev. Solid State Mater. Sci.*, **30**, 1 (2005).

[48] J. Hedström, H. Ohlsen, M. Bodegård, A. Kylner, L. Stolt, D. Hariskos, M. Ruckh and H.W. *Proc. 23rd IEEE Photovoltaic Specialists Conf.*, 364 (1993).

[49] M. S. Keshner and R. Arya, NREL/SR-520-36846; available from www.nrel.gov. Accessed 22-11-2010.

[50] C.W. Tang, *Appl. Phys. Lett.*, **48**, 183 (1986).

[51] G. Yu, J. Gao, J.C. Hummelen, F. Wudl and A.J. Heeger, *Science* **270**, 1789 (1995).

[52] M. A. Green, *Physica E*, **14**, 65 (2002).

[53] S. Guha, *Curr. Opin. Solid State Mater. Sci.*, **2**, 425 (1997).

[54] M. A. Green, *Prog. Photovolt: Res. Appl.*, **4**, 375 (1996).

[55] J. Yang, B. Yan and S.Guha, *Thin Solid Films*, **487**, 162 (2005).

[56] http://www.Sontor.com. Accessed 22-11-2010.

[57] M. Yamaguchi T. Takamoto, A. Khan, M. Imaizumi, S. Matsuda and N. J. Ekins-Daukes, *Prog. Photovolt: Res. Appl.*, **13**, 125 (2005).

[58] M. Yamaguchi, T. Takamoto, K.Araki and N. Ekins-Daukes, *Solar Energy*, **79**, 78 (2005).

[59] T. Takamoto, M. Kaneiwa, M. Imaizumi and M. Yamaguchi, *Prog. Photovolt: Res. Appl.* **13**, 495 (2005).

[60] B. O'Regan and M. Grätzel, *Nature*, **353**, 737 (1991).

[61] A. Hagfeldt and M. Grätzel, *Chem. Rev.*, **95**, 49 (1995).

[62] M. Grätzel, *Nature*, **414**, 338 (2001).

[63] M. Grätzel, *Acc. Chem. Res.*, **42**, 1788 (2009).

[64] P. Würfel, *Physics of Solar Cells: From Principles to New Concepts*, Wiley-VCH, Weinheim, 2005.

[65] R. Amadelli, R. Argazzi, C. A. Bignozzi and F. Scandola, *J. Am. Chem. Soc.*, **112**, 7099 (1990).

[66] M. K. Nazeeruddin, P. Liska, J. Moser, N. Vlachopoulos and M. Grätzel, *Helv. Chim. Acta*, **73**, 1788 (1990).

[67] M. K. Nazeeruddin, P. Pechy, T. Renouard, S. M. Zakeeruddin, R. Humphry-Baker, P. Comte, P. Liska, L. Cevey, E. Costa, V. Shklover, L. Spiccia, G. B. Deacon, C. A. Bignozzi and M. Grätzel, *J. Am. Chem. Soc.* **123**, 1613 (2001).

[68] Y. Chiba, A. Islam, Y. Watanabe, R. Komiya, N. Koide and L. Y. Han, *Jpn. J. Appl. Phys., Part 2*, **45**, L638 (2006).

[69] F. O. Lenzmann, and J. M. Kroon, *Adv. Optoelectron.*, 65073/1-65073/10 (2007).

[70] M. Grätzel, in *Nanostructured Materials for Electrochemical Energy Production and Storage*, R. E. Leite (Ed.), Springer, New York, 2008, Chapter 1.

[71] L. Kavan and M. Grätzel, *Electrochim. Acta*, 40, 643 (1995).

[72] P. J. Cameron and L. M. Peter, *J. Phys. Chem. B*, 107, 14394 (2003).

[73] S. A. Sapp, C. M. Elliott, C. Contado, S. Caramori and C. A. Bignozzi, *J. Am. Chem. Soc.*, 124, 11215 (2002).

[74] H. Nusbaumer, S. M. Zakeeruddin, J. E. Moser and M. Grätzel, *Chem. Eur. J.*, 9, 3756 (2003).

[75] P. J. Cameron, L. M. Peter, S. M. Zakeeruddin and M. Grätzel, *Coord. Chem. Rev.*, 248, 1447 (2004).

[76] U. Bach, D. Lupo, P. Comte, J. E. Moser, F. Weissörtel, J. Salbeck, H. Spreitzer and M. Grätzel, *Nature*, 395, 583 (1998).

[77] J. Krüger, R. Plass, M. Grätzel, P. J. Cameron and L. M. Peter, *J. Phys. Chem. B*, 107, 7536 (2003).

[78] N. Robertson, *Angew. Chem., Int. Ed.*, 45, 2338 (2006).

[79] S. Ito, H. Miura, S. Uchida, M. Takata, K. Sumioka, P. Liska, P. Comte, P. Pechy and M. Grätzel, *Chem. Commun.*, 5194 (2008).

[80] G. L. Zhang, H. Bala, Y. M. Cheng, D. Shi, X. J. Lv and Q. J. Yu, P. Wang, *Chem. Commun.*, 2198 (2009).

[81] J. H. Yum, D. P. Hagberg, S. J. Moon, K. M. Karlsson, T. Marinado, L. Sun, A. Hagfeldt, M. K. Nazeeruddin and M. Grätzel, *Angew. Chem., Int. Ed.*, 48, 1576 (2009).

[82] M. F. Xu, R. Z. Li, N. Pootrakulchote, D. Shi, J. Guo, Z. H. Yi, S. M. Zakeeruddin, M. Grätzel and P. Wang, *J. Phys. Chem. C*, 112, 19770 (2008).

[83] M. F. Xu, S. Wenger, H. Bala, D. Shi, R. Z. Li, Y. Z. Zhou, S. M. Zakeeruddin, M. Grätzel and P. Wang, *J. Phys. Chem. C*, 113, 2966 (2009).

[84] H. Choi, C. Baik, S. O. Kang, J. Ko, M. S. Kang, M. K. Nazeeruddin and M. Grätzel, *Angew. Chem., Int. Ed.*, 47, 327 (2008).

[85] A. Mishra, M. K. R. Fischer and P. Bäuerle, *Angew. Chem., Int. Ed.*, 48, 2474 (2009).

[86] G. Oskam, B. V. Bergeron, G. J. Meyer and P. C. Searson, *J. Phys. Chem. B*, 105, 6867 (2001).

[87] P. Wang, S. M. Zakeeruddin, J.-E. Moser, R. Humphry-Baker and M. Grätzel, *J. Am. Chem. Soc.*, 126, 7164 (2004).

[88] B. V. Bergeron, A. Marton, G. Oskam and G. J. Meyer, *J. Phys. Chem. B*, 109, 937 (2005).

[89] Z. Zhang, P. Chen, T. N. Murakami, S. M. Zakeeruddin and M. Grätzel, *Adv. Funct. Mater.*, 18, 341 (2008).

[90] B. A. Gregg, F. Pichot, S. Ferrere and C. L. Fields, *J. Phys. Chem. B*, 105, 1422 (2001).

[91] J. Kruger, R. Plass, L. Cevey, M. Piccirelli, M. Grätzel and U. Bach, *Appl. Phys. Lett.*, 79, 2085 (2001).

[92] K. Tennakone, G. Kumara, A. R. Kumarasinghe, K. G. U. Wijayantha and P. M. Sirimanne, *Semicond. Sci. Technol.*, 10, 1689 (1995).

[93] B. O'Regan and D. T. Schwartz, *J. Appl. Phys.*, 80, 4749 (1996).

[94] T. W. Hamann, R. A. Jensen, A. B. F. Martinson, H. Van Ryswyk and J. T. Hupp, *Energy Environ. Sci.*, 1, 66 (2008).

[95] M. Pagliaro, G. Palmisano, R. Ciriminna and V. Loddo, *Energy Environ. Sci.*, 2, 838 (2009).

[96] R. Jose, V. Thavasi and S. Ramakrishna, *J. Am. Ceram. Soc.*, **92**, 289 (2009).
[97] B. C. O'Regan and J. R. Durrant, *Acc. Chem. Res.*, **42**, 1799 (2009).
[98] S. Ardo and G. J. Meyer, *Chem. Soc. Rev.*, **38**, 115 (2009).
[99] L. M. Peter, *Acc. Chem. Res.*, **42**, 1839 (2009).
[100] J. B. Asbury, R. J. Ellingson, H. N. Ghosh, S. Ferrere, A. J. Nozik and T. Q. Lian, *J. Phys. Chem. B*, **103**, 3110 (1999).
[101] G. Ramakrishna, D. A. Jose, D. K. Kumar, A. Das, D. K. Palit and H. N. Ghosh, *J. Phys. Chem. B*, **109**, 15445 (2005).
[102] D. Kuang, S. Ito, B. Wenger, C. Klein, J.-E. Moser, R. Humphry-Baker, S. M. Zakeeruddin and M. Grätzel, *J. Am. Chem. Soc.*, **128**, 4146 (2006).
[103] G. Benkö, J. Kallioinen, J. E. I. Korppi-Tommola, A. P. Yartsev and V. Sundstrom, *J. Am. Chem. Soc.*, **124**, 489 (2002).
[104] A. Hagfeldt and M. Grätzel, *Acc. Chem. Res.*, **33**, 269 (2000).
[105] S. E. Koops, B. O'Regan, P. R. F. Barnes and J. R. Durrant, *J. Am. Chem. Soc.*, **131**, 4808 (2009).
[106] J. E. Moser and M. Grätzel, *Chem. Phys.*, **176**, 493 (1993).
[107] S. A. Haque, Y. Tachibana, D. R. Klug and J. R. Durrant, *J. Phys. Chem. B*, **102**, 1745 (1998).
[108] D. Kuciauskas, M. S. Freund, H. B. Gray, J. R. Winkler and N. S. Lewis, *J. Phys. Chem. B*, **105**, 392 (2001).
[109] B. O'Regan, J. Moser, M. Anderson and M. Grätzel, *J. Phys. Chem.*, **94**, 8720 (1990).
[110] S. G. Yan and J. T. Hupp, *J. Phys. Chem.*, **100**, 6867 (1996).
[111] L. M. Peter, *J. Phys. Chem. C*, **111**, 6601 (2007).
[112] K. Zhu, N. Kopidakis, N. R. Neale, J. van de Lagemaat and A. J. Frank, *J. Phys. Chem. B*, **110**, 25174 (2006).
[113] S. Nakade, Y. Saito, W. Kubo, T. Kitamura, Y. Wada and S. Yanagida, *J. Phys. Chem. B*, **107**, 8607 (2003).
[114] G. Boschloo and A. Hagfeldt, *Acc. Chem. Res.*, **42**, 1819 (2009).
[115] B. O'Regan, J. Moser, M. Anderson and M. Grätzel, *J. Phys. Chem.*, **94**, 8720 (1990).
[116] S. Södergren, A. Hagfeldt, J. Olsson and S.-E. Lindquist, *J. Phys. Chem.*, **98**, 5552 (1994).
[117] A. Solbrand, H. Lindström, H. Rensmo, A. Hagfeldt, S.-E. Lindquist and S. Södergren, *J. Phys. Chem. B*, **101**, 2514 (1997).
[118] N. Kopidakis, E. A. Schiff, N. G. Park, J. van de Lagemaat and A. J. Frank, *J. Phys. Chem. B*, **104**, 3930 (2000).
[119] D. Nistér, K. Keis, S.-E. Lindquist and A. Hagfeldt, *Sol. Energy Mater. Sol. Cells*, **73**, 411 (2002).
[120] F. Cao, G. Oskam, G. J. Meyer and P. C. Searson, *J. Phys. Chem.*, **100**, 17021 (1996).
[121] L. Dloczik, O. Ileperuma, I. Lauermann, L. M. Peter, E. A. Ponomarev, G. Redmond, N. J. Shaw and I. Uhlendorf, *J. Phys. Chem. B*, **101**, 10281 (1997).
[122] J. van de Lagemaat and A. J. Frank, *J. Phys. Chem. B*, **104**, 4292 (2000).
[123] J. Bisquert and V. S. Vikhrenko, *J. Phys. Chem. B*, **108**, 2313 (2004).
[124] A. C. Fischer, L. M. Peter, E. A. Ponomarev, A. B. Walker and K. G. U. Wijayantha, *J. Phys. Chem. B*, **104**, 949 (2000).
[125] J. Bisquert, *J. Phys. Chem. B*, **108**, 2323 (2004).
[126] J. Bisquert and V. S. Vikhrenko, *J. Phys. Chem. B*, **108**, 2313 (2004).

[127] J. R. Jennings and L. M. Peter, *J. Phys. Chem. C*, **111**, 16100 (2007).

[128] H. K. Dunn and L. M. Peter, *J. Phys. Chem. C*, **113**, 4726 (2009).

[129] F. Fabregat-Santiago, J. Bisquert, E. Palomares, L. Otero, D. B. Kuang, S. M. Zakeeruddin and M. Grätzel, *J. Phys. Chem C*, **111**, 6550 (2007).

[130] Q. Wang, J. E. Moser and M. Grätzel, *J. Phys. Chem. B*, **109**, 14945 (2005).

[131] B. C. O'Regan, K. Bakker, J. Kroeze, H. Smit, P. Sommeling and J. R. Durrant, *J. Phys. Chem. B*, **110**, 17155 (2006).

[132] A. Mihi, M. E. Calvo, J. A. Anta and H. Míguez, *J. Phys. Chem. C.*, **112**, 13 (2008).

[133] J. Bisquert, *J. Phys. Chem. C*, **111**, 17163 (2007).

[134] E. Hendry, M. Koeberg, B. O'Regan and M. Bonn, *Nano Lett.*, **6**, 755 (2006).

[135] S. Södergren, A. Hagfeldt, J. Olsson and S.-E. Lindquist, *J. Phys. Chem.*, **98**, 5552 (1994).

[136] S.-E. Lindquist, B. Finnstrom and L. Tegner, *J. Electrochem. Soc.*, **130**, 351 (1983).

[137] J. Halme, G. Boschloo, A. Hagfeldt and P. Lund, *J. Phys. Chem. C*, **112**, 5623 (2008).

[138] P. R. F. Barnes, A. Y. Anderson, S. E. Koops, J. R. Durrant and B. C. O'Regan, *J. Phys. Chem. C*, **113**, 1126 (2009).

[139] K. Westermark, A. Henningsson, H. Rensmo, S. Södergren, H. Siegbahn and A. Hagfeldt, *Chem. Phys.*, **285**, 157 (2002).

[140] K. Sayama, H. Sugihara and H. Arakawa, *Chem. Mater.*, **10**, 3825 (1998).

[141] E. S. Kwak, W. Lee, N.-G. Park, J. Kim and H. Lee, *Adv. Funct. Mater.*, **19**, 1093 (2009).

[142] A. Fujishima and K. Honda, *Nature*, **238**, 37 (1972).

[143] X. Chen and S. S. Mao, *Chem. Rev.*, **107**, 2891 (2007).

[144] M. R. Hoffmann, S. T. Martin, W. Choi and D. W. Bahnemann, *Chem. Rev.*, **1**, 69 (1995).

[145] A. P. Alivisatos, *Science*, **271**, 933 (1996).

[146] R. Asahi, Y. Taga, W. Mannstadt and A. J. Freeman, *Phys. Rev. B*, **61**, 7459 (2000).

[147] N.-G. Park, J. van de Lagemaat and A. J. Frank, *J. Phys. Chem. B*, **104**, 8989 (2000).

[148] H. Tang, K. Prasad, R. Sanjinès, P. E. Schmid and F. Lévy, *J. Appl. Phys.*, **75**, 2042 (1994).

[149] D. S. Boudreaux, F. Williams and A. J. Nozik, *J. Appl. Phys.*, **51**, 2158 (1980).

[150] L. Forro, O. Chauvet, D. Emin, L. Zuppiroli, H. Berger and F. Lévy, *J. Appl. Phys.*, **75**, 633 (1994).

[151] A. Eucken and U. A. Büchner, *Z. Phys. Chem.*, **B27**, 321 (1934).

[152] S. Roberts, *Phys. Rev.*, **76**, 1215 (1949).

[153] R. A. Parker, *Phys. Rev.*, **124**, 1719 (1961).

[154] J. C. Hulteen and C. R. Martin, *J. Mater. Chem.*, **7**, 1075 (1997).

[155] R. Tenne and C. N. R. Rao, *Philos. Trans. R. Soc. London, Ser. A*, **362**, 2099 (2004).

[156] R. Jose, V. Thavasi and S. Ramakrishna, *J. Am. Ceram. Soc.*, **92**, 289 (2009).

[157] G. Cao and D. Liu, *Adv. Colloid Interface Sci.*, **136**, 45 (2008).

[158] S. K. Deb, *Sol. Energy Mater. Sol. Cells*, **88**, 1 (2005).

[159] M. Zukalová, A. Zukal, L. Kavan, M. K. Nazeeruddin, P. Liska and M. Grätzel, *Nano Lett.*, **5**, 1789 (2005).

[160] P. Yang, D. Zhao, D. I. Margolese, B. F. Chmelka and G.D. Stucky, *Chem. Mater*, **11**, 2813 (1999).

[161] H. Lindström, E. Magnusson, A. Holmberg, S. Södergren, S.-E. Lindquist and A. Hagfeldt, *Sol. Energy Mater. Sol. Cells*, **73**, 91 (2002).

[162] G. Boschloo, H. Lindström, E. Magnusson, A. Holmberg and A. Hagfeldt, *J. Photochem. Photobiol., A,* **148,** 11 (2002).

[163] K. Shankar, J. I. Basham, N. K. Allam, O. K. Varghese, G. K. Mor, X. Feng, M. Paulose, J. A. Seabold, K.-S. Choi and C. A. Grimes, *J. Phys. Chem. C,* **113,** 6327 (2009).

[164] G. K. Mor, O. K. Varghese, M. Paulose, K. Shankar and C. A. Grimes, *Sol. Energy Mater. Sol. Cells,* **90,** 2011 (2006).

[165] K. C. Popat, L. Leoni, C. A. Grimes and T. A. Desai, *Biomaterials,* **28,** 3188 (2007).

[166] C. A. Grimes, *J. Mater. Chem.,* **17,** 1451 (2007).

[167] V. Zwilling, M. Aucouturier and E. Darque-Ceretti, *Electrochim. Acta,* **44,** 921 (1999).

[168] D. Gong, C. A. Grimes, O. K. Varghese, W. C. Hu, R. S. Singh, Z. Chen and E. C. Dickey, *J. Mater. Res.,* **16,** 3331 (2001).

[169] J. M. Macák, H. Tsuchiya and P. Schmuki, *Angew. Chem. Int. Ed.,* **44,** 2100 (2005).

[170] J. M. Macák, H. Tsuchiya, L. Taveira, S. Aldabergerova and P. Schmuki, *Angew. Chem. Int. Ed.,* **44,** 7463 (2005).

[171] M. Paulose, H. E. Prakasam, O. K. Varghese, L. Peng, K. C. Popat, G. K. Mor, T. A. Desai and C. A. Grimes, *J. Phys. Chem. C,* **111,** 14992 (2007).

[172] H. E. Prakasam, K. Shankar, M. Paulose, O. K. Varghese and C. A. Grimes, *J. Phys. Chem. C,* **111,** 7235 (2007).

[173] T. Stergiopoulos, A. Ghicov, V. Likodimos, D. S. Tsoukleris, J. Kunze, P. Schmuki and P. Falaras, *Nanotechnology,* **19,** 235602 (2008).

[174] K. Zhu, N. R. Neale, A. Miedaner and A. J. Frank, *Nano Lett.,* **7,** 69 (2007).

[175] G. K. Mor, K. Shankar, M. Paulose, O. K. Varghese and C. A. Grimes, *Nano Lett.,* **6,** 215 (2006).

[176] D.-B. Kuang, J. Brillet, P. Chen, M. Takata, S. Uchida, H. Miura, K. Sumioka, S. M. Zakeeruddin and M. Grätzel, *ACS Nano,* **2,** 1113 (2008).

[177] K. Y. Chun, B. W. Park, Y. M. Sung, D. J. Kwak, Y. T. Hyun and M. W. Park. *Thin Solid Films,* **517,** 4196 (2009).

[178] Z. Liu, V. Subramania and M. Misra, *J. Phys. Chem. C,* **113,** 14028 (2009).

[179] K. Shankar, G. K. Mor, H. E. Prakasam, O. K. Varghese and C. A. Grimes, *Langmuir,* **23,** 12445 (2007).

[180] K. Shankar, G. K. Mor, H. E. Prakasam, S. Yoriya, M. Paulose, O. K. Varghese and C. A. Grimes, *Nanotechnology,* **18,** 065707 (2007).

[181] S. Ito, N. C. Ha, G. Rothenberger, P. Liska, P. Comte, S. M. Zakeeruddin, P. Pechy, M. K. Nazeeruddin and M. Gratzel, *Chem. Commun.,* 4004 (2006).

[182] K. Shankar, J. Bandara, M. Paulose, H. Wietasch, O. K. Varghese, G. K. Mor, T. J. LaTempa, M. Thelakkat and C. A. Grimes, *Nano Lett.,* **8,** 1654 (2008).

[183] G. K. Mor, K. Shankar, M. Paulose, O. K. Varghese and C. A. Grimes, *Nano Lett.,* **6,** 215 (2006).

[184] M. Paulose, K. Shankar, O. K Varghese, G. K Mor, B. Hardin and C. A. Grimes, *Nanotechnology,* **17,** 1446 (2006).

[185] G. Schlichthörl, S. Y. Huang, J. Sprague and A. J. Frank, *J. Phys. Chem. B,* **101,** 8141 (1997).

[186] J. van de Lagemaat, N. G. Park and A. J. Frank, *J. Phys. Chem. B,* **104,** 2044 (2000).

[187] J. R. Jennings, A. Ghicov, L. M. Peter, P. Schmuki and A. B. Walker, *J. Am. Chem. Soc.,* **130,** 13364 (2008).

[188] N. N. Bwana, *Nano Res.,* **1,** 483 (2008).

[189] J. Liu, Y.-T. Kuo, K. J. Klabunde, C. Rochford, J. Wu and J. Li, *ACS Appl. Mater. Interfaces,* **1,** 1645 (2009).

[190] J. S. King, E. Graugnard and C. J. Summers, *Adv. Mater.*, **17**, 1010 (2005).

[191] C.-Y. Kuo and S.-Y. Lu. *Nanotechnology*, **19**, 095705 (2008).

[192] T. W. Hamann, A. B. F. Martinson, J. W. Elam, M. J. Pellin and J. T. Hupp, *J. Phys. Chem. C*, **112**, 10303 (2008).

[193] A. Tsukazaki, A. Ohtomo, T. Onuma, M. Ohtani, T. Makino, M. Sumiya, K. Ohtani, S. F. Chichibu, S. Fuke, Y. Segawa, H. Ohno, H. Koinuma and M. Kawasaki, *Nat. Mater.*, **4**, 42 (2005).

[194] H. Gerischer and H. Tributsch, *Ber. Bunsenges. Phys. Chem.*, **72**, 437 (1968).

[195] H. Tributsch and H. Gerischer, *Ber. Bunsenges. Phys. Chem.*, **73**, 850 (1969).

[196] H. Tributsch and H. Gerischer, *Ber. Bunsenges. Phys. Chem.*, **73**, 251 (1969).

[197] H. Tsubomura, M. Matsumura, Y. Nomura and T. Amamiya, *Nature*, **261**, 402 (1976).

[198] K. Keis, E. Magnusson, H. Lindström, S.-E. Lindquist and A. Hagfeldt, *Sol. Energy Mater. Sol. Cells*, **73**, 51 (2002).

[199] M. Saito and S. Fujihara, *Energy Environ. Sci.*, **1**, 280 (2008).

[200] K. Kakiuchi, E. Hosono and S. Fujihara, *J. Photochem. Photobiol.*, *A*, **179**, 81 (2006).

[201] K. Hara, T. Horiguchi, T. Kinoshita, K. Sayama, H. Sugihara and H. Arakawa, *Chem. Lett.*, **29**, 316 (2000).

[202] W. J. E. Beek, M. M. Wienk, M. Kemerink, X. N. Yang and R. A. J. Janssen, *J. Phys. Chem. B*, **109**, 9505 (2005).

[203] S. Rani, P. Suri, P. K. Shishodia and R. M. Mehra, *Sol. Energy Mater. Sol. Cells*, **92**, 1639 (2008).

[204] T. Yoshida, M. Iwaya, H. Ando, T. Oekermann, K. Nonomura, D. Schlettwein, D. Wohrle and H. Minoura, *Chem. Commun.*, 400 (2004).

[205] A. Otsuka, K. Funabiki, N. Sugiyama, T. Yoshida, H. Minoura and M. Matsui, *Chem. Lett.*, **35**, 666 (2006).

[206] C. Lévy-Clément, R. Tena-Zaera, M. A. Ryan, A. Katty and G. Hodes, *Adv. Mater.*, **17**, 1512 (2005).

[207] M. Law, L. E. Greene, J. C. Johnson, R. Saykally and P. D. Yang, *Nat. Mater.*, **4**, 455 (2005).

[208] E. Hosono, S. Fujihara, I. Honna and H. S. Zhou, *Adv. Mater.*, **17**, 2091 (2005).

[209] Q. Zhang, T. P. Chou, B. Russo, S. A. Jenekhe and G. Cao, *Angew. Chem. Int. Ed.*, **47**, 2402 (2008).

[210] A. Ranga Rao and V. Dutta, *Nanotechnology*, **19**, 445712 (2008).

[211] A. B. Martinson, J. W. Elam, J. T. Hupp and M. J. Pellin, *Nano Lett.*, **7**, 2183 (2007).

[212] K. Kakiuchi, M. Saito and S. Fujihara, *Thin Solid Films*, **516**, 2026 (2008).

[213] W. Chen, H. Zhang, I. M. Hsing and S. Yang, *Electrochem. Commun.*, **11**, 1057 (2009).

[214] C. Y. Jiang, X. W. Sun, G. Q. Lo, D. L. Kwong and J. X. Wang, *Appl. Phys. Lett.*, **90** 263501, (2007).

[215] U. Ozgur, Y. I. Alivov, C. Liu, A. Teke, M. A. Reshchikov, S. Dogan, V. Avrutin, S. J. Cho and H. Morkoc, *J. Appl. Phys.*, **98**, 041301 (2005).

[216] Q. Zhang, C. S. Dandeneau, X. Zhou and G. Cao, *Adv. Mater.*, **21**, 4087 (2009).

[217] I. Gonzalez-Valls and M. Lira-Cantu, *Energy Environ. Sci.*, **2**, 19 (2009).

[218] G. Boschloo, T. Edvinsson and A. Hagfeldt, *in Nanostructured Materials for Solar Energy Conversion*, T. Soga (Ed.), Elsevier BV, Amsterdam, 2006, Chapter 8.

[219] S. B. Zhang, S.-H. Wei and A. Zunger, *Phys. Rev. B*, **63**, 075205 (2001).

[220] D. M. Hofmann, A. Hofstaetter, F. Leiter, H. Zhou, F. Henecker and B. K. Meyer, *Phys. Rev. Lett.* **88**, 045504-045501 (2002).

[221] S. J. Jokela and M. D. McCluskey, *Phys. Rev. B*, 72, 113201 (2005).

[222] M. D. McCluskey, S. J. Jokela, K. K. Zhuravlev, P. J. Simpson and K. G. Lynn, *Appl. Phys. Lett.*, 81, 3807 (2002).

[223] C. G. Van de Walle, *Phys. Rev. Lett.*, 85, 1012 (2000).

[224] M. A. Butler and D. S. Ginley, *J. Electrochem. Soc.*, 125, 1 (1978).

[225] G. Redmond, A. O'Keeffe, C. Burgess, C. MacHale and D. Fitzmaurice, *J. Phys. Chem.*, 97, 11081 (1993).

[226] L. Kavan, M. Grätzel, S. E. Gilbert, C. Klemenz and H. J. Scheel, *J. Am. Chem. Soc.*, 118, 6716 (1996).

[227] L. Blok and P. L. de Bruyn, *J. Colloid Interface Sci.*, 32, 518 (1970).

[228] S. Roberts, *Phys. Rev.*, 76, 1215 (1949).

[229] D. R. Lide (Ed.), *Handbook of Chemistry and Physics*, CRC Press, Boca Raton, 2004–2005.

[230] C. L. Dong, C. Persson, L. Vayssieres, A. Augustsson, T. Schmitt, M. Mattesini, R. Ahuja, C. L. Chang and J.-H. Guo, *Phys. Rev. B*, 70, 195325-195321 (2004).

[231] L. Brus, *J. Phys. Chem.*, 90, 2555 (1986).

[232] L. Brus, *J. Chem. Phys.*, 80, 4403 (1984).

[233] U. Koch, A. Fojtik, H. Weller and A. Henglein, *Chem. Phys. Lett.*, 122, 507 (1985).

[234] D. W. Bahnemann, C. Kormann and M. R. Hoffmann, *J. Phys. Chem.*, 91, 3789 (1987).

[235] M. Haase, H. Weller and A. Henglein, *J. Phys. Chem.*, 92, 482 (1988).

[236] L. Spanhel and M. A. Anderson, *J. Am. Chem. Soc.*, 113, 2826 (1991).

[237] M. Hilgendorff, L. Spanhel, C. Rothenhausler and G. Muller, *J. Electrochem. Soc.*, 145, 3632 (1998).

[238] C. Pacholski, A. Kornowski and H. Weller, *Angew. Chem. Int. Ed.*, 41, 1188 (2002).

[239] T. Trindade, J. D. Pedrosa de Jesus and P. O'Brien, *J. Mater. Chem.*, 4, 1611 (1994).

[240] K. Keis, L. Vayssieres, S. E. Lindquist and A. Hagfeldt, *Nanostruct. Mater.*, 12, 487 (1999).

[241] M. Castellano and E. Matijevic, *Chem. Mater.*, 1, 78 (1989).

[242] H. Rensmo, K. Keis, H. Lindström, S. Södergren, A. Solbrand, A. Hagfeldt, S. E. Lindquist, L. N. Wang and M. Muhammed, *J. Phys. Chem. B*, 101, 2598 (1997).

[243] B. Liu and H. C. Zeng, *J. Am. Chem. Soc.*, 125, 4430 (2003).

[244] P. O'Brien, T. Saeed and J. Knowles, *J. Mater. Chem.*, 6, 1135 (1996).

[245] L. Vayssieres, K. Keis, A. Hagfeldt and S. E. Lindquist, *Chem. Mater.*, 13, 4395 (2001).

[246] L. Vayssieres, *Adv. Mater.*, 15, 464 (2003).

[247] L. Vayssieres, K. Keis, S.-E. Lindquist and A. Hagfeldt, *J. Phys. Chem. B*, 105, 3350 (2001).

[248] R. B. Peterson, C. L. Fields and B. A. Gregg, *Langmuir*, 20, 5114 (2004).

[249] T. Pauporte and D. Lincot, *Appl. Phys. Lett.*, 75, 3817 (1999).

[250] R. Könenkamp, K. Boedecker, M. C. Lux-Steiner, M. Poschenrieder, F. Zenia, C. Levy-Clement and S. Wagner, *Appl. Phys. Lett.*, 77, 2575 (2000).

[251] B. O'Regan, D. T Schwartz, S. M. Zakeeruddin and M. Grätzel, *Adv. Mater.*, 12, 1263 (2000).

[252] B. O'Regan, V. Sklover and M. Grätzel, *J. Electrochem. Soc.*, 148, C498 (2001).

[253] T. Yoshida, M. Tochimoto, D. Schlettwein, D. Wöhrle, T. Sugiura and H. Minoura, *Chem. Mater.*, 11, 2657 (1999).

[254] T. Oekermann, T. Yoshida, C. Boeckler, J. Caro and H. Minoura, *J. Phys. Chem. B*, 109, 12560 (2005).

[255] T. Oekermann, S. Karuppuchamy, T. Yoshida, D. Schlettwein, D. Wohrle and H. Minoura, *J. Electrochem. Soc.*, **151**, C62 (2004).

[256] T. Yoshida, K. Terada, D. Schlettwein, T. Oekermann, T. Sugiura and H. Minoura, *Adv. Mater.*, **12**, 1214 (2000).

[257] T. Yoshida and H. Minoura, *Adv. Mater.*, **12**, 1219 (2000).

[258] Z. Zhang, H. Yu, X. Shao and M. Han, *Chem. Eur. J.*, **11**, 31493154 (2005).

[259] M. H. Huang, Y. Wu, H. Feick, N. Tran, E. Weber and P. Yang, *Adv. Mater.*, **13**, 113 (2001).

[260] M. H. Huang, S. Mao, H. Feick, H. Yan, Y. Wu, H. Kind, E. Weber, R. Russo and P. Yang, *Science*, **292**, 1897 (2001).

[261] S. Chappel and A. Zaban, *Sol. Energy Mater. Sol. Cells*, **71**, 141 (2002).

[262] B. V. Bergeron, A. Marton, G. Oskam and G. J. Meyer, *J. Phys. Chem. B*, **109**, 937 (2005).

[263] B. Onwona-Agyeman, S. Kaneko, A. Kumara, M. Okuya, K. Murakami, A. Konno and K. Tennakone, *Jpn. J. Appl. Phys.*, **44**, L731 (2005).

[264] S. Gubbala, H. B. Russell, H. Shah, B. Deb, J. Jasinski, H. Rypkema and M. K. Sunkara, *Energy Environ. Sci.*, **2**, 1302 (2009).

[265] A. Kay and M. Grätzel, *Chem. Mater.*, **14**, 2930 (2002).

[266] V. Thavasi, V. Renugopalakrishnan, R. Jose and S. Ramakrishna, *Mater. Sci. Eng.*, **63**, 81 (2009).

[267] J. Bandara and U.W. Pradeep, *Thin Solid Films*, **517**, 952 (2008).

[268] N. Papageorgiou, W. F. Maier and M. Grätzel, *J. Electrochem. Soc.*, **144**, 876 (1997).

[269] T. N. Murakami and M. Grätzel, *Inorg. Chim. Acta*, **361**, 572 (2008).

[270] P. Li, J. Wu, J. Lin, M. Huang, Z. Lan and Q. Li, *Electrochim. Acta*, **53**, 4161 (2008).

[271] A. Hauch and A. Georg, *Electrochim. Acta*, **46**, 3457 (2001).

[272] A. J. Bard and L. R. Faulkner, *Electrochemical Methods: Fundamentals and Applications*, 2nd Edn, John Wiley & Sons, Ltd, New York, 2001.

[273] T. Ma, X. Fang, M. Akiyama, K. Inoue, H. Noma and E. Abe, *J. Electroanal. Chem.*, **574**, 77 (2004).

[274] Q. Wang, S. Ito, M. Grätzel, F. Fabregat-Santiago, I. Mora-Sero, J. Bisquert, T. Bessho, and H. Imai, *J. Phys. Chem. B*, **110**, 25210 (2006).

[275] S. Ito, N. L. Cevey Ha, G. Rothenberger, P. Liska, P. Comte, S. M. Zakeeruddin, P. Péchy, Md. K. Nazeeruddin and M. Grätzel, *Chem. Commun.*, 4004 (2006).

[276] F. Cai, J. Liang, Z. Tao, J. Chen and R. Xu, *J. Power Sources*, **177**, 631 (2008).

[277] S. S. Kim, K. W. Park, J. H. Yum and Y. E. Sung, *Sol. Energy Mater. Sol. Cells*, **90**, 283 (2006).

[278] G. Khelashvili, S. Behrens, C. Weidenthaler, C. Vetter, A. Hinsch, R. Kern, K. Skupien, E. Dinjus and H. Bönnemann, *Thin Solid Films*, **511–512**, 342 (2006).

[279] T. C. Wei, C. C. Wan and Y. Y. Wang, *Appl. Phys. Lett.*, **88**, 103122 (2006).

[280] A. Kay and M. Grätzel, *Sol. Energy Mater. Sol. Cells*, **44**, 99 (1996).

[281] K. Imoto, K. Takahashi, T. Yamaguchi, T. Komura, J. Nakamura and K. Murata, *Sol. Energy Mater. Sol. Cells*, **79**, 459 (2003).

[282] F. Cai, J. Chen and R. Xu, *Chem. Lett.*, **35**, 1266 (2006).

[283] S. Gagliardi, L. Giorgi, R. Giorgi, N. Lisi, Th. Dikonimos Makris, E. Salernitano and A. Rufoloni, *Superlattice Microstr.*, **46**, 205 (2009).

[284] K. Suzuki, M. Yamamoto, M. Kumagai and S. Yanagida, *Chem. Lett.*, **32**, 28 (2003).

[285] T. N. Murakami, S. Ito, Q. Wang, Md. K. Nazeeruddin, T. Bessho, I. Cesar, P. Liska, R. Humphry-Baker, P. Comte, P. Péchy and M. Grätzel, *J. Electrochem. Soc.*, **153**, A2255 (2006).

[286] T. Hoshikawa, M. Yamada, R. Kikuchi and K. Eguchi, *J. Electrochem. Soc.*, **152**, E68 (2005).

[287] Y. Saito, T. Kitamura, Y. Wada and S. Yanagida, *Chem. Lett.*, **31**, 1060 (2002).

[288] T. Muto, M. Ikegami, K. Kobayashi and T. Miyasaka, *Chem. Lett.*, **36**, 804 (2007).

[289] J. Xia, N. Masaki, K. Jiang and S. Yanagida, *J. Mater. Chem.*, **17**, 2845 (2007).

[290] K.-M. Lee, C.-Y. Hsu, P.-Y. Chen, M. Ikegami, T. Miyasaka and K.-C. Ho, *Phys. Chem. Chem. Phys.*, **11**, 3375 (2009).

[291] T. Kitamura, M. Maitani, M. Matsuda, Y. Wada and S. Yanagida, *Chem. Lett.*, **30**, 1054 (2001).

[292] Y. Saito, W. Kubo, T. Kitamura, Y. Wada and S. Yanagida, *J. Photochem. Photobiol., A*, **164**, 153 (2004).

[293] N. Fukuri, Y. Saito, W. Kubo, G. K. R. Senadeera, T. Kitamura, Y. Wada and S. Yanagida, *J. Electrochem. Soc.*, **151**, A1745 (2004).

[294] R. Senadeera, N. Fukuri, Y. Saito, T. Kitamura, Y. Wada and S. Yanagida, *Chem. Commun.*, 2259 (2005).

[295] N. Ikeda, K. Teshima and T. Miyasaka, *Chem. Commun.*, 1733 (2006).

[296] J. He, H. Lindström, A. Hagfeldt and S.-E. Lindquist, *Sol. Energy Mater. Sol. Cells*, **62**, 265 (2000).

[297] Y. Hu, Z. Zheng, H. Jia, Y. Tang and L. Zhang, *J. Phys. Chem. C*, **112**, 13037 (2008).

[298] K. Tennakone, A. R. Kumarasinghe, P. M. Sirimanne and G. R. R. A. Kumara, *Thin Solid Films*, **261**, 307 (1995).

[299] K. Tennakone, M. Kahanda, C. Kasige, P. Abeysooriya, R. H. Wijayanayaka and P. Kaviratna, *J. Electrochem. Soc.*, **131**, 1574 (1984).

[300] B. O'Regan and D. T. Schwartz, *Chem. Mater.*, **7**, 1349 (1996).

[301] J. He, H. Lindström, A. Hagfeldt and S.-E. Lindquist, *J. Phys Chem. B*, **103**, 8940 (1999).

[302] S. C. Choi, K. Koumoto and H. Yanagida, *J. Mater. Sci.*, **21**, 1947 (1986).

[303] G. Boschloo and A. Hagfeldt, *J. Phys. Chem. B*, **105**, 3039 (2001).

[304] D. M. Tench and E. Yeager, *J. Electrochem. Soc.*, **120**, 164 (1973).

[305] A. Nakasa, H. Usami, S. Sumikura, S. Hasegawa, T. Koyama and E. Suzuki, *Chem Lett.*, **34**, 500 (2005).

[306] A. Nattestad, M. Ferguson, R. Kerr, Y.-B Cheng and U. Bach, *Nanotechnology*, **19**, 295304 (2008).

[307] S. Sumikura, S. Mori, S. Shimizu, H. Usami and E. Suzuki, *J. Photochem. Photobiol., A*, **199**, 1 (2008).

[308] F. Vera, R. Schrebler, E. Muñoz, C. Suarez, P. Cury, H. Gómez, R. Córdova, R. E. Marotti and E. A. Dalchiele, *Thin Solid Films*, **490**, 182 (2005).

[309] H. Zhu, A. Hagfeldt and G. Boschloo, *J. Phys. Chem. C*, **111**, 17455 (2007).

[310] M. Borgström, E. Blart, G. Boschloo, E. Mukhtar, A. Hagfeldt, L. Hammartström and F. Odobel, *J. Phys. Chem.*, **109**, 22928 (2005).

[311] Y. Mizoguchi and S. Fujihara, *Electrochem. Solid-State Lett.*, **11**, 8 [Note(s): K78-K80] (2008).

[312] P. Qin, H. Zhu, T. Edvinsson, G. Boschloo, A. Hagfeldt and L. Sun, *J. Am. Chem. Soc.*, **130**, 8570 (2008).

[313] A. Morandeira, J. Fortage, T. Edvinsson, L. Le Pleux, E. Blart, G. Boschloo, A. Hagfeldt, L. Hammarström and F. Odobel, *J. Phys. Chem. C*, **112**, 1721 (2008).

[314] S. Mori, S. Fukuda, S. Sumikura, Y. Takeda, Y. Tamaki, E. Suzuki and T. Abe, *J. Phys. Chem. C*, **112**, 16134 (2008).

[315] P. Qin, M. Linder, T. Brink, G. Boschloo, A. Hagfeldt and Licheng Sun, *Adv. Mater.*, **21**, 2993 (2009).

[316] L. Li, E. A. Gibson, P. Qin, G. Boschloo, M. Gorlov, A. Hagfeldt and L. Sun, *Adv. Mater.*, **22**, 1759 (2009).

[317] A. Morandeira, G. Boschloo, A. Hagfeldt and L. Hammarström, *J. Phys. Chem. B*, **109**, 19403 (2005).

[318] X. Wang and D. M. Stanbury, *Inorg. Chem.*, **45**, 3415 (2006).

[319] S. A. Sapp, M. Elliott, C. Contado, S. Caramori and C. A. Bignozzi, *J. Am. Chem. Soc.*, **124**, 11215 (2002).

[320] S. Nakade, Y. Makimoto, W. Kubo, T. Kitamura, Y. Wada and S. Yanagida, *J. Phys. Chem. B*, **109**, 3488 (2005).

[321] D. Martineau, M. Beley, P. C. Gros, S. Cazzanti, S. Caramori and C. A. Bignozzi, *Inorg. Chem.*, **46**, 2272 (2007).

[322] B. M. Klahr and T. W. Hamann, *J. Phys. Chem. C*, **113**, 14040 (2009).

[323] H. Nusbaumer, S. M. Zakeeruddin, J.-E. Moser and M. Grätzel, *Chem. Eur. J.*, **9**, 3756 (2003).

[324] E. A. Gibson, A. L. Smeigh, L. Le Pleux, J. Fortage, G. Boschloo, E. Blart, Y. Pellegrin, F. Odobel, A. Hagfeldt and L. Hammarström, *Angew. Chem. Int Ed.*, **48**, 4402 (2009).

[325] S. Sheng, G. Fang, C. Li, S. Xu and X. Zhao, *Phys. Status Solidi A*, **203**, 1891 (2006).

[326] K. Tennakone, C. A. N. Fernando, M. Dewasurendra and M. S. Kariappert, *Jpn. J. Appl. Phys.*, **26**, 561 (1987).

[327] S. Nakabayashi, N. Ohta and A. Fujishima, *Phys. Chem. Chem. Phys.*, **1**, 3993 (1999).

[328] S. Sumikura, S. Mori, S. Shimizu, H. Usami and E. Suzuki, *J. Photochem. Photobiol.*, A, **194**, 143 (2008).

[329] A. Kay and M. Grätzel, *Sol. Energy Mater. Sol. Cells*, **44**, 99 (1996).

[330] H. Pettersson, T. Gruszecki, R. Bernhard, L. Häggman, M. Gorlov, G. Boschloo, T. Edvinsson, L. Kloo and A. Hagfeldt, *Prog. Photovolt: Res. Appl.*, **15**, 113 (2007).

[331] H. Han, U. Bach, Y.-B. Cheng, R. A. Caruso and C. MacRae, *Appl. Phys. Lett.*, **94**, 103102 (2009).

[332] M. K. Nazeeruddin, P. Péchy, T. Renouard, S. M. Zakeeruddin, R. Humphry-Baker, P. Comte, P. Liska, L. Cevey, E. Costa, V. Shklover, L. Spiccia, G. B. Deacon, C. A. Bignozzi and M. Grätzel, *J. Am. Chem. Soc.*, **123**, 1613 (2001).

[333] A. Ehret, L. Stuhl and M. T. Spitler, *J. Phys. Chem. B*, **105**, 9960 (2001).

[334] J.-J. Cid, J.-H. Yum, S.-R. Jang, M. K. Nazeeruddin, E. Martínez-Ferrero, E. Palomares, J. Ko, M. Grätzel and T. Torres, *Angew. Chem. Int. Ed.* **46**, 8358 (2007).

[335] D. Kuang, P. Walter, F. Nüesch, S. Kim, J. Ko, P. Comte, S. M. Zakeeruddin, M. K. Nazeeruddin and M. Grätzel, *Langmuir*, **23**, 10906 (2007).

[336] F. Inakazu, Y. Noma, Y. Ogomi and S. Hayase, *Appl. Phys. Lett.*, **93**, 093304 (2008).

[337] K. Lee, S. W. Park, M. J. Ko, K. Kim and N.-G. Park, *Nat. Mater.*, **8**, 665 (2009).

[338] J. N. Clifford, E. Palomares, M. K. Nazeeruddin, R. Thampi, M. Grätzel and J. R. Durrant, *J. Am. Chem. Soc.*, **126**, 5670 (2004).

[339] H. Choi, S. Kim, S. O. Kang, J. Ko, M.-S. Kang, J. N. Clifford, A. Forneli, E. Palomares, M. K. Nazeeruddin and M. Grätzel, *Angew. Chem. Int. Ed.*, **47**, 8259 (2008).

[340] S. Nishimura, N. Abrams, B. A. Lewis, L. I. Halaoui, T. E. Mallouk, K. D. Benkstein, J. van de Lagemaat and A. J. Frank, *J. Am. Chem. Soc.*, **125**, 6306 (2003).

[341] A. Mihi and H Míguez, *J. Phys. Chem. B*, **109**, 15968 (2005).

[342] A. Mihi, F. J. López,-Alcaras and H. Míguez, *Appl. Phys. Lett.*, **88**, 193110 (2006).

[343] Q. Wang, J.-E. Moser and M. Grätzel, *J. Phys. Chem. B*, **109**, 14945 (2005).

[344] S. Colodrero, A. Mihi, L. Häggman, M. Ocaña, G. Boschloo, A. Hagfeldt and H. Miguez, *Adv. Mater.*, **21**, 764 (2009).

[345] S. P. Bremner, M. Y. Levy and C. B. Honsberg, *Prog. Photovolt: Res. Appl.*, **16**, 225 (2008).

[346] W. Kubo, A. Sakamoto, T. Kitamura, Y. Wada and S. Yanagida, *J. Photochem. Photobiol.*, *A*, **164**, 33 (2004).

[347] M. Dürr, A. Bamedi, A. Yasuda and G. Nelles, *Appl. Phys. Lett.*, **84**, 3397 (2004).

[348] T. Yamaguchi, Y. Uchida, S. Agatsuma and H. Arakawa, *Sol. Energy Mater. Sol. Cells*, **93**, 733 (2009).

[349] J. Usagawa, S. S. Pandey, S. Hayase, M. Kono and Y. Yamaguchi, *Appl. Phys. Express*, **2**, 062203 (2009).

[350] P. Liska, K. R. Thampi, M. Grätzel, D. Brémaud, D. Rudmann, H. M. Upadhyaya and A. N. Tiwari, *Appl. Phys. Lett.*, **88**, 203103 (2006).

[351] S. Wenger, S. Seyrling, A. N. Tiwari and M. Grätzel, *Appl. Phys. Lett.*, **94**, 173508 (2009).

[352] F. Gao, Y. Wang, D. Shi, J. Zhang, M. K. Wang, X. Y. Jing, R. Humphry-Baker, P. Wang, S. M. Zakeeruddin and M. Grätzel, *J. Am. Chem. Soc.*, **130**, 10720 (2008).

[353] I. Kaiser, K. Ernst, Ch.-H. Fischer, R. Könenkamp, C. Rost, I. Sieber and M. Ch. Lux-Steiner, *Sol. Energy Mater. Sol. Cells*, **67**, 89 (2001).

[354] R. Tena-Zaera, M. A. Ryan, A. Katty, G. Hodes, S. Bastide and C. Lévy-Clément, *C. R. Chim.*, **9**, 717 (2006).

[355] K. Taretto and U. Rau, *Prog. Photovoltaics*, **12**, 573 (2004).

[356] R. Könenkamp, L. Dloczik, K. Ernst and C. Olesch, *Physica E*, **14**, 219 (2002).

[357] K. Tennakone, G. R. R. A. Kumara, I. R. M. Kottegoda, V. P. S. Perera and G. M. L. P. Aponsu, *J. Phys. D: Appl. Phys.*, **31**, 2326 (1998).

[358] D. Kieven, T. Dittrich, A. Belaidi, J. Tornow, K. Schwarzburg, N. Allsop and M. Lux-Steiner, *Appl. Phys. Lett.*, **92**, 153107 (2008).

[359] A. Belaidi, T. Dittrich, D. Kieven, J. Tornow, K. Schwarzburg and M. Lux-Steiner, *Phys. Status Solidi (RRL)*, **2**, 172 (2008).

[360] M. Nanu, J. Schoonman and A. Goossens, *Adv. Funct. Mater.*, **15**, 95 (2005).

[361] R. Könenkamp, P. Hoyer and A. Wahi, *J. Appl. Phys.*, **79**, 7029 (1996).

[362] S. J. C. Irvine, V. Barrioz, A. Stafford and K. Durose, *Thin Solid Films*, **480–481**, 76 (2005).

[363] K. Ernst, R. Engelhardt, K. Ellmer, C. Kelch, H.-J. Muffler, M.-Ch. Lux-Steiner and R. Könenkamp, *Thin Solid Films*, **387**, 26 (2001).

[364] G. Larramona, C. Choné, A. Jacob, D. Sakakura, B. Delatouche, D. Péré, X. Cieren, M. Nagino and R. Bayón, *Chem. Mater.*, **18**, 1688 (2006).

[365] A. Belaidi, R. Bayón, L. Dloczik, K. Ernst, M.Ch. Lux-Steiner and R. Könenkamp, *Thin Solid Films*, **431–432**, 488 (2003).

[366] M. Page, O. Niitsoo, Y. Itzhaik, D. Cahen and G. Hodes, *Energy Environ. Sci.*, **2**, 220 (2009).

[367] Y. Itzhaik, O. Niitsoo, M. Page and G. Hodes, *J. Phys. Chem. C*, **113**, 4254 (2009).

[368] K. Ernst, A. Belaidi and R. Könenkamp, *Semicond. Sci. Technol.*, **18**, 475 (2003).

[369] M. Biancardo and F. C. Krebs, *Sol. Energy Mater. Sol. Cells*, **91**, 1755 (2007).

[370] R. Tena-Zaera, A. Kattya, S. Bastidea, C. Lévy-Clément, B. O'Regan and V. Muñoz-Sanjosé, *Thin Solid Films*, **483**, 372 (2005).

[371] J. E. Murphy, M. C. Beard, A. G. Norman, S. P. Ahrenkiel, J. C. Johnson, P. Yu, O. I. Mićić, R. J. Ellingson and A. J. Nozik, *J. Am. Chem. Soc.*, **128**, 3241 (2006).

[372] P. V. Kamat, *J. Phys. Chem. C*, **112**, 18737 (2008).

[373] A. J. Nozik, *Physica E*, **14**, 115 (2002).

[374] D. Liu and P. V. Kamat, *J. Phys. Chem.*, **97**, 10769 (1993).

[375] D. Liu and P. V. Kamat, *J. Electroanal. Chem. Interfacial Electrochem.*, **347**, 451 (1993).

[376] A. J. Nozik, *Chem. Phys. Lett.*, **457**, 3 (2008).

[377] R. D. Schaller and V. I. Klimov, *Phys. Rev. Lett.*, **92**, 186601 (2004).

[378] R. J. Ellingson, M. C. Beard, J. C. Johnson, P. Yu, O. I. Micic, A. J. Nozik, A. Shabaev and A. L. Efros, *Nano Lett.*, **5**, 865 (2005).

[379] R. Vogel, P. Hoyer and H. Weller, *J. Phys. Chem.*, **98**, 3183 (1994).

[380] S. Hotchandani and P. V. Kamat, *J. Phys. Chem.*, **96**, 6834 (1992).

[381] A. Kongkanand, K. Tvrdy, K. Takechi, M. K. Kuno and P. V. Kamat, *J. Am. Chem. Soc.*, **130**, 4007 (2008).

[382] I. Robel, M. Kuno and P. V. Kamat. *J. Am.Chem. Soc.*, **129**, 4136 (2007).

[383] I. Robel, V. Subramanian, M. Kuno and P. V. Kamat, *J. Am. Chem. Soc.*, **128**, 2385 (2006).

[384] Q. Shen, M. Yanai, K. Katayama, T. Sawada and T. Toyoda, *Chem. Phys. Lett.*, **442**, 89 (2007).

[385] W. W. Yu, L. Qu, W. Guo and X. Peng, *Chem. Mater.*, **15**, 2854 (2003).

[386] I. Mora-Seró, S. Giménez, T. Moeh, F. Fabregat-Santiago, T. Lana-Villareal, R. Gómez and J. Bisquert, *Nanotechnology*, **19**, 424007 (2008).

[387] N. Guijarro, T. Lana-Villarreal, I. Mora-Seró, J. Bisquert and R. Gómez, *J. Phys. Chem. C*, **113**, 4208 (2009).

[388] K. S. Leschkies, R. Divakar, J. Basu, E. Enache-Pommer, J. E. Boercker, C. B. Carter, U. R. Kortshagen, D. J. Norris and E. S. Aydil, *Nano Lett.*, **7**, 1793 (2007).

[389] R. Plass, S. Pelet, J. Krueger M. Gratzel and U. Bach, *J. Phys. Chem. B*, **106**, 7578 (2002).

[390] P. Yu, K. Zhu, A. G. Norman, S. Ferrere, A. J. Frank and A. J. Nozik, *J. Phys. Chem. B*, **110**, 25451 (2006).

[391] H. J. Lee, J.-H. Yum, H. C. Leventis, S. M. Zakeeruddin, S. A. Haque, P. Chen, S. I. Seok, M. Grätzel and M. K. Nazeeruddin, *J. Phys. Chem. C*, **112**, 11600 (2008).

4

Hydrogen Adsorption on Metal Organic Framework Materials for Storage Applications

K. Mark Thomas

Sir Joseph Swan Institute for Energy Research and School of Chemical Engineering and Advanced Materials, Newcastle University, Newcastle upon Tyne, UK

4.1 INTRODUCTION

The current interest in hydrogen as a sustainable fuel is due to strategic factors relating to security of supply with concerns that world oil production is close to reaching a peak[1] and environmental concerns of global warming due to carbon dioxide emissions from fossil fuel combustion. The development of a hydrogen-based economy produces major scientific and engineering challenges, which need to be overcome before widespread use of hydrogen can be achieved. The technologies for handling and storing hydrogen in an industrial situation where safety issues can be closely controlled are well established. However, the use of hydrogen for transport applications is much more difficult because of the direct involvement of the general public in the use of hydrogen and economic, scientific and engineering constraints.

Energy Materials   Edited by Duncan W. Bruce, Dermot O'Hare and Richard I. Walton
© 2011 John Wiley & Sons, Ltd.

Hydrogen storage requirements for vehicles are very challenging and are determined by safety issues, refuelling range and overall system storage weight and volume considerations. The US DOE hydrogen on-board storage targets for vehicles are 6.0 wt% (45 g l^{-1}) for 2010 and 9.0 wt% (81 g l^{-1}) for 2015 under near ambient conditions.[2] The target for 2015 is a hydrogen powered vehicle with storage of 5–13 kg of hydrogen and a refuelling range >300 miles (483 km). The current US DOE targets are for the current US weighted average corporate vehicle and it is likely that more fuel efficient smaller and lighter cars will be used in the future leading to slightly less demanding hydrogen storage requirements. These targets are system requirements and therefore the weight of the storage tank and associated cooling, pressure and delivery control equipment must be included in the calculation for real vehicle applications.

Hydrogen has a higher combustion enthalpy than hydrocarbon fuels on a weight basis but a much lower value on a volume basis. As a consequence, when hydrogen is used as a fuel for transport applications, a much larger tank is required than currently used for hydrocarbon fuelled vehicles. Safety issues and convenience are paramount and these lead to severe practical limitations due to factors such as tank size, refuelling range, *etc.* Fuel cell vehicles provide greater efficiency thereby reducing the hydrogen storage requirements compared with internal combustion engines but increased storage volumes are still required compared with current vehicles. The durability of fuel cells and their current reliance on precious metal catalyst are issues. There are also possible environmental benefits provided that hydrogen can be produced in an environmentally sustainable manner. The development of a hydrogen economy has potential strategic and environmental benefits by reducing reliance on oil and air pollution.

The major scientific challenge to overcome before the technology necessary for change from petroleum to sustainable hydrogen as an energy carrier for transport applications can be realised is the development of a suitable storage method. The methods currently under consideration include high pressure gas, liquid hydrogen, adsorption on porous materials at relatively low pressure, complex hydrides and hydrogen intercalation in metals.[3] The requirements for vehicles are constrained by the volume required to store sufficient hydrogen for acceptable distances between refuelling. Although hydrogen storage tanks up to 70 MPa have been used there is still a limitation on the amount of hydrogen that can be delivered from a given size of tank and the energy required to compress the gas also needs to be considered. Compressed hydrogen gas storage tanks do not meet all the criteria for use in transport applications. Also, there are safety issues for vehicles with tanks containing very high hydrogen pressures. Liquid hydrogen provides a higher hydrogen storage density ($\sim \times 2$) than that of a compressed hydrogen gas tank but there are limitations due to the extremely low temperatures required (critical temperature 32.98 K)[4] resulting in significant losses during storage due to evaporation. Liquefaction costs are also significant. Currently, criteria for hydrogen storage for vehicles cannot be satisfied completely by compressed gas and liquid hydrogen storage methods currently available. Therefore, storage methods are being considered which involve storing hydrogen in a high-pressure vessel containing porous materials, metal hydrides or other materials. This combines the advantages of both compressed gas and solid state material storage to increase the total hydrogen storage capacity at lower pressures than compressed gas alone.

Porous materials have large internal surface areas due to the presence of nanometre-sized pores. They are widely used for gas storage,[3,5] purification and separation,[6,7] removal of trace pollutant species,[8,9] and as catalysts and catalyst supports.[10,11] Physisorption of hydrogen on porous materials is one of the main materials storage methods being investigated for use for transport applications.[12–20] Hydrogen adsorption has been investigated for a wide range of porous materials including, carbons, aluminosilicates, aluminophosphates, silicas, microporous polymers, covalent organic frameworks (COFs) and porous metal organic frameworks (MOFs). Comparison of hydrogen adsorption characteristics for a wide range of porous materials shows good correlations with pore structural characteristics determined by gas adsorption methods.[13,17,19]

MOFs have diverse surface chemistry,[21] pore architectures,[22] framework flexibility[12] and very large internal surface areas and pore volumes.[23] Their design is capable of being tailored in terms of pore structure and surface chemistry for specific applications. These materials are synthesised in solution from a wide range of multidentate ligands and metals or metal clusters [Secondary Building Units (SBUs)] to give materials with a very wide range of surface chemistry and porous structure characteristics.[22,24,25] Combinations of inorganic and organic linker moieties give a huge number of possible structures and many are known based on triangular, square, tetrahedral and octahedral SBUs, which lead to default structural nets. MOF structures are mainly dependent on the coordination geometry of the SBUs, which are framework joints that combine with linker units to form the framework skeleton. MOF structures with the same net can often be made by functionalising the organic linker units and this allows surface chemistry to be modified for specific interactions while maintaining the net structure. Similarly, lengthening the linker ligand may also give structures with the same net leading to a series of structures with the same crystallographic space group, but with a range of surface areas, pore sizes and pore volumes. However, lengthening the linkers to increase the distance between nodes often leads to interpenetration and this may be a limitation for synthetic studies. MOFs with zeolite type structures have also been synthesised and these may have applications for gas storage.[26,27]

The synthesis of MOFs with optimum pore size and shape, and surface chemistry for hydrogen adsorption are possible using rational design strategies. The current status of studies of hydrogen adsorption and storage on porous MOFs is reviewed and future prospects are discussed.

4.2   HYDROGEN ADSORPTION EXPERIMENTAL METHODS

Accurate measurements of hydrogen adsorption are difficult requiring stringent experimental protocols and validation using extensive reproducibility studies.[13,24,28] Hydrogen adsorption measurements for porous materials can be acquired using either (1) gravimetric instruments, which provide a direct measurement of adsorption or (2) volumetric (or manometric) measurements (Sievert's Apparatus), which is an indirect method involving adding known amounts of gas to a sample. Volumetric instruments have a simpler design, which is more economic to construct for high pressure instruments. Adsorption/desorption is driven by pressure differences in volumetric instruments and therefore the instruments are not isobaric. Gravimetric measurements are usually carried out under isobaric conditions (constant chemical potential) and this method can provide accurate adsorption/desorption kinetics.[6,29] Both methods require an ultra-clean high vacuum system with all metal seals with diaphragm and turbo pumps, which can be evacuated to 10^{-10} bar. A purification system is also needed to remove trace impurities in the hydrogen.

Gravimetric measurements require corrections to account for buoyancy effects while volumetric measurements must be corrected for the sample volume. It is often not realised that these corrections are equivalent.[28] Helium is used to determine both corrections since it is assumed not to be adsorbed. The degas procedure for gravimetric instruments involves continuous monitoring of mass in relation to time, which allows outgas rate to be specified and the equilibrium state and any sample mass losses to be quantified. Establishing that the sample is completely degassed is more difficult in the case of volumetric measurements and errors associated with introduction of doses of gas are cumulative for each isotherm pressure step and increase with increasing pressure.[28,30] Also, the importance of the amount remaining unadsorbed in the deadspace increases with increasing pressure.[31] The propagation of errors is a major source of uncertainty in determining the accuracy of high pressure hydrogen adsorption measurements because of the relative magnitudes of the corrections and amounts adsorbed. Broom and Moretto have provided a detailed discussion of sources of errors for hydrogen adsorption measurements.[28,32] Some reproducibility studies carried out by various laboratories have been published[33,34] and in these cases reasonable agreement has been obtained.

Adsorption isotherms are presented on either surface excess or absolute basis in mol g^{-1} of outgassed adsorbent.[31,35] Sometimes, for gas storage applications, total wt% loading is quoted.[36-38] Typical values for the density of adsorbed hydrogen estimated from the maximum amount of hydrogen adsorbed with the total pore volumes or crystallographic pore volumes determined using the PLATON software package, are in the range 0.045–0.07 g cm^{-3}.[39-41] The adsorbed hydrogen phase is similar to an incompressible fluid and the adsorbate density can be compared with the values for liquid hydrogen (0.0708 g cm^{-3} at 20.28 K and 0.077 g cm^{-3} at 13.8 K).[4] The density of liquid hydrogen probably represents an upper limit for the density of hydrogen adsorbed in pores at 77 K and higher temperatures.

4.3 ACTIVATION OF MOFS

Although the structure of the first coordination polymer was reported[42] in 1959, it was 1989 when the structure of another similar material was reported by Hoskins and Robson.[43] The first report of gas adsorption studies on a porous MOF was published in 1997.[44] The stability,[37] purity[45] and activation procedures[46] for MOFs are sometimes issues resulting in difficulties in obtaining repeatable adsorption results. Comparison of X-ray powder diffraction data with single crystal data allows the purity of MOF samples prior to activation to be confirmed. Ideally, the crystalline structure of the porous MOF material formed by the activation procedure is needed to establish that the framework is intact.

MOFs are synthesised in solution by a variety of methods such as slow diffusion techniques and hydrothermal and solvothermal heating in an autoclave under autogeneous pressure control to temperatures up to $\sim$473 K. The guests (templates, solvent or coordinated solvent trapped inside the framework) need to be removed to form the porous structure, which is then accessible for guest adsorption. Framework-forming bonds, which form the porous structure, are sometimes relatively weak and therefore, the activation procedure needs to be as mild as possible to avoid structural change and optimise pore volume. Comparison of crystallographic and total pore volumes obtained from gas adsorption studies also allows the activation procedure and purity of the bulk material to be assessed. The pore volumes obtained from gas adsorption studies may be lower than the corresponding values determined from crystallographic studies as a result of pore collapse or blockage on solvent removal. Removal of guests from porous structures can be achieved by heating in ultra high vacuum, exchanging with a more volatile solvent to facilitate removal and extraction with supercritical carbon dioxide. Comparison of the methods for four MOFs shows that supercritical carbon dioxide drying leads to large increases in the surface area as determined by gas adsorption measurements.[47] It was proposed[47] that supercritical carbon dioxide extraction inhibits mesopore collapse leaving micropores accessible to gases. It is apparent that supercritical carbon dioxide drying has advantages in activating the porous structures of MOFs.

The removal of coordinated solvent may lead to the formation of unsaturated or 'open' metal centres, which are high energy sites for adsorption. Therefore, thermal and chemical stability of desolvated MOF porous material structures and activation procedures are important considerations when comparing materials.

Pores are classified by the International Union of Pure and Applied Chemistry (IUPAC) by pore size as micropores ($<$2 nm), mesopores (2–50 nm) and macropores ($>$50 nm).[31] Micropores are sometimes divided into ultramicropores ($<$0.7 nm) and supermicropores (1.4–2.0 nm). The terms 'nanopore' and 'nanoporosity' are not defined precisely but refer to nanometre-sized pores. Characterisation of the porous structures of materials is difficult because some MOF materials are flexible. A variety of isotherm equations and adsorptives have been used to characterise porous structures using gas adsorption techniques. Porous structures are characterised by surface areas [determined using Langmuir, Brunauer–Emmett–Teller (BET), Dubinin–Radushkevich (DR), *etc.*, equations], pore volumes [total, micropore (DR), *etc.*] and pore size distributions.

Parameters obtained such as pore volumes, surface areas and pore size distributions are estimates for relative comparisons.

Measurements of BET nitrogen surface area at 77 K are frequently used as a standard procedure for characterisation of porous materials.[48] Studies suggested that BET surface area correlates with MOF surface areas determined from crystallographic data.[49] However, heterogeneous surfaces are found in MOFs and it is best to describe the BET surface area as an 'apparent' surface area. If a Type I isotherm has a plateau at high relative pressure, the micropore volume is given by the amount adsorbed (converted to a liquid volume) since the mesopore volume and external surface are both relatively small. Porous structure characterisation parameters (Langmuir and BET surface areas and pore volumes) obtained from gas adsorption studies of MOFs have interrelated correlations.[19]

4.4 HYDROGEN ADSORPTION ON MOFS

Figure 4.1 shows high pressure hydrogen isotherms on a surface excess and total hydrogen basis for polyhedral framework material $[Cu_3(L) \cdot (H_2O)_3)] \cdot 8DMSO \cdot 15DMF \cdot 3H_2O$ where $H_6L = 1,3,5$-tris(3′,5′-dicarboxy[1,1′-biphenyl]-4-yl)benzene (NOTT-112) over the pressure range 0–77 bar at 77 K.[50] The maximum surface excess is 7.07 wt% at 35–45 bar while the total hydrogen adsorption was 10 wt% at 77 bar. The surface excess is the amount of hydrogen present in the interfacial layer over the gas concentration present at the same equilibrium gas pressure, in which the gas phase concentration is constant up to the Gibbs surface.[31] The surface excess adsorption isotherm goes through a maximum with increasing pressure. The total amount present is calculated taking into account the high pressure gas present in the pores at equilibrium. The overall structure has three types of polyhedra: cage A with 12 open copper sites and diameter 13 Å; cage B with a diameter of 13.9 Å; and cage C with a diameter of 20 Å. MOFs with large pore volumes and BET surface areas usually have relatively low uptakes at 1 bar pressure. However, this MOF also has an appreciable uptake of 2.3 wt% at 1 bar, which is attributed to open Cu(II) centres present in the relatively small Cage A.

Isoreticular metal organic framework (IRMOF) materials were part of the series of MOF materials used in the identification of the correlation between hydrogen saturation amount and Langmuir surface area determined from nitrogen adsorption at 77 K.[51] Hydrogen adsorption on

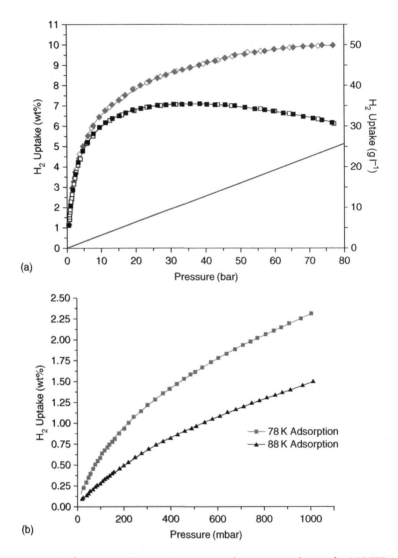

Figure 4.1 High pressure H_2 sorption excess adsorption isotherms for NOTT-112. (a) Surface excess (squares), total uptake (circles) and density of gas (line). Filled symbols (adsorption); open symbols (desorption). (b) Gravimetric total H_2 uptakes at 78 K and 88 K up to 1 bar. Reprinted with permission from Yan *et al.*, 2009[50]. Copyright (2009) Royal Society of Chemistry (RSC)

MOFs has been investigated for two series where there is a systematic change in ligand length to modify the pore volume without changing the structural topology: (1) IRMOFs of Yaghi and co-workers[5,51–57] based on Zn_4OL_3 stoichiometry (where L = various dicarboxylate linkers) (see

Figure 4.2); and (2) the copper tetracarboxylate homologous series with stoichiometry Cu_2L' (where L' = tetracarboxylate linkers) investigated by Lin *et al.* (see Figure 4.3).[38,39] Figure 4.4 shows the variation in the amount of hydrogen adsorbed at 1 bar and also high pressure (at or close to saturation) with BET and Langmuir surface areas, and pore volumes for the dicarboxylate IRMOF series. It is apparent that there are good correlations between the hydrogen uptake at high pressure and the pore characterisation parameters. No correlation was observed between hydrogen uptake at 1 bar and BET surface area and pore volume. Figure 4.5 shows the variation in the amount of hydrogen adsorbed at 1 bar and high pressure with BET and total pore volume for the tetra-carboxylate (NOTT) series. Good correlations were observed between the hydrogen uptake at high pressure and the pore characterisation parameters. This tetracarboxylate series of high surface area MOFs also does not show any correlation between the hydrogen uptakes at 1 bar pressure and pore structure characterisation parameters. Both these series show a progressive increase in hydrogen surface excess adsorption at or close to saturation at high pressure with increasing pore size, pore volume and surface area characterisation parameters.

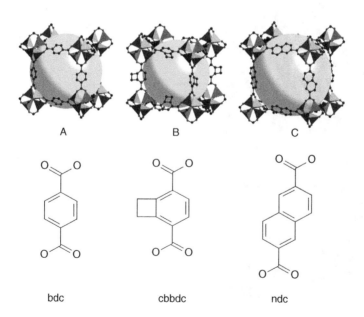

Figure 4.2 Structure of some of the IRMOF series of MOFs and the corresponding ligands: A = IRMOF-1; B = IRMOF-6; C = IRMOF-8. Reprinted with permission from Rosi *et al.*, 2003[52]. Copyright (2003) AAAS

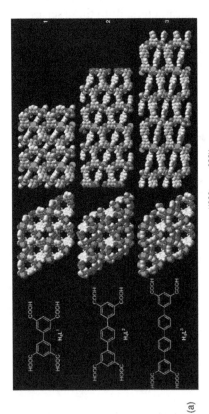

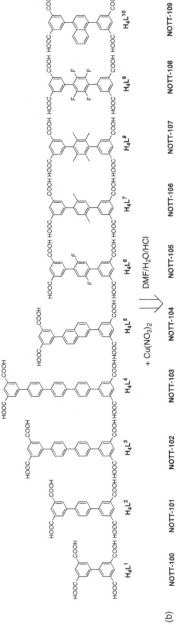

Figure 4.3 (a) Structures of some of the NOTT series of Cu_2L (where L = tetracarboxylate ligand) MOFs (L^1 = NOTT-100; L^2 = NOTT-101; L^3 = NOTT102): (middle) view along the c-axis; (right) view along the a-axis. Reprinted with permission from Lin et al., 2006[39]. Copyright (2006) Wiley-VCH Verlag GmbH & Co. KGaA. (b) Structures of the series of tetracarboxylic acid ligands in the NOTT series of MOFs. Reprinted with permission from Lin et al., 2009[38]. Copyright (2009) American Chemical Society

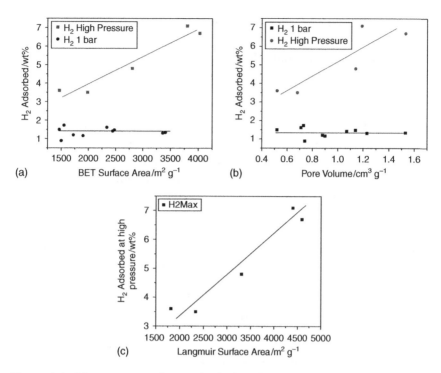

Figure 4.4 The variation of H_2 adsorbed at high pressure at $77\,K$ with pore characterisation parameters for the IRMOF series of MOFs[20,37,51,54,60,145,147]: (a) BET surface area; (b) pore volume; (c) Langmuir surface area

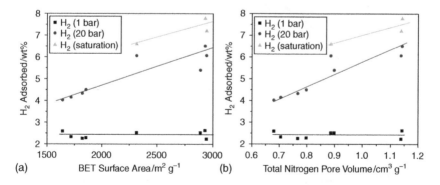

Figure 4.5 The variation of H_2 adsorbed at high pressure at $77\,K$ with pore characterisation parameters for the NOTT series of MOFs[38,39]: (a) BET surface area; (b) pore volume

Similar correlations were not observed for hydrogen adsorption at 1 bar pressure.

The adsorption potential is higher in narrow pores because of overlap of potential energy fields from the pore walls. This results in increased adsorption at low pressure. The hydrogen interactions with the surface can also be enhanced by changing the surface chemistry, for example, inclusion of open metal centres.[21,58,59] The Henry's law constant (K_H) and isosteric enthalpies of adsorption at zero surface coverage are measures of hydrogen–surface interactions. Larger pores and total pore volumes provide increased adsorption capacity at high pressures.[60] The influence of these factors can be seen in the cross-over of isotherms observed with changing pore size for the copper tetracarboxylate series of MOFs (see Figure 4.6). $Cu_2(bptc)$ (where bptc = 3,3′,5,5′-biphenylte-tracarboxylate) has a higher affinity with hydrogen than $Cu_2(tptc)$ (where tptc = terphenyl-3,3″,5,5″-tetracarboxylate) at low pressure, but the lower pore volume limits the maximum hydrogen adsorption to 4.02 wt% at 20 bar while $Cu_2(tptc)$ has higher hydrogen adsorption at >1.2 bar and 6.06 wt% at 20 bar. At low pressure and surface coverage

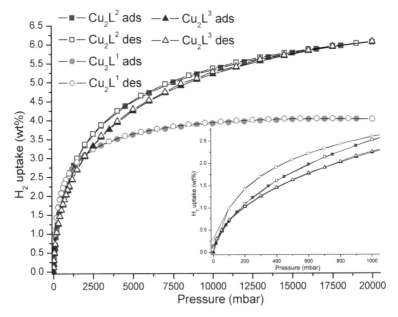

Figure 4.6 Adsorption isotherms for Cu_2L (where L^1 = bptc, L^2 = tptc and L^3 = qptc). Filled symbols (adsorption); open symbols (desorption). Reprinted with permission from Lin *et al.*, 2006[39]. Copyright (2006) Wiley-VCH Verlag GmbH & Co. KGaA

(or loadings) the enthalpy of adsorption is important whereas at high pressure the amount adsorbed involves pore filling and the pore volume provides a limit for hydrogen loading. Therefore, a compromise has to be made for the adsorption characteristics of porous materials between adsorption saturation capacity and improved adsorption at low pressure.

4.4.1   Hydrogen Adsorption Capacity Studies

Hydrogen adsorption has been studied for a wide range of porous carbons,[40,61–89] silicas,[63,90] aluminas,[63] zeolites,[91–97] porous polymers,[98–107] COFs[108–112] and MOFs.[21,27,36,38,39,50,51,54,57,58,113–189] Hydrogen adsorption measurements for porous materials carried out at pressures up to 1 bar provide data for calculating the isosteric enthalpies of adsorption at zero surface coverage, which is a measure of the hydrogen surface interaction. The smallest micropores contribute more to the hydrogen adsorption uptake at 1 bar than larger pores.[40,77,190] Figure 4.7 shows the variation of hydrogen uptakes at 1 bar and 77 K vs BET surface area for carbons,[40,63,65,67,68,70,71,74,76,78,83,89] silicas, aluminas, zeolites,[63,70,92–94,96] porous polymers,[98–100,104] COFs[108,111,112] and MOFs.[21,27,36,38,39,51,57,58,113–142,186–188] A spread of results is observed but

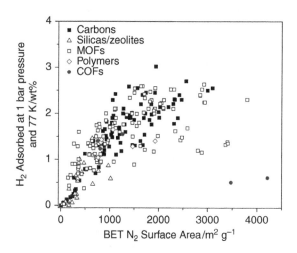

Figure 4.7   The variation of selected H_2 amounts adsorbed at 1 bar and 77 K with BET surface area for porous adsorbents: carbon materials,[40,63,65,67,68,74,76–78,81,89] silicas/zeolites,[63,92,94,96] polymers,[98–100,104] COFs[108–112] and MOFs[21,27,36,38,39,50,51,57,58,113–142,163,164,166,186,187]

similarities exist for hydrogen adsorption at 1 bar for all porous materials with surface areas below ~2000 m^2 g^{-1}. However, there is a much wider range of results for high (>2000 m^2 g^{-1}) surface area materials. No correlations were observed between hydrogen uptake at 1 bar and porous structure characterisation parameters for the IRMOF and tetracarboxylate series discussed earlier. Grand canonical Monte Carlo (GCMC) simulation calculations for hydrogen adsorption on COFs with high surface areas (3500–4500 m^2 g^{-1}) indicate that COFs have low uptakes (0.5–0.6 wt%) at 1 bar, which is consistent with larger pore sizes in these materials. As discussed later, GCMC simulations predict that these materials have the highest uptakes at high pressure (100 bar). It is apparent that the amount of hydrogen adsorbed at 1 bar pressure may not be a good guide to the maximum amount adsorbed at high pressure.

The amounts of hydrogen adsorbed under high pressure at 77 and 298 K and corresponding surface areas and pore volumes have been investigated for MOFs,[27,38,50,51,54,57,60,97,134,145–169,184] carbons,[80,83,86–88,191] zeolites,[92–95,97] silicas,[90] polymers[98,100,103,104,106] and COFs[108] and correlations between hydrogen saturation (or close to saturation capacity at 77 K) and Langmuir surface area, BET surface area and total pore volumes are shown in Figures 4.8–4.10. It is evident that there are correlations between the hydrogen uptakes at high pressure and the porous structure parameters for the porous MOF materials. The correlation for hydrogen high pressure uptake for MOFs with BET surface area is also observed for other porous materials with Types I and

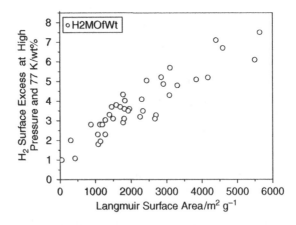

Figure 4.8   The variation of H$_2$ adsorbed (wt%) at saturation at 77 K with Langmuir surface area (m^2 g^{-1}) for porous MOF materials[21,27,37,39,46,51,54,57,60,119,127,134,136, 145–147,150,153,160–162,165,167–169,173,186,187]

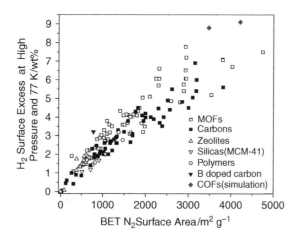

Figure 4.9 The variation of H$_2$ adsorbed (wt%) at saturation at 77 K with BET surface area (m^2 g^{-1}) for porous MOF materials,[21,36-39,41,46,50,51,54,57,60,127, 134-137,145-147,150,158,159,162-166,168,173,186,187] carbons,[68,78,80,83-87] boron doped carbon,[191] MCM-41,[90] zeolites,[92,93] polymers[98,100,104,106] and COFs[108]

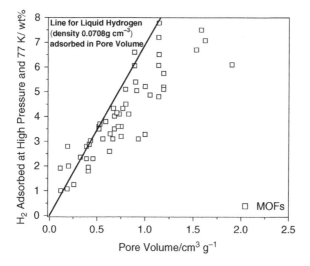

Figure 4.10 The variation of H$_2$ adsorbed (wt%) at saturation at 77 K with pore volume (cm^3 g^{-1}) for porous MOF materials[27,36,38,39,41,50,51,54,57,60,119,127,134,137, 145-147,150,153,154,158,160-162,164-167,199]

II nitrogen adsorption isotherms and therefore, the correlation is more general (see Figure 4.9). Figure 4.10 also shows a comparison of the hydrogen uptake data at or close to saturation with total pore volume and the line corresponding to liquid hydrogen filling the pores. The hydrogen saturation uptake values are usually lower than the line for liquid hydrogen filling the pores of the MOFs. This is consistent with an adsorbed phase at 77 K being similar to an incompressible fluid with a density slightly lower than liquid hydrogen (critical temperature 32.98 K).[4] MOFs have a narrower range of pore sizes and are close to the line for liquid hydrogen.

4.4.2   Temperature Dependence of Hydrogen Physisorption

Weak hydrogen–surface interactions limit hydrogen adsorption since physisorption decreases markedly with increasing temperature. This is the major obstacle to the use of hydrogen physisorption on porous materials for storage applications. Figure 4.11 shows normalised hydrogen adsorption isobars for 1 bar pressure for an activated carbon (AC), MOF $Ni_3(btc)_2(3\text{-pic})_6(pd)_3$ (C) (where btc = 1,3,5-benzenetricarboxylate, 3-pic = 3-picoline and pd = 1,2-propanediol) and the ethanol (E) and methanol (M) templated polymorphs of $Ni_2(bpy)_3(NO_3)_4$ (where bpy = 4,4′-bipyridine).[154] The differences in the shapes of the isobars in the low pressure region for E and M compared with the chiral framework, $Ni_3(btc)_2(3\text{-pic})_6(pd)_3$ and AC are

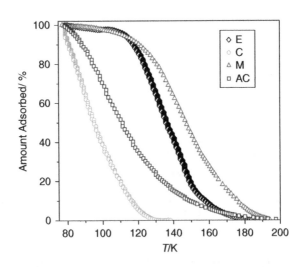

Figure 4.11   Isobars for desorption of H_2 at 1 bar with heating rate 0.3 K min^{-1} (only every 45th point is included for clarity) from: (AC) activated carbon; (C) $Ni_3(btc)_2(3\text{-pic})_6(pd)_3$; (E) E polymorph of $Ni_2(bpy)_3(NO_3)_4$; and (M) M polymorph of $Ni_2(bpy)_3(NO_3)_4$. Reprinted from Zhao *et al.*, 2004[154]. Copyright (2004) AAAS

due to hysteretic adsorption in **E** and **M** and this is discussed later. There is little conclusive evidence for significant quantities of hydrogen being physisorbed above a temperature of 195 K for a hydrogen pressure of 1 bar. The amounts adsorbed at ambient conditions and high pressure reported[36,54,60,119,127,137,145,146,150–155, 158–160,165,168,173] in the literature are usually <1 % and average ∼0.5 wt %.[19] Hydrogen surface interactions and surface chemistry are major factors under ambient temperature conditions but experimental uncertainties in these high pressure measurements make it difficult to establish correlations with characterisation parameters. Low uptakes under ambient temperature conditions are consistent with low enthalpies of adsorption for hydrogen on porous materials.[13,19,154]

4.4.3 Hydrogen Surface Interactions in Pores

Adsorption is enhanced by interactions with specific surface sites and in narrow pores by overlap of potential energies from pore walls. Smaller pore sizes enhance adsorption at low pressures whereas larger pores are necessary for increased hydrogen capacity. Porous MOFs have a greater range of surface chemistry available than in other porous materials. The surface chemistry may vary from open metal centres to hydrophobic surfaces associated with the linker ligands and also include functional groups on the ligands. The interaction between hydrogen and surface sites is critical to improving hydrogen adsorption at temperatures above cryogenic temperatures.

In 1984 Kubas *et al.* reported the synthesis and structure of the first stable organometallic complex with hydrogen coordinated side-on (η^2) to the metal.[192] Many similar complexes, mainly for second and third row transition metals, have subsequently been reported and some are stable under ambient conditions.[193,194] Hydrogen coordinated side-on (η^2) to metals involves nonclassical two-electron, three-centre bonds with a longer H-H bond than in gaseous hydrogen. Kubas[193,194] has divided the η^2 hydrogen coordination to metals into the following categories:

(1) true hydrogen complexes (H-H, 0.8–1.0 Å);
(2) elongated hydrogen complexes (H-H, 1.0–1.3 Å);
(3) compressed dihydrides (H-H, 1.3–1.6 Å);
(4) true dihydrides (H-H, ∼1.6 Å).[193]

Reversibility of hydrogen binding is often a key feature in the Kubas complexes. Studies of the variation of substitution on diphosphine ligands in [RuCl $(\eta^2$-H$_2$)(pp)$_2$][BF$_4$] (where pp $=$ bis-1,2(diarylphosphino) ethane) shows that the H-H bond length increases systematically from 0.97 to 1.03 Å with increasing electron-donor ability of the substituent.[195] This indicates that varying the substitution in ligands influences the coordination of hydrogen to metals.

These η^2 hydrogen-metal complexes are of interest in relation to a strategy of changing surface chemistry in MOFs by the inclusion of open or unsaturated metal centres for coordination of hydrogen in a similar manner. In principle, the strength of the interaction between hydrogen and metals in MOFs can be varied from very weak interactions as observed for physisorption, through elongated coordinated hydrogen to dihydrides by changing the metal, ligand and substitution on the ligand. Adsorption of hydrogen on open or unsaturated metal centres in MOFs *via* nondissociative chemisorption on metal centres may lead to significant hydrogen adsorption well above cryogenic temperatures. The objective is to bind hydrogen weakly to metals so that hydrogen adsorption occurs at higher temperatures. If the hydrogen–metal interaction is sufficiently strong so that stable hydrides are formed, it will be detrimental for hydrogen storage because of diffi-culty in desorbing the hydrogen. Therefore, fine-tuning the hydrogen–metal η^2 coordination interaction may lead to materials with facile hydrogen adsorption and desorption characteristics. It has been proposed that there is a strong correlation between coordinatively unsaturated metal centres and enhanced hydrogen surface density in many MOFs.[162]

Studies of η^2 hydrogen coordination in Kubas organometallic molecules indicate that bulky ligands tend to inhibit hydrogen splitting and this suggests confinement in pores will have an effect.[194] However, the role of confinement in pores on hydrogen coordination is unknown, but is likely to influence the strength of the hydrogen–surface group interaction in pores. The interaction of the η^2 hydrogen coordinated to a metal with the opposite pore wall is likely to have a significant influence on stability and bonding. Hydrogen interaction with open metal sites in MOFs has been observed but, so far, it is much weaker than found in so-called Kubas complexes. Only a relatively small number of Kubas complexes are known for first row transition metals whereas the MOFs that have been synthesised mainly contain first row transition metals.[193,194] Further work is required to develop new MOFs for hydrogen adsorption above cryogenic temperatures.

Theoretical and modelling studies of H$_2$ adsorption on porous materials suggest that introduction of light, non-transition metal ions, for example

Li^+, Na^+ or Mg^{2+} might enhance overall adsorption by non-dissociative H_2 binding. Mulfort and Hupp observed increases in both the isosteric heat and the overall hydrogen adsorbed for a lithium doped interpenetrated MOF $Zn_2(ndc)_2(diPyNi)$ (where ndc = naphthalene-2,6-dicarboxylate and diPyNi = *N,N'*-di-(4-pyridyl)-1,4,5,8-naphthalenetetracarboxydiimide) *via* chemical reduction with lithium metal.[139] The improvements may be due to hydrogen–lithium interactions, most likely enhanced by increased ligand polarisability and framework structural changes during chemical reduction.

Hydrogen adsorption in $[Me_2NH_2][In(bptc)]$ is enhanced by exchange of $[Me_2NH_2]^+$ for Li^+ cations.[184] The Li^+-exchanged material has a lower isosteric enthalpy for hydrogen adsorption than the parent material. The increased hydrogen capacity on cation exchange is due to an increase in the accessible pore volume, while the lower adsorption enthalpy is consistent with increased pore size.[184]

4.4.3.1 Isosteric Enthalpies of Hydrogen Adsorption

The isosteric enthalpy of hydrogen adsorption at zero surface coverage is a fundamental measurement of the hydrogen surface interaction and is constant at zero surface coverage.[40,196] The influence of adsorbate–adsorbate and other interactions are factors at higher surface coverages. The isosteric enthalpy of adsorption is very sensitive to the accuracy of experimental data and models for interpolating between isotherm points. The latter is critical in determining the Henry's law constant in the low pressure region. Hydrogen adsorption is normally measured above the critical temperature where a saturated vapour pressure cannot be used to compare isotherms at different temperatures. Therefore, virial equation methods[197,198] and Langmuir–Freundlich isotherm methods have been used. Virial parameters can be used for comparisons of hydrogen–surface and hydrogen–hydrogen interactions in pores.[40,81,199] Comparison of these methods for hydrogen and deuterium adsorption on $Zn_3(bdc)_3$ $Cu(pyen)$ (where bdc = benzene 1,4-dicarboxylate and pyenH$_2$ = 5-methyl- 4-oxo-1,4-dihydro-pyridine-3-carbaldehyde) has shown that the virial methods give a more accurate description of hydrogen and deuterium isotherms at low surface coverage.[199] This was attributed to the fact that the Langmuir–Freundlich isotherm does not reduce to Henry's law at zero surface coverage.

The isosteric enthalpies of adsorption of hydrogen on a wide range of porous materials have been reported.[34,36,38,40,50,57,58,60–62,64,87,91,103, 105,106,108,119–121,127,129,130,145,146,149,150,156,160,170–182,184,191,199–205] The enthalpies for hydrogen adsorption on porous materials at cryogenic temperatures have the following ranges:

- Carbons: 1.4–12.5 kJ mol^{-1};[40,61,62,64,87,191]
- Zeolites: 5.9–18.2 kJ mol^{-1};[62,91,202–205]
- Silicas: 5.4–10.4 kJ mol^{-1};[62]
- MOFs: 5.1–13.5 kJ mol^{-1};[34,36,38,50,57,58,60,119–121,127,129–132,136, 137,141,144–146,149,150,156,170–182,184,191,199,206]

- COFs: 2.7–8.8 kJ mol^{-1} (GCMC simulation studies);[108]
- Porous polymers: 5.7–10.0 kJ mol^{-1}.[103,105,106]

All the experimental isosteric enthalpy of adsorption values obtained for hydrogen adsorption on porous materials are much higher than the enthalpy of vaporisation of 0.9 kJ mol^{-1} for hydrogen at 20.28 K.[4] MOFs have an advantage in that surface chemistry can be varied to a greater extent than in other porous materials. The presence of open metal centres in some MOFs has been proposed as sites for η^2 hydrogen coordination similar to that found in Kubas compounds. However, although there is some evidence for stronger interactions, the enthalpies of hydrogen adsorption at zero surface coverage for MOFs with open metal centres are in most cases not appreciably higher than for other porous materials as shown by values in the range 5.3–6.7 kJ mol^{-1} observed for the tetracarboxylate series. The highest values so far obtained for MOFs are 12.3 kJ mol^{-1} for Zn$_3$(bdc)$_3$.Cu(pyen) (gas adsorption method) and 13.5 kJ mol^{-1} for desolvated Ni$_2$(dhtp)(H$_2$O)$_2$·8H$_2$O (variable temperature infrared (IR) method; where dhtp = 2,5-dihydroxyterephthalate). These values for enthalpies of adsorption at cryogenic temperatures compare with ~12.5 kJ mol^{-1} for a boron doped carbon (>7%),[191] ~11 kJ mol^{-1} for carbide derived carbons,[200] <8.2 kJ mol^{-1} for other carbons,[87] 18.2 kJ mol^{-1} for zeolites,[203] 8.8 kJ mol^{-1} for COFs (GCMC simulations) and <10.0 kJ mol^{-1} for porous polymers.

While the isosteric enthalpy of adsorption values reported so far for hydrogen adsorption on porous materials are too low for significant adsorption at ambient temperatures, high values would make hydrogen desorption very difficult. An intermediate weak nondissociative η^2 coordination or other weak chemisorption with higher isosteric enthalpy of adsorption values with facile adsorption and desorption characteristics is required for storage applications.

4.4.3.2 Diffraction Studies of Hydrogen in Pores

Direct evidence of hydrogen in pores has been provided by X-ray diffraction studies.[207,208] Takamizawa and Nakata observed dense aggregates of adsorbed hydrogen in MOF $[Rh^{II}(bza)_4(pyz)]_n$ (where bza = benzoate and pyz = pyrazine) at 90 K.[208] Hydrogen molecules adsorbed in nanochannels of a $[Cu_2(pzdc)_2(pyz)]_n$ (where pzdc = pyrazine-2,3-dicarboxylate) coordination polymer were located near a corner of rectangular nanochannels and the interaction is consistent with enhanced adsorption due to overlap of potential energy fields from the pore walls.[207]

Spencer et al. used variable temperature single crystal neutron diffraction to investigate hydrogen loaded MOF-5 $(Zn_4O(bdc)_3)$.[209] The nodal regions were the sites for hydrogen adsorption with one site located over the shared vertex of the ZnO_4 units at the centre of the node and the other over the face of ZnO_4 tetrahedra. The hydrogen adsorption sites of MOF-5 were determined using deuterium loading and powder neutron diffraction methods.[210] The initial adsorption sites were in the centre of three ZnO_3 triangular faces. The sites which were occupied next were on the top of single ZnO_3 triangles. Further sites filled at higher loading were ZnO_2 sites and sites on the hexagonal linkers. The hydrogen molecules formed three-dimensional interlinked nanoclusters at high loading.

Six D_2 adsorption sites were identified in HKUST-1 $(Cu_3(btc)_2$, where btc = 1,3,5-benzenetricarboxylate) using neutron powder diffraction methods. Initial adsorption occurred on the coordinatively unsaturated axial sites of the dinuclear Cu centre $[Cu-D_2 = 2.39(1)$ Å].[211] This distance represents a significant interaction but it is much larger than observed for σ-bonded η^2-hydrogen complexes. Competitive adsorption on other D_2 sites proceeded with increasing pore size. Neutron powder diffraction studies have shown that adsorbed D_2 is associated with the Mn centre $(Mn-D_2 = 2.27$ Å) in $Mn_3[(Mn_4Cl)_3(btt)_8.(CH_3OH)_{10}]_2$ (where btt = 1,3,5-benzenetristetrazolate).[36] One of the strongest D_2 adsorption sites was 2.47 Å from the Cu^{2+} ions in the $HCu_3[(Cu_4Cl)_3(btt)_8]_2.3.5HCl$ framework.[131] Studies have shown that in $Zn_2(dhbdc)$ (MOF-74, where dhbdc = 2,5-dihydroxyl-1,4-benzenedicarboxylate), the first site occupied has a longer $Zn-D_2$ distance (2.6 Å).[162] The metal–hydrogen distances described above are in the range 2.27–2.6 Å. This range is significantly longer than the range (1.6–1.92 Å) reported[192,195,212] for Kubas compounds. Comparison of the metal–D_2 distances with corresponding isosteric enthalpy data indicates that other factors also influence hydrogen–surface interactions.

Studies of hydrogen adsorption on $Cu_3[Co(CN)_6]_2$ using high-resolution neutron powder diffraction showed that at 1, 2 and 2.3 H_2/Cu loadings, the hydrogen was absorbed on two sites.[213] The strongest adsorption site was an interstitial location within the structure. The second adsorption site was associated with exposed Cu^{2+} ion coordination sites that result from the presence of $[Co(CN)_6]^{3-}$ vacancies.

Neutron diffraction studies revealed four distinct D_2 adsorption sites in Y(btc) and these sites are progressively occupied in the framework. The strongest adsorption sites identified were linked with the aromatic rings in the organic linkers rather than the open metal sites, as reported for other MOFs.[156] At high hydrpgen loadings highly symmetric novel nanoclusters of hydrogen molecules with relatively short hydrogen–hydrogen contact distances were formed. It is apparent that the hydrogen interactions in MOF materials are complex and that other factors exist which influence hydrogen interactions when there is confinement in pores.

4.4.3.3   Spectroscopic Studies of Hydrogen–Surface Interactions

Inelastic neutron scattering (INS) measurements on porous nickel phosphate VSB-5[214] and porous hybrid inorganic/organic framework material $NaNi_3(OH)(sip)_2$[121,214] (where sip = 5-sulfoisophthalate) indicated that initial adsorption occurred on unsaturated Ni(II) sites in the latter. As the hydrogen loading was increased, binding sites consistent with physisorption were occupied and finally interactions between hydrogen adsorbate molecules occur. Other studies of unsaturated metal centres in metal cyano[59,116,171,215] and MOF materials have suggested that the hydrogen interactions with surface sites in these materials are weaker.[21,213,215]

INS studies[216] of D_2 adsorption on HKUST-1 were consistent with previous neutron powder diffraction studies. The results showed that there were three binding sites for hydrogen loading <2.0 H_2:Cu and these sites were progressively populated with increased loading. The variation of INS peaks with temperature showed that the peak that was populated initially at H_2:Cu ratio <0.5 had the highest enthalpy (6–10 kJ mol^{-1}) while the peaks corresponding to the other sites had lower enthalpies.

Zhou and Yildirim proposed[217] that the hydrogen binding in $(Mn_4Cl)_3(btt)_8$-MOF[36] is not of the expected Kubas type because there is (a) no significant charge transfer from metal to hydrogen, (b) no evidence of any H_2-σ^* Mn-d orbital hybridisation, (c) no significant H-H bond elongation, and (d) no significant shift in v(H-H) stretching vibration.[217] The short metal–hydrogen distances and relatively high binding energies can be explained by an enhanced classical Coulombic interaction. It was concluded that this effect was unlikely to be sufficient to increase the hydrogen storage temperature to ambient conditions and that inclusion of Kubas-type motifs has the most potential for hydrogen storage on porous materials under ambient conditions.

The stretching vibration v(H-H) in hydrogen becomes IR active when adsorbed on surfaces due to lowering of symmetry. The spectra are complex because of the presence of *ortho* and *para* forms of hydrogen and *in situ* evidence for a single site catalysed conversion of adsorbed hydrogen was obtained.[218] The IR bands shift to lower energy relative to the gas phase spectrum and the intensity is proportional to the strength of the hydrogen–surface interaction. IR spectroscopy studies have shown that unsaturated Cu sites in HKUST-1 are preferential adsorption sites for hydrogen and the v(H-H) band is $\sim$70 cm^{-1} lower than the gas phase band.[218] In comparison, smaller shifts (37–45 cm^{-1}) have been observed for hydrogen adsorption on a cross-linked polymer and it was proposed that this was primarily due to specific interaction of the hydrogen molecule with the electron-rich part of the polymer.[219] The hydrogen adsorption energy estimated from the temperature dependence of IR spectral features was 10 kJ mol^{-1}. This is similar to the highest obtained from INS measurements.[216]

The shifts to lower energy in low-temperature-stable η^2 hydrogen complexes are much larger with v(H-H) typically in the range 2600–3250 cm^{-1}[193] [v(H-H) for hydrogen in gas phase is 4160 cm^{-1}].[218] The IR spectrum of [W(CO)$_3$(PCy$_3$)$_2$(η^2-H$_2$)] (where Cy = C$_6$H$_{12}$) gives v(H-H) of 2690 cm^{-1} while the H-H bond length in [W(CO)$_3$(P^{i}Pr$_3$)$_2$(η^2-H$_2$)] (where ^{i}Pr = isopropyl) was 0.84 Å as determined by neutron diffraction studies compared with 0.74 Å in hydrogen gas.[192] Raman spectroscopy has also been used to study the adsorbed hydrogen phase in porous materials at room temperature and under cryogenic conditions.[220–222] The results confirm that the interaction strength for adsorption of molecular hydrogen is small and consistent with physisorption for single-walled nanotubes, MOF-5 and HKUST-1.

4.4.4 Framework Flexibility and Hysteretic Adsorption

An interesting feature of adsorption on MOFs is that some materials have framework flexibility and structural change may occur during the adsorption process. The flexibility in the MIL series is an example of large structural changes during adsorption.[12] MOF materials may be flexible and this may occur either without any bond breaking or some materials undergo bond breaking and reforming reactions. The latter are much rarer and the former have possible applications in storage applications.

Structural change without bond breaking occurs with a scissoring motion leading to an increase in crystallographic cell volume of ethanol and methanol templated phases **E** and **M** of $Ni_2(bpy)_3.(NO_3)_4$ during adsorption of the templates.[223,224] Hysteretic adsorption was observed initially for hydrogen adsorption on $Ni_2(bpy)_3.(NO_3)_4$[154] (see Figure 4.12) and several further systems have been identified.[161,183,225] This is thought to be a kinetic trapping effect. The corresponding isobars are shown in Figure 4.11. It is apparent that hydrogen desorption does not start until higher temperatures ($\sim$110 K).[154] These materials have windows and pore cavities and the diffusion through the very narrow windows and framework flexibility are probably factors in hysteretic adsorption. Hysteretic hydrogen adsorption was subsequently observed in a related system for desolvated $[Co_2(bpy)_3(SO_4)_2(H_2O)_2](bpy)(CH_3OH)$.[226]

Llewellyn *et al.* have developed the MIL series of MOFs where the changes in structure during adsorption are much larger. Thermodynamic and structural studies of nanoporous metal benzenedicarboxylate, $M(OH)(O_2CC_6H_4CO_2)$, MIL-53(Cr) suggest that a minimum hydrocarbon adsorption enthalpy of $\sim$20 kJ mol^{-1} in the initial large pore structure is required to induce the structural transition to narrow pore structure.[227] MIL-53(Al) exhibits a reversible structural transition

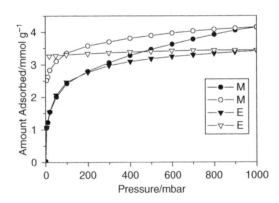

Figure 4.12 Adsorption–desorption isotherms for H_2 adsorption (filled symbols) and desorption (open symbols) isotherms on MOF adsorbents at 77 K: (M) **M** polymorph of $Ni_2(bpy)_3(NO_3)_4$; (E) **E** polymorph of $Ni_2(bpy)_3(NO_3)_4$. Reprinted from Zhao *et al.*, 2004[154]. Copyright (2004) AAAS

between an open and closed-pored structure as a function of temperature in the absence of any guest molecules.[185] Isotherm hysteresis has been observed for the high pressure hydrogen adsorption on the MIL-53 ($M =$ Al^{3+}, Cr^{3+}) at 77 K.[134] Therefore, hysteretic adsorption has been observed for MOFs with a wide degree of flexibility.

Choi et al.[161] studied hydrogen adsorption on a MOF Co(bdp) (where bdp = 1,4-benzenedipyrazolate), which showed temperature dependent broad hysteresis loops over the temperature range 50–87 K. The results suggested a pressure, temperature and gas dependent pore opening mechanism with the hysteretic behaviour governed by phase transitions with energies comparable with the hydrogen adsorption enthalpy. Some high pressure hydrogen hysteresis has also been observed for FMOF-1, a fluorous MOF material.[225] There are now several examples of hysteretic adsorption of hydrogen[134,154,161,225] and other gases.[228–230] This mechanism may have applications for hydrogen storage.

4.4.5 Comparison of Hydrogen and Deuterium Adsorption

Hydrogen and deuterium have the bond length 0.7416 Å in the gas phase and this is consistent with the classical description of isotopes.[231] However, hydrogen and deuterium have different zero point energies and small differences in physical properties such as density, boiling point, amplitudes of vibration, dissociation energy, etc., and also differences in adsorption characteristics.[40,81,199,232–234] Higher amounts of deuterium than hydrogen adsorption are observed under the same experimental conditions. Comparison of hydrogen and deuterium adsorption at 77 and 87 K shows that the molar ratios for isotherm points are in the range 1.06–1.15 for porous MOFs,[39,41,163,199] carbons[40,81] and zeolite materials.[91]

When the difference between hydrogen size and pore size is similar to the de Broglie wavelength, quantum effects occur and this could lead to molecular sieving of hydrogen and deuterium.[235–237] Quantum effects give rise to higher amounts of deuterium adsorbed compared with hydrogen.[40,41,81,199,233,234,237] The first experimental evidence for kinetic quantum molecular sieving was observed for adsorption and desorption for a carbon molecular sieve and microporous carbon.[81] Similarly, faster deuterium adsorption compared with hydrogen was reported subsequently for Zn$_3$(bdc)$_3$.Cu(pyen)[199] and zeolites.[96] Analysis of the adsorption

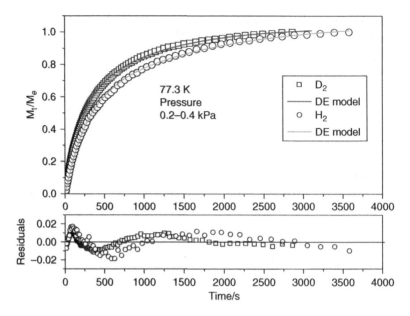

Figure 4.13 Comparison of H$_2$ and D$_2$ adsorption kinetic profiles and the corresponding fit for the double exponential kinetic model with residuals for adsorption on Zn$_3$(bdc)$_3$.Cu(pyen). Pressure increment 0.2–0.4 kPa at 77.3 K. Reprinted with permission from Chen et al., 2008[199]. Copyright (2008) American Chemical Society

isotherms using a virial equation shows that the D$_2$-D$_2$ interactions are smaller than H$_2$-H$_2$ interactions. This is consistent with the smaller amplitude of vibration of deuterium compared with hydrogen.[81,199]

Typical kinetic profiles for adsorption of deuterium and hydrogen on Zn$_3$(bdc)$_3$.Cu(pyen) are shown in Figure 4.13. Zn$_3$(bdc)$_3$.Cu(pyen) has two types of pores present in the material and independent quantum effects were observed for these pores.[199] The barriers to diffusion of hydrogen were slightly greater than of deuterium for both types of pores. Also, it was only necessary for the difference between the adsorbate and one pore dimension to be close to the de Broglie wavelength for quantum effects to be observed. The faster adsorption kinetics for deuterium compared with hydrogen is contrary to that expected on the basis of mass. This is attributed to the larger amplitude of vibration of hydrogen atoms leading to an effective larger cross-section for hydrogen resulting in slower kinetics when the pore size is similar to the de Broglie wavelength. Theoretical studies of quantum kinetic molecular sieving have been reported.[238,239] Kinetic quantum molecular sieving has possible applications for isotope separation. However, these effects are only observable at low temperature and would not be an issue for storage under ambient conditions.

4.5 CONCLUSIONS

Research on hydrogen adsorption and desorption on porous materials has focused on developing materials with capacities at cryogenic temperatures that meet US DOE targets. Several groups have reported experimental measurements showing maximum excess adsorption of 6–8 wt% of hydrogen at 77 K and <100 bar on MOF and porous carbon materials. A total hydrogen uptake of 10 wt% at 77 bar has been observed for a MOF NOTT-112 at 77 K. Hydrogen maximum surface excess capacity and BET nitrogen surface area at 77 K show a strong correlation for a wide range of porous materials. Similar correlations are observed for maximum surface excess hydrogen adsorbed with other pore structure characterisation parameters, such as pore volume, for all porous materials. These parameters indicate the importance of available surface area and pore volume in hydrogen adsorption. The hydrogen adsorbate is similar to an incompressible fluid with density similar to liquid hydrogen and the maximum amount adsorbed is limited by the available total pore volume in the adsorbent porous structure. The amounts of hydrogen adsorbed on MOFs under high pressure and ambient temperature conditions are very small.

The design of porous materials with improved adsorption capacity under ambient conditions involves enhancing hydrogen surface site interactions. The maximum enthalpy of hydrogen adsorption obtained experimentally for MOFs is 12–13 kJ mol^{-1} while for boron doped carbon it is 12.5 kJ mol^{-1} and for zeolites it is ~18 kJ mol^{-1}. The MOFs with the highest isosteric enthalpies of adsorption obtained currently are $Zn_3(bdc)_3.Cu(pyen)$[199] and $Ni_2(dhtp)$[179] and both of these materials have open metal centres. The adsorption thermodynamic results, and structural and spectroscopic observations of hydrogen adsorbed in pores with open metals centres in MOFs show that the interactions observed so far are much weaker than those observed for η^2 hydrogen binding in Kubas compounds. However, higher enthalpies of adsorption are required to obtain increased adsorption at higher temperatures for storage applications.

The stability of η^2 type hydrogen metal complexes suggest that MOFs with open metal centres acting as sites for η^2 type hydrogen coordination have scope for further development as a route to materials with enhanced hydrogen surface interactions. Modification of the hydrogen–metal interaction by varying the metal, ligand and ligand substitution is suggested by the wide variation in interaction strength between hydrogen and metals shown by structural studies of Kubas (η^2-H_2) coordination compounds. The ability to fine-tune the hydrogen–metal interaction involving weak nondissociative chemisorption rather than nonspecific physisorption is important. However, open metal centres represent a small fraction of MOF internal surface areas. Therefore strategies are required for designing materials with open metal centres where coordination of multiple hydrogen molecules per metal can occur. However, only a few stable solid bis-H_2 organometallic complexes are known.[193,194] Multiple coordination of hydrogen to Cu centres has been proposed for $Zn_3(bdc)_3$.Cu (pyen) based on the variation in enthalpy of adsorption with surface coverage. Interaction of hydrogen molecules with both sides of Cu in the Cu(pyen) pillar gives a greater influence of open metal centres.[199] In principle, incorporation of stronger interactions into MOFs using open metal centres is possible. However, interactions from other pore surfaces due to confinement in pores may influence the primary hydrogen interactions with open metal sites. Secondary interactions influence framework flexibility and window opening in MOFs, which control hysteretic adsorption characteristics and these may also be useful in the design strategies for storage applications. Progressive improvement in the understanding of surface chemistry and framework flexibility will lead to MOFs with improved hydrogen adsorption characteristics for storage applications.

ACKNOWLEDGEMENTS

The author would like to thank collaborators for many useful discussions and the Carbon Trust and Leverhulme Trust for financial support.

REFERENCES

[1] R.L. Hirsch, R. Bezdek and R. Wendling, http://www.netl.doe.gov/publications/others/pdf/Oil_Peaking_NETL.pdf, (2005) (accessed on 19 November 2010).
[2] http://www1.eere.energy.gov/hydrogenandfuelcells/storage (accessed on 19 November 2010).
[3] L. Schlapbach and A. Zuttel, Nature, 414, 353 (2001).

[4] D.R. Lide (Ed.), *CRC Handbook of Chemistry and Physics*, 74th Edn, The Chemical Rubber Co., Boca Raton, 1993.

[5] M. Eddaoudi, J. Kim, N. Rosi, D. Vodak, J. Wachter, M. O'Keeffe and O.M. Yaghi, *Science*, **295**, 469 (2002).

[6] C.R. Reid and K.M. Thomas, *J. Phys. Chem. B*, **105**, 10619 (2001).

[7] H.K. Chagger, F.E. Ndaji, M.L. Sykes, and K.M. Thomas, *Carbon*, **33**, 1405 (1995).

[8] S.Y. Jiang, K.E. Gubbins and P.B. Balbuena, *J. Phys. Chem.*, **98**, 2403 (1994).

[9] D. Britt, D. Tranchemontagne and O.M. Yaghi, *Proc. Natl. Acad. Sci. USA*, **105**, 11623 (2008).

[10] M.E. Davis, *Nature*, **417**, 813 (2002).

[11] A. Corma, *Chem. Rev.*, **97**, 2373 (1997).

[12] A.J. Fletcher, K.M. Thomas and M.J. Rosseinsky, *J. Solid State Chem.*, **178**, 2491 (2005).

[13] K.M. Thomas, *Catal. Today*, **120**, 389 (2007).

[14] M. Felderhoff, C. Weidenthaler, R.V. Helmolt and U. Eberle, *Phys. Chem. Chem. Phys.*, **9**, 2643 (2007).

[15] R.E. Morris and P.S. Wheatley, *Angew. Chem. Int. Ed.*, **47**, 4966 (2008).

[16] A.W.C. van den Berg and C.O. Arean, *Chem. Commun.*, 668 (2008).

[17] D.J. Collins and H.C. Zhou, *J. Mater. Chem.*, **17**, 3154 (2007).

[18] M. Dinca and J.R. Long, *Angew. Chem. Int. Ed.*, **47**, 6766 (2008).

[19] K.M. Thomas, *Dalton Trans.*, 1487 (2009).

[20] X. Lin, J. Jia, P. Hubberstey, M. Schroder and N.R. Champness, *Cryst. Eng. Commun.*, **9**, 438 (2007).

[21] B. Chen, N.W. Ockwig, A.R. Millward, D.S. Contreras and O.M. Yaghi, *Angew. Chem. Int. Ed.*, **44**, 4745 (2005).

[22] S. Kitagawa, R. Kitaura and S. Noro, *Angew. Chem., Int. Ed.*, **43**, 2334 (2004).

[23] G. Férey, C. Mellot-Draznieks, C. Serre, F. Millange, J. Dutour, S. Surble and I. Margiolaki, *Science*, **309**, 2040 (2005); **310**, 1119 (2005).

[24] G. Férey, *Chem. Soc. Rev.*, **37**, 191 (2008).

[25] S. Kitagawa and R. Matsuda, *Coord. Chem. Rev.*, **251**, 2490 (2007).

[26] X.-C. Huang, Y.-Y. Lin, J.-P. Zhang and X.-M. Chen, *Angew. Chem. Int. Ed.*, **45**, 1557 (2006).

[27] K.S. Park, Z. Ni, A.P. Cote, J.Y. Choi, R. Huang, F.J. Uribe-Romo, H.K. Chae, M. O'Keeffe and O.M. Yaghi, *Proc. Natl. Acad. Sci. USA*, **103**, 10186 (2006).

[28] D.P. Broom, *Int. J. Hydrogen Energy*, **32**, 4871 (2007).

[29] A.W. Harding, N.J. Foley, P.R. Norman, D.C. Francis and K.M. Thomas, *Langmuir*, **14**, 3858 (1998).

[30] E.L. Fuller, J.A. Poulis, A.W. Czanderna and E. Robens, *Thermochim. Acta*, **29**, 315 (1979).

[31] K.S.W. Sing, D.H. Everett, R.A.W. Haul, L. Moscou, R.A. Pierotti, J. Rouquerol and T. Siemieniewska, *Pure Appl. Chem.*, **57**, 603 (1985).

[32] D.P. Broom and P. Moretto, *J. Alloys Compd.*, **446–447**, 687 (2007).

[33] A. Anson, M. Benham, J. Jagiello, M.A. Callejas, A.M. Benito, W.K. Maser, A. Zuettel, P. Sudan and M.T. Martinez, *Nanotechnology*, **15**, 1503 (2004).

[34] H. Furukawa, M.A. Miller and O.M. Yaghi, *J. Mater. Chem.*, **17**, 3197 (2007).

[35] S. Sircar, *Ind. Eng. Chem. Res.*, **38**, 3670 (1999).

[36] M. Dinca, A. Dailly, Y. Liu, C.M. Brown, D.A. Neumann and J.R. Long, *J. Am. Chem. Soc.*, **128**, 16876 (2006).

[37] S.S. Kaye, A. Dailly, O.M. Yaghi and J.R. Long, *J. Am. Chem. Soc.*, **129**, 14176 (2007).
[38] X. Lin, I. Telepeni, A.J. Blake, A. Dailly, C.M. Brown, J.M. Simmons, M. Zoppi, G.S. Walker, K.M. Thomas, T.J. Mays, P. Hubberstey, N.R. Champness and M. Schröder, *J. Am. Chem. Soc.*, **131**, 2159 (2009).
[39] X. Lin, J. Jia, X. Zhao, K.M. Thomas, A.J. Blake, G.S. Walker, N.R. Champness, P. Hubberstey and M. Schröder, *Angew. Chem. Int. Ed.*, **45**, 7358 (2006).
[40] X.B. Zhao, B. Xiao, A.J. Fletcher and K.M. Thomas, *J. Phys. Chem. B*, **109**, 8880 (2005).
[41] B. Xiao, P.S. Wheatley, X. Zhao, A.J. Fletcher, S. Fox, A.G. Rossi, I.L. Megson, S. Bordiga, L. Regli, K.M. Thomas and R.E. Morris, *J. Am. Chem. Soc.*, **129**, 1203 (2007).
[42] Y. Kinoshita, I. Matsubara, T. Hibuchi and Y. Saito, *Bull. Chem. Soc. Jpn.*, **32**, 1221 (1959).
[43] B.F. Hoskins and R. Robson, *J. Am. Chem. Soc.*, **111**, 5962 (1989).
[44] M. Kondo, T. Yoshitomi, K. Seki, H. Matsuzaka and S. Kitagawa, *Angew. Chem. Int. Ed.*, **36**, 1725 (1997).
[45] J. Hafizovic, M. Bjorgen, U. Olsbye, P.D.C. Dietzel, S. Bordiga, C. Prestipino, C. Lamberti and K.P. Lillerud, *J. Am. Chem. Soc.*, **129**, 3612 (2007).
[46] J. Liu, J.T. Culp, S. Natesakhawat, B.C. Bockrath, B. Zande, S.G. Sankar, G. Garberoglio and J.K. Johnson, *J. Phys. Chem. C*, **111**, 9305 (2007).
[47] A.P. Nelson, O.K. Farha, K.L. Mulfort and J.T. Hupp, *J. Am. Chem. Soc.*, **131**, 458 (2009).
[48] F. Rouquerol, J. Rouquerol and K. Sing, *Adsorption by Powders and Porous Solids*. Academic Press, London, 1999.
[49] K.S. Walton and R.Q. Snurr, *J. Am. Chem. Soc.*, **129**, 8552 (2007).
[50] Y. Yan, X. Lin, S.H. Yang, A.J. Blake, A. Dailly, N.R. Champness, P. Hubberstey and M. Schröder, *Chem. Commun.*, 1025 (2009).
[51] A.G. Wong-Foy, A.J. Matzger and O.M. Yaghi, *J Am. Chem. Soc.*, **128**, 3494 (2006).
[52] N.L. Rosi, J. Eckert, M. Eddaoudi, D.T. Vodak, J. Kim, M. O'Keeffe and O. M. Yaghi, *Science*, **300**, 1127 (2003).
[53] O.M. Yaghi, M. O'Keeffe, N.W. Ockwig, H.K. Chae, M. Eddaoudi and J. Kim, *Nature*, **423**, 705 (2003).
[54] J.L.C. Rowsell, A.R. Millward, K.S. Park and O.M. Yaghi, *J. Am. Chem. Soc.*, **126**, 5666 (2004).
[55] J.L.C. Rowsell, E.C. Spencer, J. Eckert, J.A.K. Howard and O.M. Yaghi, *Science*, **309**, 1350 (2005).
[56] J.L.C. Rowsell and O.M. Yaghi, *Angew. Chem. Int. Ed.*, **44**, 4670 (2005).
[57] J.L.C. Rowsell and O.M. Yaghi, *J. Am. Chem. Soc.*, **28**, 1304 (2006).
[58] M. Dinca, A.F. Yu and J.R. Long, *J. Am. Chem. Soc.*, **128**, 8904 (2006).
[59] C. Prestipino, L. Regli, J.G. Vitillo, F. Bonino, A. Damin, C. Lamberti, A. Zecchina, P.L. Solari, K.O. Kongshaug and S. Bordiga, *Chem. Mater.*, **18**, 1337 (2006).
[60] M. Latroche, S. Suble, C. Serre, C. Mellot-Draznieks, P.L. Llewellyn, J.H. Lee, J.S. Chang, S.H. Jhung and G. Férey, *Angew. Chem. Int. Ed.*, **45**, 8227 (2006).
[61] W.V. Dingenen and A.V. Itterbeek, *Physica (The Hague)*, **6**, 49 (1939).
[62] I.D. Basmadjian, *Can. J Chem.*, **38**, 141 (1960).
[63] M.G. Nijkamp, J.E.M.J. Raaymakers, A.J.V. Dillen and K.P.D. Jong, *Appl. Phys. A*, **72**, 619 (2001).

[64] P. Benard and R. Chahine, *Langmuir*, **17**, 1950 (2001).

[65] H.G. Schimmel, G.J. Kearly, M.G. Nijkamp, C.T. Visser, K.P. de Jong and F.M. Mulder, *Chem. Eur. J.*, **9**, 4764 (2003).

[66] G. Gundiah, A. Govindaraj, N. Rajalakshmi, K.S. Dhathathreyan and C.N.R. Rao, *J. Mater. Chem.*, **13**, 209 (2003).

[67] H.G. Schimmel, G. Nijkamp, G.J. Kearley, A. Rivera, K.P.D. Jong and F.M. Mulder, *Mat. Sci. Eng., B*, **108**, 124 (2004).

[68] N. Texier-Mandoki, J. Dentzer, T. Piquero, S. Saadallah, P. David and C. Vix-Guterl, *Carbon*, **42**, 2744 (2004).

[69] Y. Zhou, K. Feng, Y. Sun and L. Zhou, *Chem. Phys. Lett.*, **380**, 526 (2003).

[70] H. Takagi, H. Hatori, Y. Soneda, N. Yoshizawa and Y. Yamada, *Mater. Sci. Eng B*, **108**, 143 (2004).

[71] H. Takagi, H. Hatori, Y. Yamada, S. Matsuo and M. Shiraishi, *J. Alloys Compd.*, **385**, 257 (2004).

[72] S.B. Kayiran, F.D. Lamari and D. Levesque, *J. Phys. Chem. B*, **108**, 15211 (2004).

[73] M. Shiraishi, T. Takenobu, H. Kataura and M. Ata, *Appl. Phys. A*, **78**, 947 (2004).

[74] J.B. Parra, C.O. Ania, A. Arenillas, F. Rubiera, J.M. Palacios and J.J. Pis, *J. Alloys Compd.*, **379**, 280 (2004).

[75] L. Zhou, Y. Zhou and Y. Sun, *Int. J. Hydrogen Energy*, **29**, 475 (2004).

[76] J. Pang, J.E. Hampsey, Z. Wu, Q. Hu and Y. Lu, *Appl. Phys. Lett.*, **85**, 4887 (2004).

[77] Y. Gogotsi, R.K. Dash, G. Yushin, T. Yildirim, G. Laudisio and J.E. Fischer, *J. Am. Chem. Soc.*, **127**, 16006 (2005).

[78] R. Gadiou, S.E. Saadallah, T. Piquero, P. David, J. Parmentier and C. Vix-Guterl, *Micropor. Mesopor. Mar.*, **79**, 121 (2005).

[79] E. Poirier, R. Chahine, A. Tessier and T.K. Bose, *Rev. Sci. Instrum.*, **76**, 055101 (2005).

[80] B. Panella, M. Hirscher and S. Roth, *Carbon*, **43**, 2209 (2005).

[81] X. Zhao, S. Villar-Rodil, A.J. Fletcher and K.M. Thomas, *J. Phys. Chem. B*, **110**, 9947 (2006).

[82] A. Pacula and R. Mokaya, *J. Phys. Chem. C*, **112**, 2764 (2008).

[83] F. Cheng, J. Liang, J. Zhao, Z. Tao and J. Chen, *Chem. Mater.*, **20**, 1889 (2008).

[84] A. Pacula and R. Mokaya, *Micropor. Mesopor. Mat.*, **106**, 147 (2007).

[85] Z.X. Yang, Y.D. Xia, X.Z. Sun and R. Mokaya, *J. Phys. Chem. B*, **110**, 18424 (2006).

[86] Q.Y. Hu, Y.F. Lu and G.P. Meisner, *J. Phys. Chem.*, **112**, 1516 (2008).

[87] Z. Yang, Y. Xia and R. Mokaya, *J. Am. Chem. Soc.*, **129**, 1673 (2007).

[88] M. Jorda-Beneyto, F. Suarez-Garcia, D. Lozano-Castello, D. Cazorla-Amoros and A. Linares-Solano, *Carbon*, **45**, 293 (2007).

[89] L. Zubizarreta, E.I. Gomez, A. Arenillas, C.O. Ania, J.B. Parra and J.J. Pis, *Adsorption*, **14**, 557 (2008).

[90] D.A. Sheppard and C.E. Buckley, *Int. J. Hydrogen Energy*, **33**, 1688 (2008).

[91] F. Stephanie-Victoire, A.M. Goulay and E.C.D. Lara, *Langmuir*, **14**, 7255 (1998).

[92] H.W. Langmi, A. Walton, M.M. Al-Mamouri, S.R. Johnson, D. Book, J.D. Speight, P.P. Edwards, I. Gameson, P.A. Anderson and I.R. Harris, *J. Alloys Compd.*, **356–357**, 710 (2003).

[93] H.W. Langmi, D. Book, A. Walton, S.R. Johnson, M.M. Al-Mamouri, J.D. Speight, P.P. Edwards, I.R. Harris and P.A. Anderson, *J. Alloys Compd.*, **404–406**, 637 (2005).

[94] Y. Li and R.T. Yang, *J. Phys. Chem. B*, 110, 17175 (2006).

[95] S.B. Kayiran and F.L. Darkrim, *Surf. Interface Anal.*, 34, 100 (2004).

[96] X.Z. Chu, Y.P. Zhou, Y.Z. Zhang, W. Su, Y. Sun, and L. Zhou, *J. Phys. Chem. B*, 110, 22596 (2006).

[97] J. Dong, X. Wang, H. Xu, Q. Zhao and J. Li, *Int. J. Hydrogen Energy*, 32, 4998 (2007).

[98] J.Y. Lee, C.D. Wood, D. Bradshaw, M.J. Rosseinsky and A.I. Cooper, *Chem. Commun.*, 2670 (2006).

[99] J. Germain, J. Hradil, J.M.J. Frechet and F. Svec, *Chem. Mater.*, 18, 4430 (2006).

[100] N.B. McKeown, B. Gahnem, K.J. Msayib, P.M. Budd, C.E. Tattershall, K. Mahmood, S. Tan, D. Book, H.W. Langmi and A. Walton, *Angew. Chem. Int. Ed.*, 45, 1804 (2006).

[101] N.B. McKeown and P.M. Budd, *Chem. Soc. Rev.*, 35, 675 (2006).

[102] P.M. Budd, A. Butler, J. Selbie, K. Mahmood, N.B. McKeown, B. Ghanem, K. Msayib, D. Book and A. Walton, *Phys. Chem. Chem. Phys.*, 9, 1802 (2007).

[103] J. Germain, J.M.J. Frechet and F. Svec, *J. Mater. Chem.*, 17, 4989 (2007).

[104] B.S. Ghanem, K.J. Msayib, N.B. McKeown, K.D.M. Harris, Z. Pan, P.M. Budd, A. Butler, J. Selbie, D. Book and A. Walton, *Chem. Commun.*, 67 (2007).

[105] J.X. Jiang, F. Su, A. Trewin, C.D. Wood, H. Niu, J.T.A. Jones, Y.Z. Khimyak and A.I. Cooper, *J. Am. Chem. Soc.*, 130, 7710 (2008).

[106] C.D. Wood, B. Tan, A. Trewin, H. Niu, D. Bradshaw, M.J. Rosseinsky, Y.Z. Khimyak, N.L. Campbell, R. Kirk, E. Stoeckel and A.I. Cooper, *Chem. Mater.*, 19, 2034 (2007).

[107] J. Germain, F. Svec and J.M.J. Frechet, *Chem. Mate.*, 20, 7069 (2008).

[108] S.S. Han, H. Furukawa, O.M. Yaghi and W.A. Goddard, *J. Am. Chem. Soc.*, 130, 11580 (2008).

[109] H.M. El-Kaderi, J.R. Hunt, J.L. Mendoza-Cortes, A.P. Cote, R.E. Taylor, M. O'Keeffe and O.M. Yaghi, *Science*, 316, 268 (2007).

[110] A.P. Cote, H.M. El-Kaderi, H. Furukawa, J.R. Hunt and O.M. Yaghi, *J. Am. Chem. Soc.*, 129, 12914 (2007).

[111] R.W. Tilford, S.J. Mugavero, P.J. Pellechia and J.J. Lavigne, *Adv. Mater.*, 20, 2741 (2008).

[112] J. Weber, M. Antonietti and A. Thomas, *Macromolecules*, 41, 2880 (2008).

[113] E.Y. Lee and M.P. Suh, *Angew. Chem., Int. Ed.*, 43, 2798 (2004).

[114] D.N. Dybtsev, H. Chun, S.H. Yoon, D. Kim and K. Kim, *J. Am. Chem. Soc.*, 126, 32 (2004).

[115] D.N. Dybtsev, H. Chun and K. Kim, *Angew. Chem. Int. Ed.*, 43, 5033 (2004).

[116] K.W. Chapman, P.D. Southon, C.L. Weeks and C.J. Kepert, *Chem. Commun.*, 3322 (2005).

[117] J. Perles, M. Iglesias, M.A. Martin-Luengo, M.A. Monge, C. Ruiz-Valero and N. Snejko, *Chem. Mater.*, 17, 5837 (2005).

[118] H. Chun, D.N. Dybtsev, H. Kim and K. Kim, *Chem. Eur. J.*, 11, 3521 (2005).

[119] P.D.C. Dietzel, B. Panella, M. Hirscher, R. Blom and H. Fjellvag, *Chem. Commun.*, 959 (2006).

[120] X. Guo, G. Zhu, Z. Li, F. Sun, Z. Yang and S. Qiu, *Chem. Commun.*, 3172 (2006).

[121] P.M. Forster, J. Eckert, B.D. Heiken, J.B. Parise, J.W. Yoon, S.H. Jhung, J.S. Chang and A.K. Cheetham, *J. Am. Chem. Soc.*, 128, 16846 (2006).

[122] J. Jia, X. Lin, A.J. Blake, N.R. Champness, P. Hubberstey, L. Shao, G. Walker, C. Wilson and M. Schroeder, *Inorg. Chem.*, **45**, 8838 (2006).

[123] H.R. Moon, N. Kobayashi and M.P. Suh, *Inorg. Chem.*, **45**, 8672 (2006).

[124] J.A.R. Navarro, E. Barea, J.M. Salas, N. Masciocchi, S. Galli, A. Sironi, C.O. Ania and J.B. Parra, *Inorg. Chem.*, **45**, 2397 (2006).

[125] J.A. Rood, B.C. Noll and K.W. Henderson, *Inorg. Chem.*, **45**, 5521 (2006).

[126] S. Ma, X.S. Wang, D. Yuan and H.C. Zhou, *Angew. Chem. Int. Ed.*, **47**, 4130 (2008).

[127] X.S. Wang, S.Q. Ma, K. Rauch, J.M. Simmons, D.Q. Yuan, X.P. Wang, T. Yildirim, W.C. Cole, J.J. Lopez, A.d. Meijere and H.C. Zhou, *Chem. Mater.*, **20**, 3145 (2008).

[128] H. Chun, H. Jung, G. Koo, H. Jeong and D.K. Kim, *Inorg. Chem.*, **47**, 5355 (2008).

[129] J.Y. Lee, L. Pan, S.P. Kelly, J. Jagiello, T.J. Emge and J. Li, *Adv. Mater.*, **17**, 2703 (2005).

[130] M. Dinca and J.R. Long, *J. Am. Chem. Soc.*, **129**, 11172 (2007).

[131] M. Dinca, W.S. Han, Y. Liu, A. Dailly, C.M. Brown and J.R. Long, *Angew. Chem. Int. Ed.*, **46**, 1419 (2007).

[132] F. Nouar, J.F. Eubank, T. Bousquet, L. Wojtas, M.J. Zaworotko and M. Eddaoudi, *J. Am. Chem. Soc.*, **130**, 1833 (2008).

[133] S.B. Choi, M.J. Seo, M. Cho, Y. Kim, M.K. Jin, D.Y. Jung, J.S. Choi, W.S. Ahn, J.L.C. Rowsell and J. Kim, *Cryst. Growth Des.*, **7**, 2290 (2007).

[134] G. Férey, M. Latroche, C. Sérre, F. Millange, T. Loiseau and A. Percheron-Guegan, *Chem. Commun.*, **24**, 2976 (2003).

[135] M. Kramer, U. Schwarz and S. Kaskel, *J. Mater. Chem.*, **16**, 2245 (2006).

[136] A.G. Wong-Foy, O. Lebel and A.J. Matzger, *J. Am. Chem. Soc.*, **129**, 15740 (2007).

[137] J.Y. Lee, D.H. Olson, L. Pan, T.J. Emge and J. Li, *Adv. Funct. Mater.*, **17**, 1255 (2007).

[138] H. Chun and J. Moon, *Inorg. Chem.*, **46**, 4371 (2007).

[139] K.L. Mulfort and J.T. Hupp, *J. Am. Chem. Soc.*, **129**, 9604 (2007).

[140] H. Park, J.F. Britten, U. Mueller, J.Y. Lee, J. Li and J.B. Parise, *Chem. Mater.*, **19**, 1302 (2000).

[141] S.M. Humphrey, J.S. Chang, S.J. Jhung, J.W. Yoon and P.T. Woo, *Angew. Chem. Int. Ed.*, **46**, 272 (2007).

[142] E.Y. Lee, S.Y. Jang and M.P. Suh, *J. Am. Chem. Soc.*, **127**, 6374 (2005).

[143] J.H. Yoon, S.B. Choi, Y.J. Oh, M.J. Seo, Y.H. Jhon, T.B. Lee, D. Kim, S.H. Choi and J. Kim, *Catal. Today*, **120**, 324 (2007).

[144] D.F. Sava, V.C. Kravtsov, F. Nouar, L. Wojtas, J.F. Eubank and M. Eddaoudi, *J. Am. Chem. Soc.*, **130**, 3768 (2008).

[145] A. Dailly, J.J. Vajo and C. Ahn, *J. Phys. Chem. B*, **110**, 1099 (2006).

[146] B. Panella, M. Hirscher, H. Puetter and U.Mueller, *Adv. Funct. Mater.*, **16**, 520 (2006).

[147] B. Panella, K. Hones, U. Muller, N. Trukhan, M. Schubert, H. Putter and M. Hirscher, *Angew. Chem. Int. Ed.*, **47**, 2138 (2008).

[148] A.R. Millward and O.M. Yaghi, *J. Am. Chem. Soc.*, **127**, 17998 (2005).

[149] M. Dinca and J.R. Long, *J. Am. Chem. Soc.*, **127**, 9376 (2005).

[150] Y. Li and R.T. Yang, *Langmuir*, **23**, 12937 (2007).

[151] B. Kesanli, T. Cui, M.R. Smith, E.W. Bittner, B.C. Brockrath and W. Lin, *Angew. Chem. Int. Ed.*, **44**, 72 (2005).

[152] Q.R. Fang, G.S. Zhu, M. Xue, Q.L. Zhang, J.Y. Sun, X.D. Guo, S.L. Qiu, S.T. Xu, P. Wang, D.J. Wang and Y. Wei, *Chem. Eur. J*, **12**, 3754 (2006).

[153] B. Chen, S. Ma, F. Zapata, E.B. Lobkovsky and J. Yang, *Inorg. Chem.*, **45**, 5718 (2006).

[154] X. Zhao, B. Xiao, A.J. Fletcher, K.M. Thomas, D. Bradshaw and M.J. Rosseinsky, *Science*, **306**, 1012 (2004).

[155] L. Pan, M.B. Sander, X. Huang, J. Li, M. Smith, E. Bittner, B. Bockrath and J.K. Johnson, *J. Am. Chem. Soc.*, **126**, 1308 (2004).

[156] J. Luo, H. Xu, Y. Liu, Y. Zhao, L.L. Daemen, C. Brown, T.V. Timofeeva, S. Ma and H.C. Zhou, *J. Am. Chem. Soc.*, **130**, 9626 (2008).

[157] M. Xue, G. Zhu, Y. Li; X. Zhao, Z. Jin, E. Kang and S. Qiu, *Cryst. Growth Des.*, **8**, 2479 (2008).

[158] J. Liu, J.Y. Lee, L. Pan, R.T. Obermyer, S. Simizu, B. Zande, J. Li, S.G. Sankar and J.K. Johnson, *J. Phys. Chem. C*, **112**, 2911 (2008).

[159] T. Takei, J. Kawashima, T. Li, A. Maeda, M. Hasegawa, T. Kitagawa, T. Ohmura, M. Ichikawa, M. Hosoe, I. Kanoya and W. Mori, *Bull. Chem. Soc. Jpn.*, **81**, 847 (2008).

[160] S. Surble, F. Millange, C. Serre, T. Duren, M. Latroche, S. Bourrelly, P.L. Llewellyn and G. Férey, *J. Am. Chem. Soc.*, **128**, 14889 (2006).

[161] H.J. Choi, M. Dinca and J.R. Long, *J. Am. Chem. Soc.*, **130**, 7848 (2008).

[162] Y. Liu, H. Kabbour, C.M. Brown, D.A. Neumann and C.C. Ahn, *Langmuir*, **24**, 4772 (2208).

[163] J.H. Jia, X. Lin, C. Wilson, A.J. Blake, N.R. Champness, P. Hubberstey, G. Walker, E.J. Cussen and M. Schroder, *Chem. Commun.*, 840 (2007).

[164] M. Sabo, A. Henschel, H. Froede, E. Klemm and S. Kaskel, *J. Mater. Chem.*, **17**, 3827 (2007).

[165] W. Zhou, H. Wu, M.R. Hartman and T. Yildirim, *J. Phys. Chem. C*, **111**, 16131 (2007).

[166] W. Yang, X. Lin, J. Jia, A.J. Blake, C. Wilson, P. Hubberstey, N.R. Champness and M. Schroeder, *Chem. Commun.*, 359 (2008).

[167] Q.R. Fang, G.S. Zhu, Z. Jin, Y.Y. Ji, J.W. Ye, M. Xue, H. Yang, Y. Wang and S.L. Qiu, *Angew. Chem. Int. Ed.*, **46**, 6638 (2007).

[168] X. Lin, A.J. Blake, C. Wilson, X.Z. Sun, N.R. Champness, M.W. George, P. Hubberstey, R. Mokaya and M. Schroder, *J. Am. Chem. Soc.*, **128**, 10745 (2006).

[169] Y.G. Lee, H.R. Moon, Y.E. Cheon and M.P. Suh, *Angew. Chem. Int. Ed.*, **47**, 7741 (2008).

[170] S. Ma and H.C. Zhou, *J. Am. Chem. Soc.*, **128**, 11734 (2006).

[171] S.S. Kaye and J.R. Long, *J. Am. Chem. Soc.*, **127**, 6506 (2005).

[172] J.Y. Lee, J. Li and J. Jagiello, *J. Solid State Chem.*, **178**, 2527 (2005).

[173] M. Dinca, A. Dailly, C. Tsay and J.R. Long, *Inorg. Chem.*, **47**, 11 (2008).

[174] J.T. Culp, C. Matranga, M. Smith, E.W. Bittner and B. Bockrath, *J. Phys. Chem. B*, **110**, 8325 (2006).

[175] L. Reguera, J. Balmaseda, C.P. Krap and E. Reguera, *J. Phys. Chem. C*, **112**, 10490 (2008).

[176] L. Reguera, J. Balmaseda, L.F.d. Castillo and E. Reguera, *J. Phys. Chem. C*, **112**, 5589 (2008).

[177] S.S. Kaye and J.R. Long, *Chem. Commun.*, 4486 (2007).

[178] J.T. Culp, S. Natesakhawat, M.R. Smith, E. Bittner, C. Matranga and B. Bockrath, *J. Phys. Chem. C*, **112**, 7079 (2008).

[179] J.G. Vitillo, L. Regli, S. Chavan, G. Ricchiardi, G. Spoto, P.D.C. Dietzel, S. Bordiga and A. Zecchina, *J. Am. Chem. Soc.*, **130**, 8386 (2008).

[180] O.K. Farha, A.M. Spokoyny, K.L. Mulfort, M.F. Hawthorne, C.A. Mirkin and J.T. Hupp, *J. Am. Chem. Soc.*, **129**, 12680 (2007).

[181] Y. Liu, J.F. Eubank, A.J. Cairns, J. Eckert, V.C. Kravtsov, R. Luebke and M. Eddaoudi, *Angew. Chem. Int. Ed.*, **46**, 3278 (2007).

[182] L. Pan, B. Parker, X. Huang, D.H. Olson, J.Y. Lee and J. Li, *J. Am. Chem. Soc.*, **128**, 4180 (2006).

[183] A.D. Burrows, K. Cassar, T. Duren, R.M.W. Friend, M.F. Mahon, S.P. Rigby and T.L. Savarese, *Dalton Trans.*, 2465 (2008).

[184] S. Yang, X. Lin, A.J. Blake, K.M. Thomas, P. Hubberstey, N.R. Champness and M. Schröder, *Chem. Commun.*, 6108 (2008).

[185] Y. Liu, J.H. Her, A. Dailly, A.J. Ramirez-Cuesta, D.A. Neumann and C.M. Brown, *J. Am. Chem. Soc.*, **130**, 11813 (2008).

[186] H.J. Park and M.P. Suh, *Chem.–Eur. J.*, **14**, 8812 (2008).

[187] M. Park, D. Moon, J.W. Yoon, J.S. Chang and M.S. Lah, *Chem. Commun.*, 2026 (2009).

[188] S.Q. Ma, D.Q. Yuan, X.S. Wang and H.C. Zhou, *Inorg. Chem.*, **48**, 2072 (2009).

[189] C.Y. Gao, S.X. Liu, L.H. Xie, C.Y. Sun, J.F. Cao, Y.H. Ren, D. Feng and Z.M. Su, *CrystEngComm*, **11**, 177 (2009).

[190] M. Armandi, B. Bonelli, C.O. Arean and E. Garrone, *Micropor. Mesopor. Mat.*, **112**, 411 (2008).

[191] T.C. Chung, Y. Jeong, Q. Chen, A. Kleinhammes and Y. Wu, *J. Am. Chem. Soc.*, **130**, 6668 (2008).

[192] G.J. Kubas, R.R. Ryan, B.I. Swanson, P.J. Vergamini and H.J. Wasserman, *J. Am. Chem. Soc.*, **106**, 451 (1984).

[193] G.J. Kubas, *Chem. Rev.*, **107**, 4152 (2007).

[194] G.J. Kubas, *Proc. Natl. Acad. Sci. USA*, **104**, 6901 (2007).

[195] S. Dutta, B.R. Jagirdar and M. Nethaji, *Inorg. Chem.*, **47**, 548 (2008).

[196] S. Cerny, *Chem. Phys. Solid Surf. Heterog. Catal.*, **2**, 1 (1983).

[197] J.H. Cole, D.H. Everett, C.T. Marshall, A.R. Paniego, J.C. Powl and F. Rodriguez-Reinoso, *J. Chem. Soc. Faraday Trans.*, **70**, 2154 (1974).

[198] L. Czepirski and J. Jagiello, *Chem. Eng. Sci.*, **44**, 797 (1989).

[199] B. Chen, X. Zhao, A. Putkham, K. Hong, E.B. Lobkovsky, E.J. Hurtado, A.J. Fletcher and K.M. Thomas, *J. Am. Chem. Soc.*, **130**, 6411 (2008).

[200] G. Yushin, R. Dash, J. Jagiello, J.E. Fischer and Y. Gogotsi, *Adv. Funct. Mater.*, **16**, 2288 (2006).

[201] S.H. Jhung, H.K. Kim, J.W. Yoon and J.S. Chang, *J. Phys. Chem. B*, **110**, 9371 (2006).

[202] G.T. Palomino, M.R.L. Carayol and C.O. Arean, *J. Mater. Chem.*, **16**, 2884 (2006).

[203] C.O. Arean, G.T. Palomino and M.R.L. Carayol, *Appl. Surf. Sci.*, **253**, 5701 (2007).

[204] C.O. Arean, M.R. Delgado, G.T. Palomino, M.T. Rubio, N.M. Tsyganenko, A.A. Tsyganenko and E. Garrone, *Micropor. Mesopor. Mat.*, **80**, 247 (2005).

[205] G.T. Palomino, M.R.L. Carayol and C.O. Arean, *Catal. Today*, **138**, 249 (2008).

[206] Y.G. Lee, H.R. Moon, Y.E. Cheon and M.P. Suh, *Angew. Chem. Int. Ed.*, 47, 7741 (2008).
[207] Y. Kubota, M. Takata, R. Matsuda, R. Kitaura, S. Kitagawa, K. Kato, M. Sakata and T.C. Kobayashi, *Angew. Chem. Int. Ed.*, 44, 920 (2005).
[208] S. Takamizawa and E. Nakata, *CrystEngComm*, 7, 476 (2005).
[209] E.C. Spencer, J.A.K. Howard, G.J. McIntyre, J.L.C. Rowsell and O.M. Yaghi, *Chem. Commun.*, 278 (2006).
[210] T. Yildirim and M.R. Hartman, *Phys. Rev. Lett.*, 95, 215504 (2005).
[211] V.K. Peterson, Y. Liu, C.M. Brown and C.J. Kepert, *J. Am. Chem. Soc.*, 128, 15578 (2006).
[212] G.J. Kubas, C.J. Burns, J. Eckhert, S.W. Johnson, A.C. Larson, P.J. Vergamini, C.J. Unkefer, G.R.K. Khalsa, S.A. Jackson and O. Eisenstein, *J. Am. Chem. Soc.*, 115, 569 (1993).
[213] M.R. Hartman, V.K. Peterson, Y. Liu, S.S. Kaye and J.R. Long, *Chem. Mater.*, 18, 3221 (2006).
[214] P.M. Forster, J. Eckert, J.-S. Chang, S.-E. Park, G. Férey and A.K. Cheethem, *J. Am. Chem. Soc.*, 125, 1309 (2003).
[215] K.W. Chapman, P.J. Chupas, E.R. Maxey and J.W. Richardson, *Chem. Commun.*, 4013 (2006).
[216] Y. Liu, C.M. Brown, D.A. Neumann, V.K. Peterson and C.J. Kepert, *J. Alloys Compd.*, 446–447, 385 (2007).
[217] W. Zhou and T. Yildirim, *J. Phys. Chem. C*, 112, 8132 (2008).
[218] S. Bordiga, L. Regli, F. Bonino, E. Groppo, C. Lamberti, B. Xiao, P.S. Wheatley, R.E. Morris and A. Zecchina, *Phys. Chem. Chem. Phys.*, 9, 2676 (2007).
[219] G. Spoto, J.G. Vitillo, D. Cocina, A. Damin, F. Bonino, and A. Zecchina, *Phys. Chem. Chem. Phys.*, 9, 4992 (2007).
[220] B. Panella and M. Hirscher, *Phys. Chem. Chem. Phys.*, 10, 2910 (2008).
[221] A. Centrone, D.Y. Siberio-Perez, A.R. Millward, O.M. Yaghi, A.J. Matzger and G. Zerbi, *Chem. Phys. Lett.*, 411, 516 (2005).
[222] A. Centrone, L. Brambilla and G. Zerbi, *Phys. Rev. B*, 71, 245406 (2005).
[223] A.J. Fletcher, E.J. Cussen, T.J. Prior, M.J. Rosseinsky, C.J. Kepert and K.M. Thomas, *J. Am. Chem. Soc.*, 123, 10001 (2001).
[224] A.J. Fletcher, E.J. Cussen, D. Bradshaw, M.J. Rosseinsky and K.M. Thomas, *J. Am. Chem. Soc.*, 126, 9750 (2004).
[225] C. Yang, X.Wang and M.A. Omary, *J. Am. Chem. Soc.*, 129, 15454 (2007).
[226] D. Bradshaw, J.E. Warren and M.J. Rosseinsky, *Science*, 315, 977 (2007).
[227] P.L. Llewellyn, G. Maurin, T. Devic, S. Loera-Serna, N. Rosenbach, C. Serre, S. Bourrelly, P. Horcajada, Y. Filinchuk and G. Férey, *J. Am. Chem. Soc.*, 130, 12808 (2008).
[228] S.M. Humphrey, S.E. Oungoulian, J.W. Yoon, Y.K. Hwang, E.R. Wise and J.S. Chang, *Chem. Commun.*, 2891 (2008).
[229] S. Ma, X.S. Wang, E.S. Manis, C.C. Collier and H.C. Zhou, *Inorg. Chem.*, 46, 3432 (2007).
[230] J.T. Culp, M.R. Smith, E. Bittner and B. Bockrath, *J. Am. Chem. Soc.*, 130, 12427 (2008).
[231] B.P. Stoicheff, *Can. J. Phys.*, 35, 730 (1957).
[232] A.J. Fletcher and K.M. Thomas, *J. Phys. Chem. C*, 111, 2107 (2007).
[233] H. Tanaka, H. Kanoh, M. El-Merraoui, W.A. Steele, M. Yudasaka, S. Iijima and K. Kaneko, *J. Phys. Chem. B*, 108, 17457 (2004).

[234] D. Noguchi, H. Tanaka, A. Kondo, H. Kajiro, H. Noguchi, T. Ohba, H. Kanoh and K. Kaneko, *J. Am. Chem. Soc.*, **130**, 6367 (2008).

[235] J.J.M. Beenakker, V.D. Borman and S.Y. Krylov, *Phys. Rev. Lett.*, **72**, 514 (1994).

[236] J.J.M. Beenakker, V.D. Borman and S.Y. Krylov, *Chem. Phys. Lett.*, **232**, 379 (1995).

[237] R. Yaris and J.R.J. Sams, *J. Chem. Phys.*, **37**, 571 (1962).

[238] A.V.A. Kumar and S.K. Bhatia, *Phys. Rev. Lett.*, **95**, 245901 (2005).

[239] A.V.A. Kumar, H. Jobic and S.K. Bhatia, *J. Phys. Chem. B*, **110**, 16666 (2006).

Index

Printed and bound by CPI Group (UK) Ltd, Croydon, CR0 4YY

Printed and bound by CPI Group (UK) Ltd, Croydon, CR0 4YY

16/04/2025

14658545-0001